Martin Bartelheim – Roland Hardenberg – Irmgard Männlein –
Simone Riehl – Tobias Schade – Thomas Scholten (ed.)

ResourceCultures

How Resources Affect Societies

ResourceCultures

How Resources Affect Societies

Martin Bartelheim – Roland Hardenberg – Irmgard Männlein
– Simone Riehl – Tobias Schade – Thomas Scholten (ed.)

Reichert Verlag Wiesbaden 2025

Cover images:
left - Thomas Scholten
middle - Martin Bartelheim
right - Codex Tubingensis zu Platon, Universitätsbibliothek Tübingen

Bibliografische Information der Deutschen Nationalbibliothek
Die Deutsche Nationalbibliothek verzeichnet diese Publikation
in der Deutschen Nationalbibliografie; detaillierte bibliografische Daten
sind im Internet über http://dnb.dnb.de abrufbar.

Gedruckt auf säurefreiem Papier
(alterungsbeständig – pH 7, neutral)

Tauernstraße 11, 65199 Wiesbaden
info@reichert-verlag.de, www.reichert-verlag.de

ISBN: 978-3-7520-0868-5 (Print)
eISBN: 978-3-7520-0330-7 (Ebook)
https://doi.org/10.29091/9783752003307

Contents

Preface

This book is the result of 12 years of interdisciplinary research in the Collaborative Research Centre SFB 1070 ResourceCultures.

It is thus the product of the contributions of many people. Before wishing you an enjoyable read, the editors would like to take this opportunity to express their heartfelt thanks to the many helping hands who have been involved in the preparation and completion of this publication. These include, of course, all the authors of the volume as well as the researchers of the SFB 1070 for their contributions, comments, feedback and reviews.

In addition to the authors involved in Chapters 1 to 6 as well as the excursuses and case studies, special mention should be made of Gabriele Alex and Peter Kühn, who identified interesting topics for the excursuses and coordinated them, as well as Raffaella Da Vela, Wulf Frauen and Beat Schweizer, who coordinated Chapter 6 with the case studies.

The editors are also particularly grateful to Döbereiner Chala-Aldana for his extensive feedback and very helpful reviews of the chapters, excursuses and case studies, and to Chiara Sava Raich for her valuable support in coordinating this project. Richard Szydlak kindly edited the illustrations and created wonderful maps in a consistent style. The editors would also like to thank the editorial team, namely Uwe Müller, Henrike Srzednicki and Carolin Manzke for reading, commenting, correcting, preparing and editing the manuscript, Anke Scholz and Sandra Teuber for their valuable contributions to the further development of the resource concept, the University of Tübingen for providing the SFB 1070 with continuous support and a professional research environment, and last but not least the team at the Dr. Ludwig Reichert Verlag (Wiesbaden), namely Marie-Christine Schimpf and Sarah Reinish, for reviewing, proofreading and preparing the final version for publication.

Finally, we would like to thank the German Research Foundation (DFG) for funding the SFB 1070 from 2013 to 2025 (project number: 215859406), and a considerable number of reviewers for evaluating this extraordinary endeavour.

Martin Bartelheim, Roland Hardenberg, Irmgard Männlein,
Simone Riehl, Tobias Schade, Thomas Scholten

Tübingen, May 2025

1 Why Do Resources Matter? An Introduction to ResourceCultures

Martin Bartelheim, Tobias Schade, Thomas Scholten

Resources are omnipresent in today's discussions – but what exactly do we mean by resources and why do resources matter? These are not only the fundamental questions with which this book begins, but they also summarise the core idea of the Collaborative Research Centre SFB 1070 ResourceCultures at the Universities of Tübingen and Frankfurt, from whose research synthesis this book emerged.

In modern times, the dominant view of resources has evolved significantly, reflecting a complex interplay of economic, environmental and social factors. Traditionally, resources were seen primarily as commodities – raw materials or assets that could be extracted, processed and consumed for economic gain. These include 1) energy sources (e.g. coal, gas, oil, water) needed to fuel the economy as well as individual and collective needs, 2) precious metals (e.g. gold, silver) considered as objects of status and wealth, 3) rare earth elements (e.g. scandium) or light metals (e.g. lithium) needed in technological production and consumption processes, 4) materials for food security (e.g. soil, water, fertilisers), 5) forms of processed goods (e.g. construction materials). Added to this are 6) infrastructures associated with these materials, which are understood here as networks of people and things (e.g. production and distribution chains, knowledge transfer) and tangible material components (e.g. roads, transport vehicles). Accordingly, discussions of resources often focus on issues of exploitation, overuse and scarcity, distribution, sustainability and, more recently, recycling of resources. As such, resources and related issues of security of supply, depletion, prosperity and sustainability can be seen as highly relevant and topical in contemporary societal discussions. This is true not only in science and politics, but also in the public media and in popular cultural adaptations.

A noteworthy example is the Club of Rome's report on the 'Limits to Growth' in 1972 (Meadows et al. 1972), which addressed issues of resources, food, population development and environmental pollution in the context of overexploitation and sustainability. The report continues to influence the public debate on resources to this day and underlines the enormous importance of the issue of human resource use. Another example is the awareness of the potential depletion of limited raw materials. These are often regarded as 'systemically relevant' resources, such as oil or phosphorus. The predicted end of oil production ('peak oil'), which Hubbert (1956) anticipated and which would reach its maximum around the year 2000 and then begin to decline, would certainly have a significant impact on general consumption and production in today's oil-dependent societies. A similar situation is being discussed with regard to phosphorus production ('peak phosphorus'), as an indispensable component of crop fertilisers (Cordell et al. 2009), challenging global food security.

However, as we have moved further into the 21st century, this view has changed. Increased awareness of environmental issues, climate change and sustainability has led to a more nuanced understanding of resources. Resources are now often seen not merely as commodities, but as components of a larger ecological and social system. This perspective emphasises the interconnectedness of human activities and the environment, recognising that overexploitation and pollution can have detrimental effects on ecosystems and, ultimately, on human well-being. Current discussions on resources are linked to the concepts of 'natural capital' and 'ecosystem services', with resources being considered in the context of 'Sustainable Development Goals' (Bouma 2014). Even if, for example, questions about the limits to growth, the depletion of key resources and sustainability may reveal shifts in discourses, the modern view of resources is still often characterised by the dominance of capitalist production, consumption and valuation. This in turn means that resource management in pre-modern or non-capitalist societies and even intangible dimensions of resources are often not taken into consideration – at least rarely in isolation from our modern economic understanding.

Sustainability has emerged as a key principle in the resource debate. The concept of sustainable resource management advocates the responsible use of resources to meet the needs of the present without compromising the ability of future generations to meet their own necessities. This approach challenges the dominant paradigm of unlimited growth and resource consumption, as put on the agenda by the Club of Rome more than 50 years ago, and urges practices that prioritise conservation, efficiency and renewable alternatives. For example, the rise of renewable energy sources such as solar and wind reflects a shift towards using resources in ways that minimise environmental impact and promote long-term sustainability.

Another important aspect of the modern perspective on resources is the recognition of social equity, justice and ethics. Resources are increasingly viewed through the lens of access and distribution, with a focus on how they can be shared more equitably between different populations and continents. This has led to discussions of the 'resource curse', where countries rich in raw materials often face economic and social challenges due to mismanagement and corruption. Modern approaches advocate inclusive policies that empower local communities and ensure that the benefits of resource exploitation are distributed fairly and sustainably.

In addition, technology has played a critical role in reshaping our understanding of resources. Innovations in data analytics, artificial intelligence and biotechnology are improving our ability to manage resources more efficiently. For instance, precision agriculture uses technology to optimise the use of water and fertiliser, minimising waste while maximising yields. Similarly, advancements in recycling and materials science are paving the way for a circular (bio-)economy, where resources are reused and repurposed rather than discarded.

In summary, the dominant view of resources in modern times has been challenged by attempts to offer alternative understandings of tangible and intangible resources not only from economic and political, but also from cultural perspectives (see Chapter 2). Resources are not 'given' by nature and have no essentialist value, but are 'created' and 'valued' in socio-cultural contexts. This becoming of resources has been addressed, for example, by Richardson and Weszkalnys (2014), who mentioned gold, which was valued differently by 'Incas' and 'Spaniards' in the early modern period due to different socio-cultural perspectives. This may also be discussed for other times or spaces where a purely economic perspective was or is not dominant. Thus, not only natural raw materials such as gold, stones or oil can become resources for certain societies, but also, for example, landscapes, practices, objects or even the past. Subsequently, the use of resources by people defines identities and solidifies social groups or societies. Thus, the interaction and use of resources can lead to dependencies for societies, which can be described as 'entanglements' according to the archaeologist Hodder (2014). This can be illustrated by the example of oil (petroleum), which was known and sometimes used for a long time in the human past, but did not play a central role until it gained importance from the 19th century onwards and became even more important in the 20th century due to socio-technological developments. In this epoch, which is also referred to as the 'Age of Oil', a 'ResourceCulture Oil', so to say, developed around the resource of oil, in which the knowledge of raw material deposits and extraction, the processing and handling of oil and the associated infrastructures not only became an aspect of power and identity formation (e.g. for 'petrostates') as well as a pop-cultural symbol (e.g. as a status of wealth and progress), but also led to strong dependencies of entire societies. On the one hand, some of these dependencies have had societal, economic and political consequences in the past, such as the 'oil crisis' in 1973 and the 'energy crisis' in 1979, both with rising oil prices and the need to save energy, and subsequently the 'oil glut' in the 1980s with increased production, but lower demand and thus falling oil prices. On the other hand, these dependencies are also difficult for oil-based societies to overcome today, as oil is used not only for heating and as a fuel for transport and mobility, but also as a component of plastics, clothing, cosmetics, medical products, etc. Even infrastructure and entire professions have developed around this resource and have become entangled. Today, oil has the status of a key resource that can be considered 'systemically relevant', even if alternatives are being sought. However, it is important to stress that this is not because of any inherent value of oil, but because oil has become a resource for modern societies, and modern societies have subsequently emerged around the resource of oil and maintained this relationship economically and socio-culturally. In this context, Rogers (2015) and Watts (2005), for example, address oil as a resource, while Wilson et al. (2017) address 'petrocultures' as networks around oil, so to speak.

This example of an evolving perspective underscores the need for a holistic approach to analysing and understanding resources and to recognise the intricate socio-cultural networks within and between human societies. These result in specific ResourceCultures, which are presented and explained in this book. This new concept of ResourceCultures has been developed, applied, explained, adapted and refined over twelve years of collaborative research by scientists from archaeology, history, classical philology, social and cultural anthropology, cultural studies, economic history, theology, geography and soil sciences in the Collaborative Research Centre SFB 1070 RESOURCECULTURES. In twenty-two interdisciplinary projects, the researchers analysed cultural dynamics from the perspective of tangible and intangible resources.

The plurality and diversity of the researchers involved, including a considerable number of principal investigators, postdoctoral researchers and doctoral candidates, is reflected not only in the diversity of the research carried out over the years, but also in the structure of this book. Theoretical aspects of resource issues (Chapters 2–5) are followed by various case studies from different times and places around the world (Chapter 6). The individual chapters and case studies follow differentiated perspectives and pluralistic approaches of ResourceCultures. In concluding this interdisciplinary collaboration, it cannot be denied that it is a challenge to find a common language between different researchers from different disciplines, since each discipline and research area has its own concepts and definitions, sometimes even for superficially similar terms. Ways to understand culture and landscape, for example, are diverse. We believe that standardising such terms is neither necessary nor helpful. Rather, there is a danger of generalising findings in a way that obscures their disciplinary specificities. Instead we have developed the analytical tools of ResourceComplexes and ResourceAssemblages (Chapter 2), which allow for diachronic comparison and comparability in different geographical contexts.

Despite these disciplinary differences, all chapters, excursuses and case studies share a cross-disciplinary perspective on cultural resources and a similar terminology. They are intended not only to encourage readers to reflect on familiar narratives of resources and societies, but also to provide heterogeneous starting points for other researchers to follow from the perspective of ResourceCultures.

The new concept of ResourceCultures, from which the book takes its name, and its theoretical framework are introduced in Chapter 2 by Martin Bartelheim, Roland Hardenberg and Thomas Scholten. It not only considers the socio-cultural construction, valuation and use of particular resources, whether tangible or intangible, but also their positioning within different resource networks – so-called ResourceComplexes and ResourceAssemblages. Both terms define analytical perspectives for describing, analysing and comparing different levels and gradations of resource networks and related resource management in different spatial, temporal and socio-cultural contexts. While these concepts were subject to dynamic consideration and development in the early stages of our research, the perspectives presented here (see Chapter 2) represent the synthesis of our discussions. The focus has been on aspects of 'developments', 'movements' and 'valuations'. In the following Chapters 3–5, different aspects of these dynamic processes are illustrated with various examples from different spatial and temporal settings in order to shed light on parallels and differences in resource management in the present and the past. In doing so, we contradict the expectation of straightforward evolutionary developments through time.

Under the aspect of 'developments', Martin Bartelheim, Jens Kamlah and Simone Riehl (Chapter 3) present examples that deal with the topic of resources in processes of socio-cultural change and highlight questions of change and adaptation as well as developments in landscapes, also relating the resource concept to cultural niche theory. The examples used by Thomas Scholten, Peter Pfälzner and Jörg Baten (Chapter 4) deal with resources in the context of processes of 'movements', which are related to the question of mobilities and the settlement of spaces, highlighting, among other things, the modalities of movement, the role of pathways and processes of long-distance movement. In Chapter 5, 'valuations', Roland Hardenberg, Irmgard Männlein and Thomas Thiemeyer selected examples of value creation in relation to resources, the symbolic dimensions of resources and the sacralisation of resources, also talking about resource communities and identity-forming processes, as well as political dimensions of resources.

Chapters 2–5 are framed by short excursuses contributed by various authors, which aim to provide additional in-depth information on specific issues and key aspects discussed in the chapters. Chapter 6 presents eleven case studies that reflect different aspects of resource becoming and resource management, illustrated from ethnological, archaeological, historical and interdisciplinary perspectives. The introduction to Chapter 6 provides a detailed overview of the individual contributions to the chapter.

A large number of researchers have contributed to this joint publication with written texts, inputs, discussions, comments and feedback, making this publication also a collaborative work of a collaborative team. In addition to the main authors of Chapters 3–5, other researchers from the SFB 1070 projects provided examples from their research for these chapters: Döbereiner Chala-Aldana, Nicholas Conard, Marta Díaz-Zorita Bonilla, Harald Floss, Wulf Frauen, Sigrid Hirbodian, Sophie Hüglin, Michael Kienzle, Thomas Knopf, Christian Kübler, Peter Kühn, Natascha Mehler, Maike Melles, Jesse Millek, Karin Polit, Richard Posamentir, Elena Revert Francés and Lukas Werther, among others, contributed discussions or short texts (Chapters 3 and 4). The introduction to Chapter 5 and sections of Chapter 5.1 are slightly revised versions of texts written by Roland Hardenberg, Thomas Knopf and Beat Schweizer, with comments and additions by several colleagues. Other sections of Chapter 5.1 are based on revised versions of texts by Karl-Heinz Stanzel (Chapter 5.1.1) and texts by Beat Schweizer (Chapter 5.1.1), Peter Pfälzner (Chapter 5.1.2) and Roland Hardenberg (Chapter 5.1.3). Some parts of Chapter 5.1.3 have already been published in Hardenberg 2017. Julia Gilfert (Chapter 5.3.1), Manuel Respondek (Chapter 5.3.2) and Peter Zeller (Chapter 5.3.3) provided comments based on their research projects on the examples given in Chapter 5.3.

2 The Concept of ResourceCultures

Martin Bartelheim, Roland Hardenberg, Thomas Scholten

Modern concepts of resources have evolved from their traditional economic definition to a broader, multidimensional understanding applicable to various aspects of human life (Bridge 2009; Richardson/Weszkalnys 2014). In the 19th and 20th centuries, resources were seen primarily as natural raw materials for the production and exchange of goods within an economic framework (e.g. Apter 2005; Barbier 2011) in a functional relationship with the environment (e.g. Zimmerman 1933). This view was expanded to include not only material but also immaterial resources (Carrier 2021; Kaneff 2018; Uchibori 2011), and extended beyond economic contexts to include politics, society, culture, religion, ecology and conservation (Hausmann/Perreaux 2018). This semantic expansion has broadened the scope of the term 'resource' to include a wider range of tangible and intangible means for achieving various goals in different spheres of human existence.

2.1 Concepts of Resources

Despite this expansion from nature to culture, the economic and functional roots of the concept of resources have not been fully overcome. In other words, while the term has been applied more broadly in different fields, its origins in economics, and in particular rational choice theory, still influence its understanding and application. The term 'resource' has been used in material and non-material contexts, but has remained a rational means of achieving certain ends (Ferry/Limbert 2008). Accordingly, resources are usually considered to be 'scarce', the 'capital' of a person or group, which can be 'accumulated' and should be used 'rationally' or 'functionally' to satisfy certain 'needs' of people or systems (Harvey 1974; Franquesa 2019). For Bourdieu, for example (Bourdieu/Waquant 1992, 119), resources are known to be the basis of the different types of capital – economic, social, cultural and symbolic – which can be converted into each other and depend to a considerable extent on social networks. In ecology and conservation, resources are often understood as components of the natural environment, such as soil, plants and animals, that provide benefits to both humans and other organisms (e.g. Davidov 2014). This includes ecosystem functions and services, i.e. processes and activities performed by the components of an ecosystem, such as soil, that contribute to its overall objectives, such as stability, resilience, productivity, health and security for human needs and well-being (Millennium Ecosystem Assessment 2005). Like ecosystem functioning, ecosystem services are essential to human well-being. In line with the classical approach to resources mentioned above, they are understood as 'natural capital' (Costanza et al. 1997) and play a critical role in supporting economic activities, health and overall quality of life (Daily 1997).

These resource concepts, which operate within economic parameters, represent a modern industrial understanding and a particular way of life focused on industrial production and capitalist consumption, in which market principles are applied to all aspects of life. In societies and worldviews that prioritise other aspects of human well-being, such as dwelling in an environment according to certain social, religious or cosmological principles, the concept of resources takes on a different meaning and importance. In their view, resources are not just seen as inputs for economic production, but are recognised as essential elements for maintaining the orders and identities of a society and its relations with the environment. Such resources may change their value over time, for example gold may no longer represent purity and sacredness but stable profits, while others may seem constant over thousands of years but appear in different cultural settings and geographies. The latter are, for example, all those different means used by people around the world to establish social relationships with human and more-than-human beings in order to live according to a particular vision of the world.

By understanding resources as means for establishing relationships across time and space, the concept of resources transcends the modern economic definition and shows its embeddedness in broader notions of social order, value systems, cosmologies, hierarchies, sustainability and human well-being. This holistic understanding of resources is consistent with the idea that prosperity encompasses not only material wealth but also social, cultural and spiritual well-being. Therefore, the management and use of resources in these societies are guided by cultural principles that seek to create, maintain and change social relationships and identities. This often involves attempts to establish particular relationships with the environment or with agents of cosmological domains, and is guided by a variety of cultural values. In contemporary times, environmental conservation, social justice and cultural integrity are just some examples of such resource use. Such values of a society can be in conflict with each other, for example, when the protection of life and personal well-being is in conflict with the drive for profit. To understand this fully, it is necessary to examine these contradictions on a case-by-case basis and at different levels, i.e. at the level of ideas, rules and actions. For example, values may be contradictory at the ideological level, but people find ways of overcoming these contradictions by adopting certain norms or acting in certain ways.

In the light of such observations, social, cultural, historical and environmental sciences have developed an expanded understanding of resources in past and present societies. They show that resources, in the sense of material and immaterial means, can be the basis of a variety of social, political and cultural processes. For example, Giddens (1984, 258) distinguishes between authoritative and allocative resources. For him, the former are means and capabilities that serve the exercise of power, while the latter ensure access to the material basics of life. Lowenthal (1985) has initiated the so-called heritage studies, in which various cultural studies are involved to explore how different societies use resources to represent their history(ies) and to evaluate and make sense of contemporary processes (Sørensen/Carman 2009). In historical studies, Ash (2002, 32) has freed the concept of resources from its purely economic meaning in order to describe the reciprocal relationships between science and politics. For him, resources can also be cognitive, apparatus, personnel, institutional and rhetorical. In geography, Watts (2005) considers oil as a resource and uses the terms oil complex and oil assemblage. These approaches are currently being further discussed, including by Rogers (2015) and Schritt (2016), particularly to analyse political configurations in the context of social change and commodity use. In the natural sciences, the relationships between humans and things, or humans and the environment, are understood as complex adaptive systems (Lansing 2003). These can serve as an interdisciplinary framework for studying social and natural systems (Teuber et al. 2017) to better understand human responses to cultural and environmental challenges. Research on materialities uses the concept of resource and its social and political associations to link analytical engagement with how objects acquire meaning through their production and circulation in historical contexts. Here, the dynamics of the physical properties of materials are interconnected and contribute to their own social histories through processes such as weathering and erosion, or evaporation and condensation (Richardson/Weszkalnys 2014; Johansen/Bauer 2018).

The result of these interdisciplinary dialogues is a new concept of resources that differs from conventional economic definitions in several ways. The social significance of resources is emphasised, as resources serve to create, maintain and change social relations and identities. This social significance of resources is closely linked to cultural values and value creation. According to this understanding, raw materials only become resources when they acquire cultural value and are used in culturally specific ways. Their value is not given by the nature of things. This non-essentialist aspect emphasises the social significance of resources. Soil, for example, is not an important resource simply because it occurs in a particular geographical location and functions as a source of nutrients and water for plant growth (see excursus 'Soil'). Soil is assigned a specific benefit and value through its use by humans within cultural frameworks and in relation to particular social contexts, such as soil health and soil security. Understanding soil management requires an understanding of how this practice is embedded in wider human domains, such as food production in a particular place (Teuber et al. 2022). For example, the value of soil differs when food is hunted and gathered by foragers or produced by farmers in a specific geographical location that is considered their habitat and settlement area.

This example shows that a resource is always culturally constructed through changing attributions and conditions of use. On the one hand, this means that basically anything can become a resource, but on the other hand, it also means that the social meaning of resources is tied to specific cultural, temporal and spatial contexts. As a result, the dynamics that arise from the use of resources become an important topic of study. Particular attention should be paid to socio-cultural developments (Chapter 3), spatial and social movements (Chapter 4), conflicts, curses and blessings, and processes of valorisation and sacralisation of resources (Chapter 5).

2.2 Resources as Part of a Network

In the previous chapter it was shown that soil as a resource is an important basis for food production embedded in specific social and cultural settings. Soil quality is not simply part of the choice of a particular place to live and settle, but an expression of local values and existing cultural practices (see excursus 'Favour/Disfavour'). In addition, there may be other values associated with particular forms of food production, such as access to water, the ability to hunt, better conditions for the birth of children and protection from enemies. Soil is therefore always a resource that is embedded in a network with other elements that people consider important. These are not only material, but also immaterial, such as the knowledge of how to grow plants as a source of food and the risks of childbirth. In this way, a ResourceComplex is created around local settlement and food production.

Resources are therefore not seen as isolated elements, but as part of a planned network called ResourceComplex (Hardenberg et al. 2017). Based on this definition, different

ResourceComplexes can be analysed for different times and places in terms of their social significance and various dynamics resulting from the use of resources can be addressed (see Chapter 6). A wide range of variations can be observed in the cultural connotation of resources and how they can affect the formation, maintenance and transformation of communities in different ways (Bartelheim et al. 2015).

The cultural construction of resources described above using the example of soil and its embedding in an existing ResourceComplex is not static, but a dynamic process involving activities that lead to change and transformation. Focusing on these cultural dynamics rather than on specific objects leads to a new understanding of resources and ResourceComplexes as processes of value allocation and value creation. These processes can thus provide information about the values of a resource to the community, whether these are given, shared or contested, and the means by which they can be accessed (see Chapter 6). Focusing on values means focusing on the material and immaterial aspects of resources. This provides information about the significance of the material quality of a resource, the extent to which the same immaterial value can be effective in different resources and whether the same material resource simultaneously incorporates several socially relevant values. The latter is, for example, an important analytical approach for analysing the implementation of the operationalised Sustainable Development Goals (SDGs) (United Nations 2015; Wittwer et al. 2021). Soil as a resource plays a key role here, as it is given importance in twelve of the 17 SDGs (Lal et al. 2021) and is therefore assigned multiple values.

Another aspect for the discussion of socio-cultural dynamics is the role of overuse or even destruction of resources, such as the mechanisation of agriculture and the associated increase in soil erosion (Montgomery 2007) or factory farming and its environmental consequences due to overfertilisation (Sutton et al. 2011). ResourceComplexes are therefore elements in the system of resource value creation and no longer purely functional networks. A ResourceComplex thus becomes a heuristic of the dynamic resource network described above. As such, it is intentionally similar in different places or at different times and allows them to be analysed and compared from the same perspective, in this case value allocation.

This idea of the ResourceComplex as a specific perspective opens up the possibility of analysing and describing networks and comparing their concrete manifestations. For example, the resource metal (see excursus 'Materiality of Metals') in an anthropogenic context, goes through a chain of processes, starting with the extraction, for example of gold and copper since the early Copper Age, as a raw material, through mining, concentration, purification and alloying, transport and the associated trade of the product. Each step in the process requires not only the chemical element, but also a variety of other elements, starting with knowledge of the location of the deposit in the landscape and the craftsmanship required to extract it from the rock or sediment. Further material and immaterial elements are added during subsequent use, for example from the artistic, symbolic, political and economic spheres. By taking the perspective of a ResourceComplex, i.e. by studying the human organisation of a specific resource use and by examining the interrelationships involved in the production and handling of metal products, concrete networks emerge, such as mining settlements or the trading of precious metals on the stock exchange, which characterise a specific cultural context.

From this perspective, the dynamics resulting from the existence of such networks concern things, properties, relations, circumstances or events. These interact in a contingent way, i.e. they establish historical, often arbitrary relations. The resulting networks thus change over time as the elements of the network begin to interact with each other and acquire new values and meanings. For example, the ResourceComplex of gold mining changed with the arrival of the Spanish conquistadores in South America, as the human labour resource tied to gold mining was transferred to silver mining and the military sector (Barragán Romano/Zagalsky 2023). The network was still focused on metal extraction, but as a result of certain socio-cultural processes it shifted from a regional to an intercontinental scale. In this sense, the network has been transformed into a structure or assemblage. Such a perspective on resource networks, which emphasises these contingencies and developments arising from the interactions between elements, is called ResourceAssemblage. Assemblage (DeLanda 2006; Bennett 2010) or *agencement* (Deleuze/Guattari [1980] 1992) emphasises the historical contingency of structural parts of a network. From the perspective of assemblage theory, the components of such a network are connected to each other through different materiality and longevity. The connections express the simultaneity of the non-simultaneous, for example the interplay of different elements of different materiality, temporality and dynamics, which do not simply contradict each other, nor necessarily form a stable whole, but rather allow something new to emerge in relation to what has been before.

As such, the perspective of ResourceAssemblages opens up new spaces and makes them analysable through the decoding and reinterpretation of resource networks. A crucial point here is that the assemblage character of a network is not inevitable, but emerges from the interaction of resources, the timing, duration and outcome of which are contingent. Elements of a network interact, and through the coincidence of several events without a direct causal link, such as the intentions of actors, the elements change or are formed anew. This process is not subject to any discernible laws, such as human planning, and makes it possible to analyse the significance of the supernatural worlds of 'non-humans' or 'more-than-humans' (Tsing 2015) in the context of participatory research (Bastian et al. 2017) and of the relationships between things that focus on the efficacy of materiality (see Chapter 6).

Both perspectives, ResourceComplex and ResourceAssemblage, form a central heuristic for interpreting the role of resource networks in a socio-cultural context (fig. 1). These are fundamentally new concepts that turn resource networks into analytical categories, which in turn open up new perspectives. They shed light on the complex dynamics of resource use and offer different ways of looking at how communities use resources and integrate them culturally into a ResourceCulture. It is crucial to recognise that the two concepts of ResourceComplex and ResourceAssemblage should not be seen as opposing or exclusive, but rather as complementary perspectives that together offer a more comprehensive understanding of resource use. They show how resource use involves both structure and change, continuity and innovation, and how these aspects are intertwined and shape a community's specific ResourceCulture. On the one hand, the ResourceComplex approach emphasises more the structural aspects of resource use, including the planning, organisation and integration of different resources into a cultural context. This may relate to the long-term use of particular resources and how they are embedded in a community's way of life. The ResourceAssemblage focus, on the other hand, emphasises the processual and dynamic aspects of resource valorisation and use. It emphasises the constant adaptation, appropriation and reorganisation of resources by communities over time. It brings to the fore the diversity and flexibility of resource practices, which can change according to a community's interests and circumstances.

Object	Heuristic/Perspectives
Resource(s) (definition)	ResourceComplex (functional, intentional)
ResourceCultures (conceptual)	ResourceAssemblage (diachronic, contingent)

Fig. 1 The new concept of the dynamic interaction of resources and cultures. Resources and ResourceCultures are phenomena in time and space that can be described and understood through two heuristic perspectives, namely ResourceComplexes and ResourceAssemblages. ResourceComplexes point towards the functional, synchronic as well as intentional human use of resources, ResourceAssemblages point towards the contingent, diachronic and dynamic interactions between resources and other elements (Graphic design: Richard Szydlak).

The distinction between ResourceComplex and ResourceAssemblage can be illustrated by the example of human-made and human-used landscapes (see excursus 'Landscape'). Landscapes can provide various resources whose cultural value is manifested in planned human actions, such as agriculture, transhumance, mining, water pumping and transport (e.g. Benítez de Lugo/Mejías Moreno 2022), as well as the construction of sanctuaries and the creation of protected areas (e.g. Da Vela 2023). Here, looking at a landscape from the ResourceComplex perspective emphasises the networks that people intentionally create from material and immaterial components at specific places and times (Bartelheim et al. 2021b; Knopf et al. 2021; Chala-Aldana 2024). In contrast, analysing landscapes from the perspective of a ResourceAssemblage highlights specific historically contingent developments and ruptures, for example in the northern Mesopotamian foothills (e.g. Pfälzner et al. 2018) or the Sierra Morena of southern Spain. Since prehistoric times, the Sierra Morena has been characterised by agro-pastoral landscape use, which in some regions has changed significantly due to external influences (especially interest in external resources). In some parts, from the last pre-Christian millennium to the present day, it has been almost completely replaced by the exploitation of mineral resources as the locally dominant form of resource use, which has significantly changed the character of landscape use. In other parts, the external demand for pork ham and cork products, in particular, has led to a significant intensification of production in recent decades, which is still part of the agricultural sector, but which uses the agricultural resources of the landscape in a completely different way to the earlier, often transhumant, way of farming (Bartelheim et al. 2022; Melles 2021).

2.3 From Resource to Culture: Conceptual Framework of ResourceCultures

Following the theoretical analysis and redefinition of resources, two perspectives for analysing human-resource interactions have been presented and contrasted: the ResourceComplex approach for structural aspects of resource use and the ResourceAssemblage approach for processual and dynamic aspects of resource use. It is concluded that both have different but complementary strengths and allow for a holistic understanding of human-resource interactions as well as the processual and dynamic aspects of resource valorisation and use. From the relevant observations at different times and places and from different perspectives (see Chapter 6), comprehensive representations of resource use over time and in specific spaces can be developed. Such representations are called ResourceCultures here. As such, they are objective (fig. 1) in the sense that their description refers to concrete networks that exist in time and space and consist of materials and environments, actors, things, monuments and landscapes, knowledge, techniques and infrastructures, practices and orders. The two perspectives of ResourceComplex and ResourceAssemblage are used to identify different implications and meanings of ResourceCultures. On the one hand, these are intentional, anthropocentric and synchronic dimensions. On the other hand, there are dimensions of contingent, accidental, non-human, diachronic and dynamic action. Here, elements interact and form new relationships on the basis of their own agency. Such a dual approach opens up an understanding of socio-cultural dynamics over long periods of time and across space, and provides clues to the changing valuations and symbolic representations of things that people consider essential to their way of life. In this way, ResourceCultures enable an understanding of human responses to past and present cultural and environmental challenges, and are relevant to new ways of thinking about sustainability, resilience and long-term environmental planning for the future. They also address one of the major challenges in archaeological, anthropological and environmental research: how societies emerge, persist and change.

Thus, ResourceCultures are based on valuations of certain material and/or immaterial things in space and time, how they are handled, and the socio-cultural dynamics associated with them. A ResourceCulture develops its own inherent logic, which can be clearly distinguished from that of others, on the basis of networks around resources, which have two dimensions, the intentional and the contingent, captured by the analytical tools of ResourceComplexes and ResourceAssemblages. The socio-cultural dynamics to which ResourceCultures are subject are not purposeful or necessary, but multidimensional and open-ended.

The concept of culture underlying the concept of ResourceCultures differs from an essentialising concept of societies as homogeneous and static entities. In recent decades, the notion of self-contained cultures that are more than the sum of their parts and can be described independently has come under criticism. Studies of individual cultures were supplemented by studies of their interconnectedness (e.g. Barabási 2002; Hidasi 2017), although neither can fundamentally overcome the idea that it is 'cultures' that are interwoven. The perspective has changed, however, in that more attention has been paid to the interconnections between open and evolving social and cultural contexts, with a shift in perspective from macro-sociological issues to micro-, everyday- and cultural-historical issues, as well as infrastructures and their effects. Cultural units have been replaced by some researchers with studies on the cultural dimension (Kuada/Gullestrup 1999; see e.g. Wentink 2020), i.e. the effect of cultural attributions of meaning from different origins on the circulation of things, the attribution of identities (see excursus 'Identity'), or the legitimation of power relations (see excursus 'Power'). In this context, the earlier territorial attachment of the concept of culture to specific spaces has been supplemented by an interest in the connections of cultural elements across time and space.

For the concept of ResourceCultures, the foundations of symbolic anthropology are taken up, namely the concept of culture of 'self-spun webs of significance' going back to Weber and emphasised by Clifford Geertz (1973). Culture, in this sense, is an ongoing process of negotiating rules of coexistence and systems of meanings that are taken to be representative of a group, or that are retrospectively ascribed to a group as representative. In cultural studies, cultures are widely understood as learned and, to varying degrees, shared systems of meaning

Excursus: Religious Practices

The illustration shows a terracotta figurine from the 3rd millennium BC: a person standing in front of an altar makes an offering to a deity. This act of offering can be understood as an intangible resource (Hardenberg et al. 2017, 14) – just like the many other forms of religious practice.

Religious communities, past and present, have in common that parts of their respective religious conceptions are expressed through specific practices. The collective performance of such practices, based on shared religious ideas, strengthens communities and is an important resource for the formation, maintenance and renewal of religious identities.

Religious practices can take many different forms. Within this broad diversity of religious practices, sacrificial offerings form an important group (cf. Burkert 1972; Rappaport 1999). Various forms of sacrificial practices are essential for most ancient religious communities as well as for many contemporary religions. Food offered to goddesses or gods is often the object of such sacrifices (Kamlah 2012, 527–528). Offering rituals are acts by which members of religious communities symbolically provide their deity or deities with food. These symbolic acts are linked to divine blessings that the believers have already received or hope to receive in the future. In these religious conceptions, acts of sacrifice bridge the boundaries between two opposing spheres: that of the divine and that of the human. For such religious conceptions, the cult of sacrifice forms the basis of the human-divine relationship, and acts of sacrifice performed together strengthen the inner bond of communities.

This can be illustrated by the terracotta figurine shown in the illustration, which dates from around 2800 BC. Excavated from the Early Bronze Age town of Hirbet ez-Zeraqon in northern Jordan, it shows a person (possibly a priest) offering an animal to a deity. Hirbet ez-Zeraqon is a model example of a city of the first urban culture in the Levant (ca. 3000–2500 BC) and of Early Bronze Age Levantine religion (de Miroschedji 2011). The central importance of the religious practice represented here is evident from the urban plan of the Early Bronze Age city. The excavations have revealed a temple district in the centre of Hirbet ez-Zeraqon, at the highest point of the city, which contained two temples and a large open-air altar. This example shows that the offerings made in the temple precinct were a central resource for the inhabitants of the Early Bronze Age city in the formation and maintenance of their community.

Jens Kamlah

Fig. 2 Depiction of a religious practice ca. 2800 BC: a terracotta figurine (8.5 cm high) from northern Jordan shows a person offering an animal sacrifice to a deity (Photo: Institute for Biblical Archaeology, University of Tübingen).

that are empirically expressed in language, writing, texts, knowledge, objects, bodies, institutions, symbols and a wide variety of actions. Based on this, the concept of ResourceCultures describes an interrelationship of practices (social), valuations (ideational) and materialities (material environment) that shape identities. The elements of these worlds of meaning are interrelated and form the basis of practices (see excursus 'Religious Practices'), which in turn influence and change them. The systems of meaning emerge in relation to the material world. Therefore, a strict dichotomy between imagination and matter is inappropriate, since human beings are part of this material world. Thus, it must be considered that the contextually changing properties of resources have an impact on the emergence of meaning systems and the actions associated with them. Another aspect arises from the consideration that cultures are always contingent. Ideas and practices and their changes cannot be reduced to universal or natural conditions. They are interwoven with specific horizons of meaning and are subject to historical change. Under the condition of cultural contingency, comparisons in cultural studies make use of polythetic categories and are directed towards understanding cultural difference through contrast or towards exploring formal commonalities, such as similar processes or recurring principles of composition and structure (e.g. Ingram/Silverman 2016). The contingency thesis does not imply a denial of the importance of resources for human existential functions, but rather that resources are not viewed and handled according to universal laws. Rather, it should be emphasised that ResourceCultures interpret ways of dealing with resources as historically embedded cultural attributions of meaning. These conceptions of and ways of dealing with resources lead to dynamics involving social relations, entities or identities that are always already symbolically constructed. The relevant point is the assumption that these structured worlds of meaning, analogous to Reckwitz (2006), are contingent and deal with very different times and spaces. ResourceCultures are furthermore based on the assumption that the materiality of culture influences social reality. This is evident, for example, in the objectification of things, tools and instruments and the built and natural environments of ResourceCultures in which they exist.

ResourceCultures can thus be distinguished on the basis of resource use, with certain resources being more significant than others. Resources and their use can have meaning for the collective as a whole, for example, body and health in times of pandemics (e.g. Rodrigues et al. 2023; see excursus 'Health'). ResourceCultures, therefore, are what shape collectives through these processes that can be distinguished from other collectives. Comparisons between groups can therefore be used to identify differences or similarities between the resource practices of social groups.

A key feature of both cultures and societies is their dynamic nature, due to ongoing processes of negotiation and resulting actions and reactions. Three fundamental dimensions of human existence – time, space and imagination – have been considered as basic explanatory patterns for social and historical change. Time provides the framework within which events, processes and developments take place (Giddens 1984). Space is a physical and social construction (Harvey 1989). It organises social relationships, distributes resources and influences human interaction and its processes. Finally, imagination enables people to create alternative realities, explore new ways of living together and develop ideas for the future. It drives social movements, cultural and technological innovation and political change (Taylor 2004). Reflecting on the interactions between people and resources and how they can be analysed and explained, three dynamics can be identified that are of paramount importance to any ResourceCulture in terms of time, space and imagination. First, resources play an

Excursus: Health

Fig. 3 Health (Photo: Thambi Durai Thangavel).

This picture depicts the performance of a Hindu *puja* in June 2020, when wearing a mask was mandatory. The masks were also placed on the statues and offered to Ganesha for his blessing.

As the number of Covid-19-related deaths began to rise in March 2020, concerted public health measures were taken at local, national and global levels, with interventions in the public and private sectors. Overnight, lockdowns, curfews, quarantines and contact restrictions became the 'new normal' for citizens around the world. The effects were manifold – everyday routines were disrupted, and the resulting changes in living and working conditions and infrastructures had an impact on people's social and, above all, economic conditions (e.g. González/Marlovits 2020; Samuelsen/Toé 2021). The pandemic has made the world realise that health is a key resource on which peace and security, exchange and trade, social and personal relationships depend. Health is needed to keep society running, to provide essential services, to maintain infrastructures and to guarantee legal security. From the perspective of **ResourceComplexes**, maintaining a certain level of health during the pandemic required an elaborate network of services and material things, spatial reorganisations and technological innovations. In the case of Covid-19, this included medical and managerial staff, hospitals and care units, oxygen supplies and medicines, masks and vaccines, nurses and doctors, research and knowledge, all of which suddenly became scarce resources in the course of the pandemic. In addition, the cooperation of the population was essential to maintain a certain level of public health. Not everyone saw public health as the most important resource – in many countries, people demonstrated against Covid-19 policies and measures because they felt their constitutional rights were being violated. Individual freedom – e.g. to meet relatives, not to wear a mask or be tested, to be allowed to work – was prioritised over public health. The protests for and against the Covid-19 measurements show the multivocality in society about which resources – in this case, public health or individual freedom – are most important.

Health has also been theorised as an individual resource for everyday life (WHO 1986), a stock of biopsychosocial resources that influences people's ability to obtain sustainable livelihood options, but also enables participation in social activities and the pursuit of personal goals (Williamson/Carr 2009). Medical knowledge and healing techniques (e.g. Ayurveda, Traditional Chinese Medicine, Siddha, Unani, homeopathy or biomedicine) have always been an integral part of human societies and in all regions (Krause et al. 2012). Health is, has been and will continue to be a central resource in pandemic and non-pandemic times.

Gabriele Alex

important role in developments, second they initiate movements and third they are subject to valuation. The next step is to analyse and interpret ResourceCultures in more detail, considering these three main dynamics: developments (Chapter 3), movements (Chapter 4) and valuations (Chapter 5).

3 Developments

Martin Bartelheim, Jens Kamlah, Simone Riehl

Developments are understood here as open-ended dynamics that affect the structure of societies and social actors. They do not necessarily represent linear evolutionary changes. The focus is on socio-cultural processes in the management of specific resources, the basic thesis being that societies develop and change as a result of the management of resources, just as societal processes influence the management of resources (see also Laland et al. 2000). Important aspects include the links between resources and the social or political organisation of societies. This includes the question of whether changes in resource availability have triggered societal processes or, conversely, whether societal and political problems have led to different resource management strategies and economic structures. It also involves the necessary organisational and institutional preconditions of knowledge, technology, financial capital and practices for the extraction, distribution and consumption of resources, as well as the individuals (groups) who possess or control these preconditions.

3.1 Concepts on Developments

Structural change in societies has always been a central theme in archaeological and historical research, as well as in modern cultural studies. There has been close cooperation between these disciplines for decades. In contrast to archaeology and history, cultural anthropology is rarely able to trace processes of change over long periods of time due to a lack of sources (Sahlins 1985; 2000, 415–470). Modern cultural studies however, through direct contact with the protagonists, can capture aspects of socio-cultural systems and processes that archaeology misses due to its temporal distance from the events under study and the often strong filtering of its information base. In the study of periods with little or no writing, the use of analogy has been well established from the beginning (Currie 2016, 84), with two main schools of thought providing the models for these comparisons. Given the early recognition of the need to use models other than modern Western ones for the structure of societies distant in time, these are, on the one hand, the historical sciences (usually those relating to Europe and the Mediterranean) and, on the other, modern cultural studies, especially cultural anthropology. Classical studies, Egyptology and ancient Oriental studies, as disciplines with a high proportion of written historical records, have for a long time relied largely on statements in contemporary testimonies to describe social conditions and their changes. In recent decades, these disciplines have also tended to varying degrees, to adopt models of socio-political organisation or processes of social change from cultural anthropology.

In archaeology, the identification of hierarchies and their imagined development is usually the leitmotif in descriptions of societies and their change (Bartelheim 2023). For this purpose, social typologies developed in ethnology (especially Sahlins 1963; Service 1962; 1975; Fried 1967; Friedman 1982) have often been used, especially since the 1960s and 70s.

Despite their widespread acceptance, their use has not been without criticism, since on the one hand the representativeness of these types, which were developed on the basis of field research in selected regions, has been questioned and it has been noted that the definitional boundaries between them are blurred, making it difficult to transfer them to other societies (e.g. Eggert 2007, 268–269). On the other hand, it seems problematic to place such types, which should rather be seen as individual snapshots, on an imaginary evolutionary line, e.g. from 'big men'-systems to chiefdoms to states (Yoffee 1993, 63–72). The use of models based on social and political conditions in recent history in the archaeology of preliterate periods has also been controversial (Hayter 1994, 44–45). Although observations of processes with great temporal depth can be used as a basis, the question remains to what extent the transferability of conditions from one epoch to another is possible and meaningful (Freeman 1968, 262; Eggert 2007, 255–257). The basic tendency to assume a diachronic linear development of social structures from egalitarian relations to the formation of complexly organised hierarchies, which explicitly or implicitly prevails in large parts of research, has been confronted with alternative perspectives in recent years. For example, Kristiansen argued for more wave-like ups and downs in the formation of hierarchical structures during the Bronze and Iron Ages in Europe (Kristiansen 1998), or Zimmermann for cyclical movements in the cultural development of prehistoric societies (Zimmermann 2012). A plea for a broader view of the diversity of social relations in ancient societies than a fixation on vertical structures can be found in the conference volume 'Beyond Elites' edited by Kienlin and Zimmermann (2012).

The study of social structures is closely related to research on the foundations of their emergence, maintenance or change. In archaeology, the focus is usually on the economic and environmental conditions for the functioning of social communities. In particular, the use of resources, especially raw materials (e.g. Kenoyer 1998), agricultural possibilities (e.g. Boserup 1965; Clark/Haswell 1964; Nelson 1996; Renfrew 1978; Zvelebil/Lillie 2000) or exchange systems (e.g. Blau 1964; Bouchaud/Mézard 2000; Massa/Palmisano 2018) are often seen in close connection with the formation and preservation of hierarchies. A particular role is played by metal, the extraction, processing and distribution of which, due to the technological and organisational complexity of the processes required, as well as the desirability of the material, is widely assumed to have required centralised organisational structures and thus promoted the formation of hierarchies (summarised by Bartelheim 2007; Kienlin 2010; see excursus 'Materiality of Metals'). The intensification of extensive exchange networks supported the tendency to form vertical social structures by simultaneously limiting access to them within the participating societies (Kristiansen/Larsson 2005; Hansen/Müller 2011; Kristiansen/Suchowska-Ducke 2015). Analyses of socio-political processes, including those focusing on interactions with agricultural intensification, point in a similar direction. While no significant social hierarchies have yet been identified for the beginnings of agricultural activities in the context of human sedentarisation in the Old World (e.g. Lüning 2005; 2012; Bellwood 2022), this is often postulated for the subsequent development (e.g. Müller 2012). This extends to the formation of complexly organised and technologically sophisticated irrigation systems in riverine landscapes in arid environments, where the foundations are assumed for the emergence of larger, steeply hierarchically organised ruling units, e.g. states and empires, such as those of the Ancient Near Eastern and Egyptian empires, the Indus civilisation, or the state systems on the west coast of South America (the so-called 'hydraulic societies' of Wittfogel [1981]). Alternative models discussed for reconstructing the organisation of resource use are mainly forms of community associations, in the sense of kinship or neighbourhood systems, which do not necessarily presuppose centralised governance struc-

Excursus: Materiality of Metals

The materiality of metal as a resource becomes tangible through a manufacturing process that requires a network of other resources to make metal usable for humans. In addition to ore as the basic raw material, this network includes knowledge about the discovery, extraction and processing of metals, but also material things such as wood, leather and stone for the practical implementation of the manufacturing process. In addition, depending on the landscape and production conditions, there is the satisfaction of the basic needs of the producers (especially housing, clothing and food), the organisation of distribution and strategies for securing the infrastructure for transporting the raw material to the end consumer. These elements around the resource metal, their developments and contingencies, shape the group's **ResourceCulture**.

Such complex processes can be observed particularly well in prehistoric Europe with regard to the use of copper. During the Copper, Bronze and Iron Ages (4th to 1st millennium BC), it was increasingly and intensively used as a raw material to produce objects with practical uses (mainly as tools or weapons) or for ornamental and symbolic purposes. Due to its wide range of uses, the material was apparently soon valued as an important resource, as evidenced above all by its wide distribution, with sometimes long transport routes from the original deposit, and the often large quantities found (fig. 4). In addition, the material and the forms seem to have had a symbolic meaning. Copper was often mined near the surface in open-cast mines, but prehistoric underground mines are also known, such as the Arthurstollen at the Mitterberg (Austria), where ore was mined at a depth of about 200 m below the surface in the Bronze Age (Stöllner et al. 2006). The management of such a complex mining operation reveals extensive knowledge, which, as an intangible resource, was an essential component of the **ResourceCulture** around the metal copper. The technological and organisational diversity of the processes required to produce copper, as well as the attractiveness of the material, is widely believed to have necessitated centralised management structures and thus promoted the formation of hierarchies, with clear implications for socio-economic and political developments (Turck et al. 2019).

The valuation of copper as a resource in Central Europe is likely to have changed from the middle of the last millennium BC onwards, when iron (mainly in the form of steel) increasingly became the dominant material for the manufacture of tools and weapons, as its greater hardness made it more suitable for this purpose. Copper or bronze (alloyed with tin) was now used mainly for jewellery and other luxury items.

Martin Bartelheim

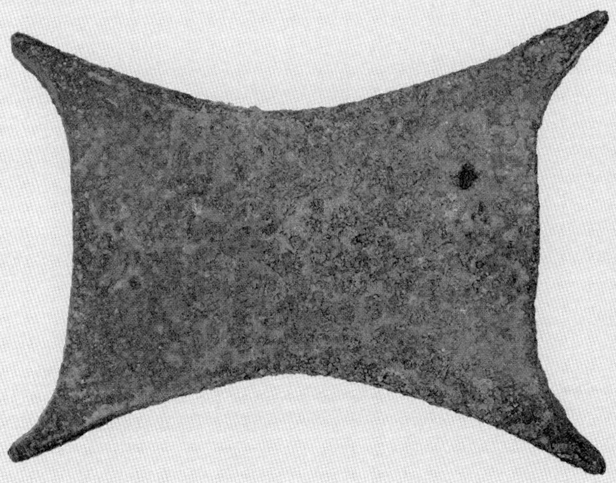

Fig. 4 A so-called 'ox-hide ingot' from the shipwreck of Ulu Burun, Turkey (14th century BC). Length: 54 cm; width: 33 cm. This is a shape with a weight of ca. 30 kg into which raw copper was cast to be transported from the production site to the final consumer (Photo: Martin Bartelheim).

tures. This highlights the need to understand the complexity of societies not only in terms of creating the most extensive power relations, but also to look for motivations of social interaction other than mere prestige and personal influence (Rowlands 1971; Roscoe 2000; Kohring/Wynne-Jones 2007; Kienlin 2012; Bartelheim 2022).

Differential access to resources is thus widely seen in ancient, historical and modern cultural studies as a cause of cultural and social change. Resources can provide a reciprocal category of analysis based on the aforementioned basic thesis that societies develop and change as a consequence of the use of resources, just as social processes affect the use of resources. In doing so, they offer a broader perspective on socio-cultural processes and their interpretation, and illustrate that such developments are not straightforward, linear and goal-oriented, but multidimensional.

By analysing the use of resources in the context of development, it is possible to move beyond a narrow focus on ecologically deterministic perspectives, narratives of 'crisis' and evolutionary understandings of culture, and to focus on other perspectives. In particular, the analysis of diachronic development trends can illustrate the interplay between resource use and the formation, preservation and transformation of cultural and social identities. By combining the methodologies of archaeology, history, cultural studies and the natural sciences, it is possible to shed light on the influence of both social and natural factors on these processes. In the following, therefore, the relationship between resource use and social development will be examined primarily from the perspectives of 'social dynamics', 'multidimensionality of historical processes' and 'development in landscapes' in order to understand resource-oriented developments.

3.2 Social Dynamics

Social dynamics are understood here as processes that lead to transformations of social states and orders in response to internal and external changes. These include, above all, processes of change in social structures and institutions, in business communities and in the differentiation of social positions and lifestyles of individuals, both in the short and long term (Smelser 1992). The causes of change are mostly political, economic, cultural and natural factors. Networks play an important role in social dynamics (White 2003), i.e. structures of actors whose individual behaviour is influenced by interactions with others and their behaviour.

Cultural change is closely related to social change, which can include the elements mentioned above, but also describes changes in perceptions, ideas, values and ideologies that can feed back into social processes. These interdependencies are expressed below using the term socio-cultural to illustrate that perceptions and the use of resources create dynamics that affect social relationships, entities or identities that have always been symbolically constructed.

3.2.1 Constancy and Change in Patterns of Use

Constant patterns of resource use make social and cultural change particularly visible. The analysis of resource use patterns allows us to both identify persistent resource use strategies and examine the related societal dynamics from a *longue durée* perspective, enabling us to describe long-term social development trends. In many cases, a close correlation can be observed between continuous developments in resource use and socio-economic dynamics.

One such constant pattern of resource use can be observed in the south of the Iberian Peninsula in the lower Guadalquivir Valley and the neighbouring Sierra Morena during prehistory (3rd to 1st millennium BC). There is an assemblage of regionally typical agricultural, faunal and inorganic resources that, despite some modifications at the transition from the Chalcolithic to the Bronze Age (end of the 3rd millennium BC), indicates an associated long-lasting agropastoral ResourceCulture (Bartelheim et al. 2022). It is characterised by the use of fertile arable land in the valley, pasture in the Guadalquivir Valley and the mountainous areas of the Sierra Morena, all practised with varying intensity. In addition, there are intensive linkages within the region and the use of the rich ore deposits of the Sierra Morena as a resource in temporally varying forms and intensities (Bartelheim 2007; Arboledas Martínez et al. 2022; Chala-Aldana 2022; 2024). The establishment of the forest pasture system of 'Dehesas' – a traditional form of managed woodland for agropastoral use in south-western Iberia and the northern Maghreb – in the region since the 3rd/2nd millennium BC (Stevenson/Harrison 1992) allowed an effective use of the landscape in the form of a combination of field cultivation and animal husbandry, or the keeping of different animal species such as pigs, sheep/goats and cattle. Particularly for the more recent prehistoric, Roman and medieval periods, seasonal mobility can also be assumed to have ensured an adequate supply of livestock and landscape use (Bartelheim et al. 2022).

The archaeological expression of the agropastoral ResourceCulture is, on the one hand, the uniformity of the find record since the Chalcolithic over long periods of time of this use pattern specifically oriented to local ecological conditions. On the other hand, it is the long continuities in the settlement areas as an expression of constantly evolving socio-cultural dynamics. Some changes can be observed in the course of the arrival of the 'Phoenicians' on the southern coast of the Iberian Peninsula during the 1st millennium BC (Aubet 2001; López-Ruiz/Doak 2019). Given their economic and cultural practices, which were very different from those of the local population, and their markedly different interest in raw materials and agricultural products, an intensification of agricultural and metallurgical production can be observed (González Wagner/Alvar Ezquerra 1989; Pérez Macías 1996). In some places, there is also evidence of a shift in economic emphasis, for example from agricultural to metallurgical (Bartelheim 2007). Most importantly, however, the continuity of settlement patterns and settlement locations at mostly fortified elevations argues for a relatively small impact of these transformations on the social dynamics of the region (Escacena Carrasco et al. 2018; Ferrer Albelda et al. 2022). It was not until the establishment of Roman rule in the region in the last two centuries of the last millennium BC, with the subsequent profound changes in the settlement and economic system (Schattner 2025), that the southern Iberian Peninsula underwent a significant caesura and was incorporated into a significantly altered socio-economic structure of external origin.

3.2.2 Resources as the Basis of Socio-Cultural Dynamics

Provided that at least some information (usually in written form) on the political structure of a society is available, it is possible to examine the role of resource use in socio-cultural development processes. In most cases, it is possible to identify resources that play a central role as foundations and catalysts in such developments. These resources can be mobile, immobile or even intangible. For example, the excavation of Bronze Age palaces (2000–1200 BC) in Syria yielded many metal artefacts, primarily made of gold, silver and copper or bronze. In-depth studies characterising this material corpus have shown that precious metal objects, especially gold, functioned to represent the self of the royal elite and to shape their relationships and practices (Puljiz 2021). Gold served primarily as a means of social demarcation and as an element in the creation and maintenance of power. Objects made of this material were involved in a wide range of contextual references and social practices. These cover the whole range of courtly and elite activities. Gold was a means for actors to gain access to the courtly elite and to document it for the outside world. This points to social practices such as 'conspicuous consumption' (Veblen 1899), which were of great importance for the socio-political function and mode of action of the ResourceCulture of gold during the 2nd millennium BC.

Gold can be seen as part of a ResourceCulture with numerous components, which was structured, controlled and modified by the actors according to their needs. The primary importance of gold in the Syrian palaces of the 2nd millennium BC was to express social hierarchies and to consolidate social identities through a specific cultural valuation of the material as part of the royal sphere (Puljiz 2021). In this role, the resource served to differentiate and manifest the social order and perpetuate socio-political hierarchies. When political conditions changed, as they did at the transition to the Late Bronze Age, when the Syrian kingdoms lost their autonomy and fell into vassalage relationships with the Mitanni and Egyptian empires, the ResourceCulture of gold changed as well. Restricted political access to gold led to a shortage of this socially and symbolically valuable resource. New techniques were developed to compensate for this scarcity, such as the alloying of gold with silver or the diffusion gilding of silver objects (Schwab/Pernicka 2021; Puljiz/Schwab 2021), which led to a modification of the gold ResourceCulture, since silver and the corresponding working techniques to gild it also had to be incorporated.

In another example, medieval castles in south-west Germany, as immobile material goods, represented a resource for the formation of rule by the lower nobility as crystallisation points for the exercise of rule from the first possession of castles by ministerials in the 12th century to the establishment of independent dominions (Froehlich/Weidenbacher 2020; Froehlich 2023). Thus, a complex landscape of castles of lower nobility developed around the centres of power of the higher nobility, which enabled the owners of ministerial castles to participate in the lifestyle of the higher nobility. This was based on the possibility to use a set of resources, which can be seen as a ResourceComplex, ranging from the castles themselves, the income from economic activities, the securing of property to the exercise of power.

Especially in connection with monasteries, castles were part of the aristocratic ResourceCulture in the creation and development of high aristocratic rule. While from the 11th century onwards the nobility began to use the castle as the focal point of spatially manifested rule and at the same time to focus their understanding of dynastically defined ruling families from here, the monasteries founded in the vicinity of the ancestral castle that gave the dynasty its name took on the function of family burial grounds and thus as a focal point of dynastic *memoria* (Dendorfer 2017). In many respects, the castle and the monastery functioned together as a manorial unit, the parts of which can be analysed as a ResourceComplex. Castles served not only as the nucleus of aristocratic rule, but also for military control of an area, for securing and developing economic resources, as an administrative hub, as a centre for trade and crafts, as a symbol of rule, as a stage for aristocratic habitus and also as a sacred centre.

In addition to their spiritual function, monasteries were often used to secure land and lordship, not least as an investment in the founder's salvation. Land was protected from encroachment by neighbours, and the donating noble family was able to rule the area more effectively. Land was often permanently linked to a main line of the dynasty (Andermann/Bünz 2019). In terms of personnel, the appointment of family members to leadership positions provided them with useful and powerful positions and secured their own clientele access to spiritual benefices and thus also economic resources (Spieß 1993; cf. e.g. Geiß 2018; Diekjobst/Hohlstein 2018). This significance as a ResourceCulture is the only way to understand the large number of monasteries founded by the nobility and the enormous estates associated with them.

3.2.3 Network Formation

Networks are relevant to the discussion of the development of societies primarily because such constellations of knowledge, techniques, people, places and objects/media must interlock in a certain way in order to make resources usable for a community (see Chapter 2; Klocke-Daffa 2017, 255; Da Vela et al. 2023). Functional and intentional relationships in networks can be described using the analytical perspective of ResourceComplexes. However, both networks and ResourceComplexes are only partially planned by humans and under human control. What is considered and used as a resource is subject to constant change, new human and material contextualisations and contingency-based attributions of value and meaning. In a diachronic perspective, such processes are described by the term ResourceAssemblage, which is intended to analytically capture the unpredictable or unstable, also in the development of networks (Bartelheim et al. 2021a).

Network structures around the use of stones (especially chlorite) can be observed in connection with crafts, trade and agriculture in south-eastern Iran. There, they ensured the extraction and processing of stones as well as the functioning of interregional exchange systems and the nutrition of the population in changing ways and over longer periods of time since at least the 4th millennium BC. In particular, from the 3rd millennium BC onwards, the extraction and distribution of chlorite can be traced from the Bagh-e Borj Mountains across the Jiroft Plain to the ports of the Persian Gulf (Pfälzner/Alidadi Soleimani 2017; Pfälzner et al. 2019). From there, the stones reached Mesopotamia via the Arabian Gulf, where they were apparently valued and used primarily as material for high-quality art objects. The Early Bronze Age extraction and transport of the stones is associated with the so-called Jiroft culture, which maintained a complex system of exchange relations in south-eastern Iran, consisting of river valley agriculture and pastoralism, characterised by regional mobility, especially between the Gulf coast and the Iranian highlands, as far as Central Asia and even Baluchistan and India (Karami et al. 2020). Understanding this as a ResourceComplex allows us to analyse the social dynamics associated with it, most notably the formation of the settlement centre at Konar Sandal in Jiroft and a dense accumulation of settlements in the 3rd millennium (Madjidzadeh 2008).

The network of stone use within the Jiroft culture collapsed completely at the end of the 3rd millennium BC. The reasons for this collapse are still unknown. This was followed by a period of 1,500 years during which the region remained uninhabited. In the Achaemenid period (6th to 4th century BC), the reconstruction of the network in the Jiroft region begins. Structures similar to those of the Early Bronze Age emerge, with similar settlement locations and densities, as well as similar routes and economic activities. Chlorite also seems to have become popular again during this period and circulated within the Achaemenid

Empire. In conjunction with the network, socio-economic dynamics comparable to the Early Bronze Age are evident in south-eastern Iran, as evidenced by similar settlement densification and the establishment of fortification systems (Pfälzner et al. 2019). The newly established networks and settlement systems show strong continuity through the Hellenistic, Parthian and Sassanid periods and into the early Islamic period. In the early Islamic period, an important route from Jiroft to the Persian Gulf, marked by numerous settlements, explains the supraregional importance of the medieval port of Hormoz (Pfälzner et al. 2019).

From a diachronic perspective, the resource-use related ResourceCulture of the Jiroft region is characterised by discontinuities, even if it is stable over long periods of time. The activation or reactivation of networks follows different modalities. However, the outcome is strikingly similar in the cases observed. This can be deduced from the basic structural similarity of the settlement systems, the similar path connections and the similar resources. Settlements and pathways are important resources in themselves, representing the nodes and linkages of the activating human networks. They are linked to other resources such as mineral raw materials, various other trade goods and agricultural products. Water is also an important resource in this system. This illustrates the complexity and high degree of integration of the ResourceCulture that developed here at different times.

Similarly, a comparison of Bronze Age, historic and modern ways of dealing with the landscape in the agriculturally favourable valley of the Guadalquivir and the neighbouring arid mountain zone of the Sierra Morena in southern Spain shows that networks have been the crucial element of effective land use at all times. These operated primarily through various forms of mobility, be it transhumance, the transport of goods or migration (see Chapter 4). Even if the region's eventful history, especially in the last 2,000 years, has led to some breaks (Bartelheim et al. 2022), the connectivity that played an important role as early as the Chalcolithic and Bronze Age (3rd/2nd millennium BC) is mainly regional (Chala-Aldana 2022; 2024; Díaz-Zorita Bonilla et al. 2022). It was of major importance in combining the benefits of stationary economic activities such as agriculture or ore mining/metallurgy with mobile components such as livestock or the exchange of raw materials or finished products.

On the basis of much more extensive networks, the 'Phoenician' West developed during the period from the 9th to the 6th century BC in the central and western Mediterranean region as part of processes of spatial development and identity formation (Aubet 2001). In this context, for example, evidence from the oldest settlement layers of Carthage from the 9th and 8th centuries BC has shown that there was a widely ramified system of contacts with other 'Phoenician' settlements in both the western and eastern Mediterranean, and that the 'Phoenician' development of the West built on older, central Mediterranean, transmaritime networks of the Late Bronze Age and Early Iron Age, which were integrated from the outset into regionally differentiated ResourceCultures (Schweizer/Schön 2018; Schön/Töpfer 2016; Flügel et al. 2018).

Among the western 'Phoenician' settlements, Carthage developed into a dominant city and successively into a hegemonic power over the Punic settlements in the west during the 6th century BC. The western 'Phoenician' ResourceCulture seems to have favoured processes of centralisation and the formation of a Carthaginian aristocracy. The consolidation of Carthaginian hegemony can be seen around the city and on the Punic islands of the central Mediterranean. This was achieved on the one hand through the development of agricultural resources in the context of an intensive villa economy supported by the aristocracy, but also through the mining of ore and stone raw materials, with regions now understood as Carthaginian being secured by fortifications and new settlements (De Vincenzo 2013; Delile et al. 2019; Schäfer 2019). Within this process, the western 'Phoenician' ResourceCulture seems to have favoured the formation of a Carthaginian citizenry and aristocracy through the centralisation of older networks, whose membership now itself became an important resource for Carthaginian society. Furthermore, it is clear from the archaeological record of Carthage that the orientation of the older networks towards the metropolis triggered a series of urbanistic processes that were forced by the land-owning aristocracy and can be read in the archaeological record, for example in the expansion of the city, the densification of urban development or the construction of industrial areas, which in turn posed supply problems for the growing urban population (Bockmann et al. 2018; Flügel et al. 2018; Schön 2020a).

The integration of the formerly Carthaginian-dominated territories into the Roman Empire, which began with the conquest of Carthage during the Third Punic War in 146 BC, meant above all a realignment of the existing local social and governance systems with Rome, but by no means the end of Punic culture. The Punic language, Punic cults and magistrates persisted well into the imperial period, as did the networks established in Punic times (Hobson 2019).

The analysis of diachronic developments in the social sphere shows that constant processes are usually accompanied by similar trends in resource use. Interruptions also tend to lead to changes in the forms of use and the associated networks behind them. In the case of non-linear, contingent developments, it becomes clearer that resource-oriented processes often have a multi-dimensional character and are characterised by changes in perception and evaluation, which is why they do not follow clearly recognisable regularities.

3.3 Multidimensionality of Historical Processes

The use of resources includes the development and extraction as well as the processing, distribution and use of socially relevant resources. It triggers certain dynamics, i.e. multidimensional processes of change, which affect individual sectors of society or society as a whole. ResourceCultures are not static, but are subject to specific socio-cultural dynamics. These dynamics are not directed or necessary, but multidimensional and open. Societies develop and change through the use of resources, just as social processes affect the use of resources. 'Multidimensionality' emerges as an essential feature of historical processes of resource use, because many historical examples show that these processes are not straightforward and linear.

3.3.1 Changes of Social Relations and Identities

Research on ResourceCultures focuses on the cultural dimensions of resource practices. In this context, the cultural studies perspective on resources emphasises the cultural dimension of meanings, values and ideas that – in the context of the construction and reconstruction of identities – turn something into a resource (see Chapter 5). The meanings that make something a resource change over time or vary regionally. Resources and groups of actors thus constitute each other in specific temporal and spatial contexts. It can therefore be assumed that resources are culturally and temporally variable and thus subject to historical change.

Historical developments, in which resources have been used as means to create, maintain or change social relations and identities are characterised above all by the fact that they are multilayered and multidimensional. This is particularly evident in analyses of long-term developments in resource use, such as studies of Bronze and Iron Age agriculture in the southern Levant (Kamlah/Riehl 2021). These cover a period of almost 3,300 years (ca. 3600 to 330 BC) and provide a multifaceted picture of how the interactions between agricultural resources and peasant actors evolved over millennia. The ways in which agriculture and animal husbandry affected the social relations and identities of peasant groups in the southern Levant over time – and how these actors affected resources – can be studied through the interaction of archaeological, archaeobotanical, zooarchaeological, iconographic, epigraphical and exegetical methods (Vermeersch et al. 2021). Interdisciplinary studies such as these clearly show that the dynamics of agricultural resource management do not allow for monocausal explanations, nor do they follow linear developments. Rather, they are open-ended and multidimensional development processes.

In the different sub-regions of the southern Levant, different forms of agricultural use have developed even in phases of constant natural conditions. The temporal variability was partly influenced by cultural decisions and partly by historical-political developments. For example, viticulture, which has been continuously present in the inland hill country since the Early Bronze Age (from ca. 3600 BC) and throughout the Bronze (Kamlah/Riehl 2021, 198) and Iron Ages, was an identity-forming component of the southern Levantine family-based small-scale farming system. This is evidenced, among other things, by numerous symbolic valuations of wine attested by Old Testament texts for the Late Iron Age (Orendi 2020). On the other hand, the expansion of wine-growing areas into the southern Levantine coastal plains and into the hinterland of the port cities, as it occurred both in the Late Bronze Age (ca. 1500–1200 BC) and in the Iron Age IIB–C (ca. 800–550 BC), was influenced by historical-political and economic-historical developments, which caused significant social changes in both periods. The triggering factors were, on the one hand, Egyptian domination during the Late Bronze Age and, on the other hand, the Mediterranean wine trade and its exchange networks, which developed from the 8th century BC onwards. Egyptian domination of the southern Levant during the Late Bronze Age led to agricultural overproduction, as agricultural products – including large quantities of wine – were brought to Egypt from the southern Levant. A second wave of wine overproduction can be observed for the southern Levant in the Iron Age IIB, this time not caused by the conditions of a foreign great power, but by the dynamics of the Mediterranean wine trade (Orsingher et al. 2021).

3.3.2 Globalisation versus Regionalisation

The multidimensionality of historical processes is particularly evident when they involve diametrically opposed tendencies. Even historical periods of fundamental change, which superficially appear to be the simple collapse of an existing culture, turn out to be multifaceted processes when analysed more closely from the perspective of resource use and Resource-Assemblages.

A frequently discussed example of profound historical change is the transformation and disintegration of the Levantine state world during the transition from the Late Bronze Age to the Iron Age. Numerous explanations have been offered for this change (Cline 2014; Millek 2019, 29–87). The conventional view of this change is dominated by the idea that it was driven by the collapse of international trade, which led to the collapse of Late Bronze Age cultures in the Levant. This view proves problematic when transregional exchange is analysed as a resource and when developments in the use of this resource are examined (Kamlah 2016; Maeir 2023). The intensity of transregional exchange, which was high during the Late Bronze Age (ca. 1500–1200 BC), declined markedly at the end of this period, but long-distance trade did not collapse completely. For example, during Iron Age I (ca. 1200–1000 BC), numerous finds of Egyptian amulets and cedar imported from the Lebanon Mountains are known for the southern Levant (Kamlah 2016, 210–216).

Furthermore, it is evident that developments in the use of transregional exchange differed between sub-regions and localities in the southern Levant (Kamlah 2016, 215; Millek 2019). For example, the use of iconographic media as a resource in the southern and northern Levant shows partially opposite trends. While the importance of cylinder seal images declined after the end of the Late Bronze Age in the southern Levant, a rather different development can be observed in the ResourceAssemblage of architectural sculptures and reliefs in the northern Levant (Herrmann 2019): here, in the Syro-Hittite states of Iron Age I, the iconographic-medial recourse to the period of the Hittite Great Empire gained great importance in the sense of a resource for the formation of collective memory (see excursus 'Collective Memory'). Such resource-based analyses show that different forms of connectivity and exchange were formative factors for the various societies of the Late Bronze Age and Iron Age Levant. They reveal divergent developments for the transition period from the Late Bronze Age to the Iron Age, showing tendencies towards both globalisation and regionalisation, and thus revealing different types of Levantine ResourceCultures. As such, they attest to the multidimensionality of historical processes in the eastern Mediterranean region at the end of the 2nd and beginning of the 1st millennium BC.

3.3.3 Changes of Cultural Exchange Networks

Cultural exchange networks undergo multidimensional processes of transformation in their resource function. This can be seen, for example, in the long-term development of settlements in a large cultural region in south-eastern Iran. Over a long period of time, from the Neolithic through the Bronze Age, Classical and Islamic periods to the present day, fundamental changes in settlement patterns and resource use have occurred in the region. The area encompasses several geographically and ecologically diverse sub-areas, including the Jiroft Basin, adjacent plains and upland areas, suggesting diversified resource use. Consequently, the area provides an excellent opportunity to observe developments and changes in resource use over nearly 10,000 years. Archaeological investigations, based on a large-scale archaeological survey and accompanying cultural anthropological case studies, show how resource use in the region is still subject to multidimensional developments today. Restructuring of settlement systems and associated resources has occurred repeatedly in the area over millennia (Karami et al. 2021, 278–286). From a *longue durée* perspective, it becomes clear that several aspects play an essential role during each cultural period: the internal structuring of regional networks in the region to satisfy basic needs (i.e. resources that are fundamental to the livelihood of the society concerned), the nature of changes in the networks and the resources around which the networks are built and the region's connection to supraregional networks. Within the area, the same type of resources (stones such as chlorite, diorite and calcite; agriculture; livestock; road networks) have been important for regional socio-cultural development in completely different periods. From an analytical perspective, however, specific Resource-

	Perturbation	Relocation
Inceptive	People initiate change in their selective environment by physically altering their surroundings, e.g. clearing forests to create agricultural fields	People expose themselves to a novel selective environment by moving to a new location, e.g. moving to other places in the expectation of acquiring more resources
Counter-active	People counteract earlier environmental change by physically altering their surroundings, e.g. by building irrigation canals to prevent crop failure due to drought	People react to environmental change by moving to a more suitable location, e.g. by emigrating to another region to escape resource scarcity

Fig. 5 Examples of the four categories of niche construction adapted from Laland/O'Brien (2010) (Graphic: Simone Riehl; Graphic design: Richard Szydlak).

Cultures emerge for each period, which differ considerably in their dimension and distribution. In their multidimensional developments they can be analysed as ResourceAssemblages that make processes of social and cultural change, collapse and restitution comprehensible.

3.4 Developments in Landscapes

The perspective of continuously developing ResourceAssemblages in changing landscapes (see excursus 'Landscape'; see also 'counteractive behaviour' in fig. 5), or in other words, non-linear human socio-cultural evolution in changing environments, is reflected in the idea of 'niche construction theory' (NCT), which has been discussed since at least the 1980s.

Originally derived from evolutionary biology, NCT emphasises the ability of organisms to modify natural selection in their environment and thereby act as co-directors of their own and other species' evolution (Lewontin 1983). In recent decades, however, NCT has increasingly recognised the role of culture in human evolution (Odling-Smee et al. 2003; Laland et al. 2000; Laland/O'Brien 2010). This is particularly relevant where our species appears to have engaged in extensive environmental modification through cultural practices. Proponents of these concepts suggest that events such as the transition to novel food sources with the advent of agriculture (human-plant co-evolution), or the dispersal of human populations into new environments with different climatic regimes, were self-imposed (Borenstein et al. 2006).

Landscapes, as part of the environment, are an essential component of cultural niche construction. Considering the four categories of niche construction described by Odling-Smee et al. (2003), inceptive or counteractive perturbation, inceptive or counteractive relocation, they all have in common that they are co-determined by human perception as a societal reflection of their intentional worlds (fig. 5), in the sense of Ingold (2020). The concepts of movements (see Chapter 4) and valuations (see Chapter 5) consider the perception and valuation of landscapes in different human societies as a factor of local cultural niche construction, societal change and regional movement, including feedback mechanisms leading to mutual transformation of landscapes and societies, while the concept of developments (this chapter) emphasises the long-term trends in the development of ResourceCultures in different landscapes and their interactions within landscapes through particular Resource-

Excursus: Landscape

Landscape is a concept that is often misinterpreted. It may therefore be useful to begin by stating what landscape does not refer to, namely the natural environment (Crumley/Marquardt 1990). Landscape is not something that surrounds people, but refers to the interplay of human impacts and spatial potentials (Filippucci 2023). This difference is more fundamental than it seems at first glance: because the concept of landscape focuses on the interplay of things rather than on the thing itself, the term is relational rather than material. In other words, landscape describes a relationship that unfolds in a particular area, not the area itself. An example may illustrate this: the iconic painting 'Wanderer über dem Nebelmeer' by Caspar David Friedrich shows the back of a walker looking out over misty mountain peaks. It is an example of a human looking at nature rather than dwelling in it (Ingold 1995). This externalising understanding of nature as 'something else' has long been exposed as a cultural construct. Anthropologists such as Philippe Descola or Eduardo Viveiro de Castro, in their studies of Amazonian societies, have shown that this way of being in the world is only one possible ontology among others and by no means a universal way of perceiving the world (Castro 1998; Descola 2011). This idea may have its roots in the Neolithic period and has since been manifested in identity-constituting mass media such as the Bible: '[...] fill the earth and subdue it; and have dominion [...] over every living thing that moves upon the earth.' However, it must be recognised that this questionable way of being in the world has been globalised to such an extent that it can be seen as one of the major problems of our time, since it contributes greatly to current challenges such as climate warming, pollution, etc.: the idea of nature as something unrelated to oneself seems to make it difficult to deal with it responsibly. In the humanities, landscape is therefore understood as a space of interaction between natural-environmental possibilities and human responses to them. These responses can be diverse and reflect the values and meanings inscribed in a landscape, which influence not only how resources are used, but also what becomes a resource in a given landscape. We then interpret these relationships as different **ResourceCultures**. One example is an indigenous group of former pastoral nomads in south-eastern Iran who have recently begun to mine mineral resources for industrial purposes (Frauen/Klocke-Daffa 2023). This changed way of life represents not only a shift in resource use, but also a different way of being in the world, creating a new landscape for those in it. In another example, Da Vela (2022) conceptualises the northern Apennine Mountains during the Iron Age as a 'memoryscape' in which space, memory, culture and emotion coagulate into an identity-constituting entity. In this case, the analysed landscape itself becomes a resource.

Wulf Frauen

Fig. 6 The interplay of human impacts and spatial potentials that creates a landscape. In this case, it is not just wind turbines, but also sheep on parceled-out pastures, a small group of trees and a tractor working on the farmland (Photo: Molgreen, CC BY-SA 3.0).

Complexes and ResourceAssemblages, i.e. more specifically the analysis of different modes of niche construction.

Above all, the unifying concept was that landscapes become resources only within the framework of specific cultural perceptions of human groups and are thus strongly linked to social processes (Chapter 3.4.1). This principle permeates all modalities of landscapes, including their direct use for the extraction of raw materials (Chapter 3.4.2) or more complex processes subordinated to humans (Chapter 3.4.3).

3.4.1 Landscapes as ResourceAssemblages

Landscapes can be viewed as ResourceAssemblages, composed of specific elements such as water, soils, pathways and resource deposits, or as socially constructed entities such as agriculture, economy, exchange and trade networks that link different landscape types, often including the diversity and evolution of perceptions.

A vivid example of the role of landscapes as ResourceAssemblages is the diachronic comparison of the use of different landscape zones of the southern Iberian Peninsula (see Chapter 3.2.1). A simplistic association of specific natural conditions with analogous subsistence forms, such as the agrarian prosperity of the Guadalquivir Valley with sedentarism and agriculture, or the arid conditions of the mountainous interior with higher mobility and pastoralism, is not supported by the archaeological evidence and must be modified to include specific socio-economic dynamics that are strongly based on networks (Bartelheim 2022). In this landscape, socio-economic and socio-cultural aspects form a very specific ResourceAssemblage that evolves through individual and mutual changes in these parameters. These are not necessarily related to temporal changes, as has been demonstrated for the continuous agropastoral ResourceCultures of the southern Iberian Peninsula throughout the Chalcolithic and Early Bronze Age. Furthermore, an ethnological study of the historical development of the Dehesa-type economy shows that there are considerable regional differences in how people of different traditions perceive a very similar landscape type (Bartelheim 2022). The discrete perceptions of the inhabitants of the Dehesas of the Sierra Morena compared to those of Extremadura are based on socio-political diversity, including different intentional worlds (*sensu* Ingold 2020). While the inhabitants of the Dehesas of the Sierra Morena maintain unique systems of pig breeding and cork production that involve private economic disparities, the inhabitants of the Dehesas of Extremadura are part of a more complex system that involves collective rights of use. Such historically grown differences determine different intentional worlds and perceptions, which co-determine the measures taken to preserve the landscape.

In archaeology, visualisation is an important tool for analysing differences and developments as ResourceAssemblages at the landscape level. The parameters documented for this purpose are often geographic in nature, such as diachronic landscape models for the Jiroft region of south-eastern Iran based on settlements, resource deposits, hydrogeology and pathways, which can help to elaborate spatial relationships and networks. Another example is the study of differences in agricultural ResourceAssemblages as perspectives based on archaeobotanical evidence at different sites in the southern Levant. Such an analysis made it possible to specify the well-known phenomenon of the decline of emmer wheat cultivation from the Middle Bronze Age onwards in favour of free-threshing wheat, which dominated the wheat spectra from the Late Bronze Age onwards and can be described in terms of NCT as a hitherto unknown form of perturbation (Riehl 2009). Map visualisation suggests that, despite these general trends, emmer wheat remained an important crop species in the Jesreel Plain, a region of multiple importance as an agronomically productive landscape as well as a key route connecting Egypt and Syria (Via Maris) (Nicolì 2024). The continuous and exclusive dominance of emmer wheat over free-threshing wheat in the Jesreel Plain may emphasise a strong link with Egypt, where emmer wheat was also the preferred wheat species (Nesbitt/Samuel 1996; Samuel 1993). Furthermore, the diachronic visualisation of cereal assemblages suggests a strong relationship between climatic dynamics and agroecological requirements of the plants. Comparatively moister conditions during the Early Bronze Age II–III, in contrast to the Early Bronze Age IV, are visible in the southernmost regions with a shift from wheat to barley, whereas especially during the relatively humid Iron Age I, free-threshing wheat strongly increases north of 31.8° latitude. Such human-induced multiple spatial and diachronic developments reflect well the role of landscapes as ResourceAssemblages.

3.4.2 Changes in Techniques and Strategies of Landscape Use

Changes in techniques and strategies of landscape use as part of cultural niche construction fall mainly into the area of inceptive or counteractive perturbation (fig. 5). The use of landscapes for sustainable subsistence is closely linked to people's decisions about where and how to settle. Strategies of landscape use are linked to natural or man-made frameworks, as for example in the 3rd millennium BC in south-eastern Iran through a continuous abandonment of a microclimatically advantageous transhumant lifestyle in favour of an industrial lifestyle, a shift from pastoral landscape use to mining based on permanent settlements (Pfälzner/Alidadi Soleimani 2017; Pfälzner et al. 2019; see also Chapter 3.2.3).

Long-term developments in landscape use against the background of settlement patterns are also evident in the case studies of the Guadalquivir Valley. The archaeological evidence suggests that settlement activity shifted to higher altitudes at the beginning of the Bronze Age, possibly for several reasons, including strategic objectives as well as an increase in aridity at the end of the 3rd millennium BC, representing a form of counteractive relocation in the construction of cultural niches (Bartelheim et al. 2021c; see fig. 5). Nevertheless, the previously inhabited landscapes continued to be exploited, as suggested by stable strontium isotope analysis of human and animal bones from nearby regions (Díaz-Zorita Bonilla et al. 2022).

Intensive soil management as a characteristic of agricultural societies is particularly relevant with the introduction of the plough and its increasingly systematic use, as increased tillage sometimes leads to higher yields when perennial weed species are eliminated. Zooarchaeological evidence from archaeological sites in the southern Levant suggests that the use of cattle as part of animal husbandry increased during the Early Bronze Age I and may be related to the use of the cattle-drawn plough. This would have allowed larger areas to be cultivated, and the associated increase in agricultural production may have been an important factor in the emergence of the first urban culture in the southern Levant. For centuries after the Early Bronze Age, simple wooden ploughs drawn by cattle served as important tools for preparing the soil for sowing in the southern Levant. A change can only be observed in the later Late Bronze Age and the Early Iron Age, when the advent of iron allowed for technical innovation: the tip of the wooden plough could now be reinforced with iron ploughshares. The distribution of archaeological finds of such ploughshares shows, on the one hand, that they were used in regions where the stony terrain made them particularly useful. On the other hand, they also occur in regions where agriculture was particularly intensified by systematic cultivation during the Iron Age II, namely in the Jezreel Valley and the Shefela.

Land use dynamics are also linked to ancient people's perceptions of the landscape, which are reflected in specific land use practices within the ResourceComplex soil. Using biomarker and phytolith analysis as well as anthracology, Henkner et al. (2017) were able to show that the slopes of the Baar region (South Germany) were mainly used for animal husbandry and forest management during the Late Bronze and Early Iron Age (1300-450 BC) (see also excursus 'Colluvial Deposits'). Techniques and strategies of land use practices can thus be seen as a ResourceAssemblage that is culturally inherited and implemented in human niche construction on the one hand and produces feedback mechanisms that continuously transform the landscape on the other.

3.4.3 Changing Framework Conditions in Landscapes

Framework conditions in landscapes change at different rates, from seasonal to centennial, and are closely intertwined with human perceptions and decisions. They also contribute to the construction of cultural niches at different levels. This was particularly evident in the diachronic comparison of modes of landscape perception and use during the Bronze Age, historic and modern times (see Chapter 3.4.1). Networks, including mobility and transhumance ('inceptive and counteractive relocation'; see fig. 5), were a crucial element in coping with changing framework conditions in landscapes (see Chapter 3.2.3).

Changing framework conditions associated with catastrophic events are an important variable in the development of interregional exchange from the Late Bronze Age to the Early Iron Age in the southern Levant. Millek (2019) discussed various forms of destructive forces, including climate change, seismic storms, warfare, migration, the disruption of international trade routes, socio-political upheaval and famine, and their role in the historical development of the Levant at the end of the Bronze Age (ca. 1200 BC; Millek 2019). Given the lack of an overarching explanation for the end of the Late Bronze Age, reflected in other research that describes it as a multifaceted problem (e.g. Cline 2014), Millek attempts to develop regional theories of 'collapse' by examining evidence of destruction at 49 archaeological sites in the southern Levant. Although this study does not directly examine changing environmental conditions in landscapes, his study nicely outlines the diversity of destruction patterns at the local level, supporting the view of an overall complexity of cultural change.

The role of freshwater supply (see excursus 'Freshwater Scarcity on Islands'), both as an existing parameter and as an analytical ResourceComplex in medieval and post-medieval sites on Mediterranean islands, is particularly subject to changing environmental frameworks, including landscapes. The definition of the knowledge and management of environmental factors as a ResourceAssemblage has allowed the consequences of these aspects to be recognised far beyond the islands studied. Dierksmeier (2020) examined the effects of long-term water scarcity and found a relationship between income and access to clean drinking water, leading to social upheaval and conflict in the Canary and Balearic Islands. The study is also a good illustration of different forms of cultural niche construction and, in particular, of further consequences of inceptive perturbation (see fig. 5). The transformation of water into an economic resource leads to even greater social inequality and, consequently, to more and more complex forms of water resource management.

Excursus: Colluvial Deposits

The analysis of colluvial deposits has provided evidence of human-induced soil erosion since the Neolithic period, although not to the same devastating extent as today, where soil erosion threatens the function and viability of many ecosystems and food security (Wuepper et al. 2020). We consider colluvial deposits to be the unintended product of soil erosion caused by human activities. Colluvial deposits (fig. 7), which consist of different M horizons (M comes from the Latin *migrare*), mainly originate from upslope eroded soils (Zádorová et al. 2023). The time of formation of these M horizons can be dated to correlate phases of land use and reconstruct prehistoric and historic land use practices using biogeochemical analyses (Scherer et al. 2021a).

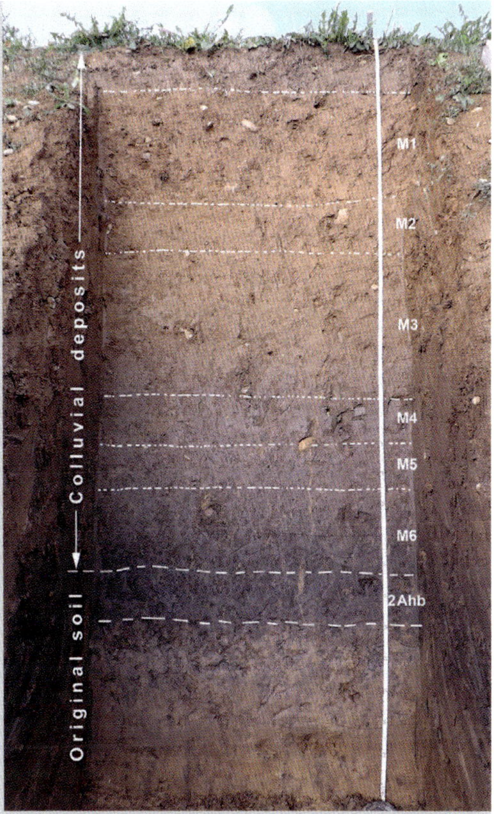

Fig. 7 Colluvial deposit with six M horizons covering the original pre-iron-age land surface. For detailed information on the analysis results for the reconstruction of Bronze Age land uses from the colluvial deposits see Scherer et al. (2021b) (Photo: Sascha Scherer).

Biogeochemical analyses (e.g. polycyclic aromatic hydrocarbons) of the M6 horizon in the figure (Scherer et al. 2021b) – dating to the Middle Bronze Age – indicate controlled burning of biomass at low combustion temperatures, which was a characteristic prehistoric practice for the maintenance of land use areas. An accumulation of faecal remains such as paleo-urease enzymes or 5ßstanols, the latter derived from omnivorous digestive processes, can be interpreted as a Bronze Age input of faeces. The presence of oak-dominated forest ecosystems, inferred from anthracological analyses, and the high levels of faecal remains are consistent with the presence of pig bones in the pit fillings of the adjacent Bronze Age settlement. These results, together with the archaeological finds, suggest a diverse land management model of a Middle Bronze Age settlement with wooded pasture, mainly used for pig husbandry and wood procurement, and an open landscape with arable land and livestock husbandry.

As farmers usually did not leave their fields due to soil erosion and downslope formation of colluvial deposits, the latter are spatio-temporally linked to other natural and socio-cultural factors and thus form a central component of a **ResourceComplex**. This allows us to capture the diversity and variability of former land use and perceptions of the land, as experience and tradition helped to make land use more controllable (Knopf 2017). From the perspective of **ResourceComplexes**, 'colluvial deposits' are embedded in a network of interactions and linkages between their components and enable a holistic view of the colluvial system from a natural and cultural science perspective. From the perspective of **ResourceAssemblages**, archaeopedological analysis of colluvial deposits can decipher changes over time and identify complex-adaptive dynamics in land-use practices, soil management, settlement dynamics and specific social valorisation processes (cf. James et al. 2021). Colluvial deposits are thus an essential archive for the analysis and interpretation of land use history and settlement dynamics at least since the Neolithic, as well as for the general perception of soils in a spatio-temporal perspective (Knopf et al. 2021; Pietsch/Kühn 2017).

Peter Kühn

Excursus: Freshwater Scarcity on Islands

As part of our research (Schön/Dierksmeier 2021; Schön 2025), we studied islands as locations with **ResourceCultures** that differ from the mainland. The black nets in the image are an example of this. The nets look like insect screens without window frames. In fact, these simple net structures are a response to extreme freshwater scarcity on Fuerteventura, one of the driest Canary Islands. The wooden frames and soft nets lead to plastic piping and allow moisture to be collected or 'harvested'; which is why this technique is known as 'fog harvesting'. Fog harvesting is one of many innovative forms to combat water scarcity on islands, which is a reoccurring attribute of insularity (see **excursus 'Islands and Resources, Islands as Resources'**).

Differences between islands and the mainland become particularly prominent when we consider the importance of water as a scarce resource within a larger **ResourceComplex**. Islanders often had a long history of coping with resource scarcity and natural disasters. Research has shown that arrangements for access to water have not only influenced food production, the use of watermills for energy and the like, but also migration and related phenomena such as the introduction of epidemics. The management of climatic environmental factors – understood here in terms of **ResourceAssemblages** – had consequences far beyond the individual island. The environmental history of island societies should therefore also be understood as a global environmental history, following the methodological results of our research.

The governance structure and the level of political corruption of a given society have a direct impact on the distribution of water resources. Particularly on islands where freshwater supplies are in private hands, limited access to water exacerbates existing income inequalities; islanders without 'hydraulic citizenship' may suffer from a diminished ability to participate in society (Anand 2017). But the picture is not all bleak. The study of island water scarcity sheds light on the coping mechanisms, adaptations and innovations of islanders over the centuries (Schön/Dierksmeier 2021). In some cases, islanders provide examples that could be replicated with the necessary modifications on other islands or on the mainland; as such, they should be seen as pioneers in coping with ongoing environmental change. For example, in the Canary Islands, a separate water police force was established between 1500 and 1800 to protect the water supply. Water-yielding plant species called 'Fountain Trees' were carefully guarded, low-cost household solutions to purify water (*la destiladera Canaria*) were implemented, extensive cisterns collected rainwater, fog moisture was harvested from pine trees and communities came together to protest their rights to water (Dierksmeier 2020).

In short, the study of water as part of a much larger **ResourceCulture** reveals the many ways in which water supply has been affected by multiple geological, geographical, social, cultural and economic challenges, and reveals a long history of solutions to one of the most significant enigmas of island life.

Laura Dierksmeier / Frerich Schön

Fig. 8 Fog harvesting nets for water collection on Fuerteventura (Photo: Laura Dierksmeier).

3.5 Synthesis

The analysis of cultural, social and economic developments shows their close correlation with trends in resource use. Like patterns of use, resource values often vary regionally, over time and between different social groups. In any ResourceCulture, effective resource use requires networks that influence social dynamics. The consequences of these social developments, in turn, influence the way resources are used. It is this entanglement that characterises the specific functional and intentional relationships of the respective networks, which can be analysed as ResourceComplexes in the first place. In most cases, however, these cannot be planned, but depend on various factors, especially natural or political ones. Such contingency in development often characterises ResourceCultures and determines the specific dynamics of the associated society.

The processes that describe these multiple linkages between resource development and use, as well as social, political and cultural dynamics, are complex and multidimensional. While modes of resource use and associated networks are constant, social developments tend to be largely continuous – although there may still be differences. Change, on the other hand, often leads to ruptures in social, political and economic processes. Just as the emergence and functioning of networks is unpredictable, the social dynamics of resource use are not linear or monocausal. This is particularly evident in cases where supraregional relations play an important role and a variety of unpredictable external factors can act, or in the diachronic consideration of landscape use. From the analytical perspective of ResourceAssemblages, their development reflects a variety of deliberate, multiple spatial and diachronic acts of shaping human environments while adapting to needs and types of resource use. Change is often associated with shifting strategies in human responses to changing needs, practices and social, political, economic and natural conditions. Here, the theoretical model of 'cultural niche construction' proves to be a good approach to analytically capture the ResourceCultures that are specific to each society – such as the one related to water in the Balearic and Canary Islands (see excursus 'Islands and Resources, Islands as Resources') – and that vary regionally as well as diachronically. In the case of changes in resource management strategies, it is clear that networks played an important role in their implementation as well as in the restoration of earlier models of resource use, as shown in the diachronic analysis of the stone exploitation system in south-eastern Iran. Networks enabled the involvement of a greater potential of people and means to respond adequately to the complexity of needs and demands of resource use. Changes in these networks, in turn, represent important factors that can also influence the development of landscapes that are considered as ResourceAssemblages, as illustrated by a comparison of the Dehesa systems in Andalusia and the neighbouring Extremadura in southern Spain.

In the interplay between societies and resources, it has been shown that developments as social and cultural dynamics are closely linked to tendencies in the valuation and use of resources, as well as in the respective networks that enable their management. Stable socio-cultural developments are often associated with stable ResourceCultures, as illustrated by the long-lasting continuous agropastoral land use in the Spanish Sierra Morena or the establishment of castles and monasteries as key resources for the rule of noble dynasties throughout the Middle Ages in southern Germany. Sometimes, however, changes in resource use strategies are necessary to ensure such stable socio-cultural developments. This is illustrated by the way in which elite power was established in Syria in the 2nd millennium BC, where it was largely based on the ability to display wealth, particularly through the display of gold. This ResourceCulture required changes in resource use when access to gold became limited, by alloying silver with gold or gilding silver objects to maintain social positions and keep the system going.

In other cases, significant changes in socio-cultural development are triggered by the emergence of desires for the use of specific resources, as can be observed in the western Mediterranean as a consequence of the arrival of the 'Phoenicians' there in the 10th/9th centuries BC (Bartelheim 2007). Conversely, social and political developments also led to changes in resource management strategies and economic structures, as in the Bronze Age southern Levant after the Egyptian conquest and subsequent economic changes.

4 Resources and Movements

Thomas Scholten, Peter Pfälzner, Jörg Baten

The concept of movement can be applied to almost all aspects of human-nature interrelations, whether physical or conceptual. In the context of socio-cultural dynamics and ResourceCultures, not only physical entities such as people or commodities are in motion, but also ideas, knowledge and other forms of resources (Feldhay/Hertz 2019; Nail 2018; Quet 2014). They often initiate movements and profoundly influence them in many ways. In the context of ResourceCultures, movements are described and analysed not only as resource-related processes of spatial appropriation, but also as processes in themselves that drive socio-cultural developments and the symbolic dimensions of the relevant resources. These processes are understood as catalysts for socio-cultural developments. This also touches on the question of how the need for resources arises, and it is necessary to identify what resources were involved in the movements and how they were socially, politically, symbolically or religiously embedded. Were the movements directed, planned or controlled, and what part of a social group was involved? What forms of communication and contact were established? For example, did a conflict over or lack of resources drive the movements? Did this lead to internal or external conflicts between social units over material and immaterial resources?

The following chapters will address the above questions about the relationship between resources and movements. First, the theoretical background to this relationship will be discussed (Chapter 4.1). This is followed by an examination of the different modalities of movements (Chapter 4.2), which will show the different dimensions in which movement can be understood, defined and conceptualised. The modes of intraregional mobility (Chapter 4.3) show how movements operate and what role they play in the constitution of social and cultural entities. In a cross-cultural approach, this is illustrated by past and present examples from Germany, Spain, the British Isles, Mesopotamia, Iran and India. In all these examples, it becomes clear that pathways are an important element of movements in a variety of ways. Long-distance movements (Chapter 4.4) are interregional or transregional and include long-distance economic transactions as well as migrations of social groups. Examples range from the Palaeolithic through the Bronze Age and Classical Antiquity, to the Middle Ages and the Viking Age.

4.1 Theoretical Background for Understanding the Relationship between Movement and Resources

Movement can first be described as any kind of activity in space and the processes that underlie it. The representation of the type of movement is initially descriptive, for example in the form of who or what is moving from A to B. The two basic components of a movement are the process as such and the exchange between A and B associated with it (Cresswell 2006, 2). Other properties of movement are its speed, acceleration and duration, adding a temporal component to the initial spatial perspective (Cresswell 2006, 4). Analysis tends to focus on the causes of different forms of movement and the social effects they produce.

The causes and social effects of movements are complex, even when the trigger is monocausal, such as the outbreak of war or famine. Resources are always involved and are defined and valued by the participants in the course of movements in order to transport, exchange, acquire or use them. In addition to material resources such as wood, stones, precious metals, jewellery and food, as well as infrastructures, immaterial resources such as information, knowledge and ideas are relevant control variables of movement processes. A good example of the complex relationships between causes and social effects and meanings of movements is swidden agriculture. This farming system, in which the land is left fallow between cultivation periods to allow natural vegetation to return and the soil to recover, is a practice of local farming populations that are particularly adapted to the natural environment, such as in the highlands of Odisha in present-day India (Hardenberg 2018a) or in Q'eqchi' Maya villages in present-day southern Belize (Downey et al. 2020). After the autumn harvest, the products of the field are transported to the village. In this new location, they are not only food, but a culturally valued resource, used for self-sufficiency, the basis of the population's culinary diet and an important part of the religious festivals and sacrifices that shape people's relationships with cosmological forces. Immaterial resources thus structure people's movements in space and also condition mobility. Movement is thus understood as a social process that characterises the change of people and things in geographical space. It can be distinguished from mobility as the sum of social processes of movement, such as social rhythms, orders, strategies and social effects of changes in place (e.g. Beaudry/Parno 2013).

Following the concept of ResourceCultures and integrating both material and immaterial resources, which are understood as culturally constructed by the respective societies in terms of their meaning (see Chapter 2), the distinction between movement and mobility dissolves. Accordingly, the perception and evaluation of places, landscapes and actions by their inhabitants lead to social and cultural meanings that are triggers and control variables of movements in space (Adey 2017, 63; Cresswell 2006, 1–7). They lead to the development of social structures, which in turn influence the shaping of habitats and landscapes (see Chapter 3). In Bronze Age landscapes, for example, these are the construction of interaction networks through spatial corridors (Chala-Aldana 2024) and the creation of flat surfaces for fields (Scherer et al. 2021b). The consequences are spatial developments and processes of resource appropriation in terms of movement and mobility, which in turn feed back into social orders and identities.

An analysis of movements on the basis of the material and immaterial resources that drive and control them thus creates the possibility of an initially descriptive and technical analysis of the modalities of movement (see excursus 'Machine Learning') without loading them with meanings and assumptions about cause-and-effect relationships. Resources are thus the basis for the emergence, maintenance and transformation of movements as expressions of social relations, orders and identities (Bartelheim et al. 2021b). In a subsequent step, the characteristics of movements can be placed in a socio-cultural context and analysed as a ResourceComplex, including networks of knowledge, practices, techniques, people, places, objects and media, and further diachronically as a ResourceAssemblage (see Chapter 2), in order to gain new insights into the causes of movements and the social effects they produce.

Excursus: Machine Learning

The integration of Machine Learning (ML) into the study of **ResourceCultures** offers a groundbreaking perspective on the resource use of human societies. ML, a vital branch of artificial intelligence, enables systems to independently identify patterns and distil knowledge from large data sets (Bishop 2006). This enables innovative exploration of the complexities of resource management across diverse cultural landscapes. ML involves systems that learn from data without explicit programming, and dynamically refine decisions and predictions as more data is processed. Its autonomous nature in identifying patterns and extracting insights makes it indispensable for the study of **ResourceCultures**. ML illuminates complex dynamics within **ResourceComplexes** and **ResourceAssemblages**. Using algorithms, ML systems analyse data, learn and make decisions based on their learning. This continuous process facilitates constant refinement and improves understanding of the intricacies of resource management. In essence, ML is revolutionising the way we study resource management, shedding light on the complex interplay between societies and their resource use strategies (Argyrou/Agapiou 2022).

The use of ML algorithms to analyse large datasets, including archaeological evidence, environmental cues and socio-cultural records, facilitates the discovery of hidden relationships and intricate patterns in resource management across different cultural milieus (Bickler 2021). In archaeology, for example, ML techniques are invaluable for investigating the spatial distribution of artefacts (Bellat et al. 2022) and discerning temporal relationships, providing crucial insights into historical resource management strategies (Felicetti et al. 2021) and the socio-economic dynamics that shaped them (Character et al. 2024). Furthermore, ML helps to reconstruct lost artefacts and historic sites (Wang et al. 2024), thereby preserving cultural heritage and deepening understanding of past resource management methods and their contemporary relevance (Farina et al. 2024).

ML also plays a central role in exploring the societal impacts of migration, trade and infrastructure development. Through sophisticated network analysis, ML illuminates the intricate web of human interactions and their role in the evolution of resource use and the emergence of **ResourceCultures**. Using Geographic Information Systems and state-of-the-art remote sensing data (Guarino/Brucato 2024), ML enables researchers to optimise resource use in unprecedented ways and facilitates advanced spatial pattern analyses and mobility assessments (Zimmer-Dauphinee et al. 2024).

In summary, the use of ML in **ResourceCultures** research heralds a significant paradigm shift in research. It provides a comprehensive framework for studying the socio-cultural dynamics that shape resource use practices. Through interdisciplinary collaboration and methodological innovation, the potential of ML can be fully realised, leading to the development of sustainable resource management strategies that are informed by a deep understanding of the complex relationship between humans and nature. This integration not only enriches our collective knowledge, but also underscores the importance of a nuanced approach to managing the resources that sustain life on our planet.

Ruhollah Taghizadeh-Mehrjardi

Understanding resources as processes also leads to a different understanding of movements and their spatial dynamics. For example, interregional exchange, through which both objects and ideas are disseminated, determined the nature of the development of agricultural resources in Bronze and Iron Age Palestine, as well as political and economic change (Kamlah/Riehl 2021). The integration of immaterial resources in the sense of ResourceComplexes for a better understanding of everyday as well as developmental movements, which include objects as well as knowledge, information and ideas, could be clearly outlined, for example, using religious speeches as a resource in South and Central Asia (Conrad/Hardenberg 2020) and for the ResourceComplex of health and body using the example of Unani medicine in South India (Sieler 2022; see also excursus 'Health'). This opens the way to combining movements with valuations (see Chapter 5) and analysing them as a component of ResourceComplexes as well as a cause of their changes in the sense of ResourceAssemblages.

4.2 Modalities of Movement

Movements of people, objects and materials in space can be classified into linear, radial and random movements, among others (e.g. Padgham 2012). A linear, uniform motion is characterised by a constant velocity or acceleration of a given duration in one direction. This may be dictated by topographical structures such as mountain passes and valleys (Tobler 1993, 3), by boundary conditions and the return of resources such as water and food supplies (Pyke et al. 1977; Hayden 1981; Kelly 1995) or by predetermined infrastructural trajectories such as pathways and roads. Radial movements are movements from a central starting point in different spatial directions at the same time or movements within a region that point to or originate from central locations (e.g. Binford 1980). Random movements appear random from an observational perspective, in the sense of contingent, unforeseen or unsuspected activity in space. All three types of movement can be observed intraregionally as well as transregionally.

Early general models of mobility in cultural anthropology make use of such movement typologies. They derive them from the behaviour of contemporary hunter-gatherer groups. Examples are freely migrating groups with no territory, restricted migrating groups with territorial boundaries, radiating and circumpolar migrating groups, as well as center-dwelling migrating groups returning seasonally to a central settlement and semi-permanent sedentary groups occupying a settlement year-round but changing settlement every few years (Beardsley et al. 1955). Similar models have also classified groups as fully nomadic, semi-nomadic, semi-sedentary and fully sedentary (Murdock 1967), revealing a combination of spatial and temporal components linked to the frequency of movement from the home base or household (Salzman 2004, 18). This has also been considered in pastoral and animal husbandry practices. Following essentially the same combinatorics, pastoralism has been divided into four types: sedentary pastoralism (Vega Toscano et al. 1998), nomadic pastoralism (Khazanov 1984; Arnold/Greenfield 2006), semi-nomadic pastoralism (Khazanov 1984) and transhumant pastoralism, which adds a directional vector to the spatial and temporal component (Geddes 1983; Chang/Koster 1986; Cherry 1988; Nisbet et al. 1991; Halstead 1981; Sherratt 1981; 1983; Costello/Svensson 2018). More recent approaches attempt to move beyond models that classify groups into mutually exclusive typologies by identifying combinations of different mobility strategies within hunter-gatherers (Kelly 1995) and considering different types of movement within sedentary or pastoral groups (Bernbeck 2008).

These examples show that a modality that focuses on particular groups or ways of life makes assumptions about their causes and raises fundamental epistemological problems of a disciplinary perspective. Furthermore, they show that the temporal component of spatial movements is closely linked to human action, determining the timing as well as the duration, frequency and amplitude of a movement. As a result, movements are charged with meaning (Adey 2017) and acquire a specific character that varies across cultures, time or space and can be viewed in a particular psychosocial context (Cresswell 2010). An exception to this are large material movements caused by natural events, such as volcanic eruptions, tsunamis, floods and landslides, which can also have social impacts and whose causes can currently also be linked to human activities and movements, for example floods in Germany on the Elbe (2002) or in the Ahr valley (2022).

Thus, movement between places in space, whether linear, radial or random, is driven by factors such as resource availability, wealth, investment and political, religious, cultural and social interactions, coercion, disasters and scarcity. Subsistence strategies and economic opportunities are also considered as drivers of mobility in sociological studies (Awinia 2020; Gardner/Osella 2003; Geist/McManus 2008). In addition, both dynamic and long-term stable environmental factors such as climate, topography, soils and biota can play a significant role, for example by motivating or minimising movement through soil quality and associated food security (Scholten et al. 2002; see excursus 'Soil'). Extensive studies of human mobility in the Late Quaternary link culture and environment and analyse their interaction with respect to the migration of anatomically modern humans out of Africa (Stringer 2011). Many studies in this context (e.g. Litt et al. 2021) have shown that, in addition to climate and environment, the cultural context, with processes such as innovation transfer and population growth, was already crucial for human behavioural strategies and migration in the Pleistocene. Furthermore, Shennan et al. (2013) showed that sedentism in Europe was followed by a rise and fall pattern in the density of regional populations, suggesting that demographic patterns may have resulted from endogenous causes.

Another aspect of movements is power (see excursus 'Power'), which, in the context of knowledge as an intangible resource, goes beyond mere access to basic materials and natural products such as stones (Pfälzner/Alidadi Soleimani 2017) and finds expression in palace cultures (Puljiz 2021, 523–529), domains of rule in the Middle Ages (Froehlich 2023), colonisation (Schön/Schäfer 2021) or island economies (Dierksmeier et al. 2021). The connection between specific forms of knowledge and power is illustrated by images as symbols of power (Herrmann 2019) or health and medical knowledge (Sieler 2022). In addition to intangible resources, the circulation of objects, such as obsidian (Williams-Thorpe 1995), but also wood from trees (Daly/Nymoen 2008) or gold (Puljiz 2021), is also an aspect of the exercise of power, especially with regard to the connectivity and networking of human groups. As elements involved in various social and spatial interactions, objects can materialise identities and thus represent a group ideology (DeMarrais et al. 1996; Mehler 2009). They can be understood as gifts, artistic representations, relics or artefacts and have the ability to move through space and time. They can also be triggers of colonial spatial development and the conflicts and integration processes that accompany it.

Furthermore, movements can be seen as the starting point for global trade relations and networks (see Chapter 3) and as part of phenomena such as 'world systems' or the concept of 'globalisation' as described for the European and Near Eastern Bronze Age (Kristiansen

Excursus: Soil

Five kilos of carrots a year. That is what a farmer can expect to harvest from one square metre of this moorland. The carrots contrast sharply with the black, humus-rich topsoil. This layer used to be up to 2 m thick. Today, the fertility of peatland soils is threatened in many places around the world as the once nutrient-rich humus continues to be degraded.

The history of soils in much of the northern hemisphere begins with the melting of the glaciers about 10,000 years ago. Developed over long periods of time under the influence of climate, topography, rock and biota, soil is a resource that plays a vital role in supporting life on Earth. It is a living, dynamic system that provides a medium for plant growth and serves as a habitat for various organisms, including humans. As such, soil is a finite and non-renewable resource for human life that must be carefully managed to ensure its long-term sustainability. Descriptions such as soil health, soil fertility, soil productivity and soil quality, but also actions such as soil management and soil conservation, assume that soils are not only a material resource but also a valuable asset for people in a socio-cultural context. From this perspective, soils can be seen as an integral part of **ResourceCultures** and analytically understood as a **ResourceComplex** (James et al. 2021). Soils are integrated into agricultural food production, or their analysis provides insights into the dynamics of colonisation, to name but two examples.

Fig. 9 Bog soil in Bernese Witzwil between Lake Biel and Lake Neuchâtel, in the 'Great moss', the largest cultivated fen area in Switzerland (Photo: © Agroscope [Gabriela Brändle, Urs Zihlmann] LANAT [Andreas Chervet]).

Soils are subject to many changes as a result of human activities such as agriculture, urban development, mining and industrial land use. These activities can affect soil health, reduce soil fertility and limit soil productivity. Soil degradation can lead to reduced crop yields, increased erosion, loss of biodiversity and reduced water quality. As a result, today's society is taking steps to protect and conserve this valuable resource. Sustainable soil management practices include reducing soil erosion, maintaining soil organic matter levels and the responsible use of fertilisers and other inputs. Such dynamic processes to improve soil health and fertility are essential to maintain soil productivity and ensure food security for future generations. They can be analysed as **ResourceAssemblages**, using the adaptive cycle metaphor and the social-ecological system as tools to show the complexity of human-environment interactions and their evolution (Teuber et al. 2017).

Thomas Scholten

1998; 2011; Kristiansen/Larsson 2005; O'Connor 2010), for example the early tin trade in the Bronze Age (Hodos 2017).

It becomes clear that the controlling factors and processes of spatial movement, as well as the movement itself, must always be understood in a spatio-temporal context. This leads to the question of spatial and temporal dynamics and the extent to which controlling variables change. For example, concepts of scale in geology, soil science, geomorphology and climatology (Kerry/Oliver 2011; Behrens et al. 2018; 2019; among others), with timescales ranging from millennia to millions of years and global in scope, differ by orders of magnitude from those of socio-culturally driven movements. An example of the latter is the metaphor of the rhizome in geography (Burini et al. 2021) as a theoretical tool for understanding how people move through space in terms of mobility networks. Other examples include daily errands or annually recurring trade relationships with neighbouring communities of areas. This is related to the idea explored by Massey (2004) of how places are created and changed by people in a causal sense that can be understood in the context of dynamic state factors. With regard to animal mobility, a common concept is the theory of optimal foraging. It postulates that when hunters reach a point of diminishing returns in one area because resources are exhausted, they move on to the next area (Pyke et al. 1977). However, ethnographic studies show that foragers do not wait until all yields are fully exploited, but move on when the daily yield of a particular component is deemed insufficient for the needs of the group (Hayden 1981). This behavioural ecology approach suggests that, according to the theory of optimal foraging, people will seek the highest yield of something in a given area and only move on to lower ranking components when high ranking components become scarce. In this context, diminishing returns and moving on to the next location have been described in patch choice theory or marginal value theory (Kelly 1995). As transport distances increase, people will move components with higher yields back to a location. This shows how the dynamics of movement can affect resource formation and lead to an understanding of the modalities and processes of movement as an interrelationship between space and people. They represent the temporal component or momentum of movement along with their spatial perception, starting from the desire or need to acquire, transport or use material and immaterial resources. The movement of individuals or groups in a spatio-temporal context can thus be classified into three categories of human movement that represent spatio-temporal patterns and can be applied independently of assumptions about cause and effect.

Daily movements, or the regular bridging of distances to meet immediate subsistence or other basic economic needs, are, for example, regular visits to places for daily supplies of water, food, energy sources and building materials. In this way, humans have shaped tropical forests and their soils for millennia, especially since the advent of agriculture (Lewis et al. 2015).

A second category of movements are irregular but repeated movements. These include gatherings for cultural, social, political, religious or economic purposes, as well as transit movements to get from one place to another or to transport objects and people (Chapter 4.3), such as religiously motivated movements within the liminal Late Bronze Age and Early Iron Age landscape of the Heuberg in southern Germany with the Heidentor of Egesheim as a sacrificial site (Miera et al. 2019).

The third category of spatio-temporal movement patterns consists of movements to access and develop new sites, and movements of change resulting from small-scale changes in

Excursus: Power

'Castles. Symbols of Power' ('Burgen. Symbole der Macht') is the title of an influential monograph by Joachim Zeune (1996). But what is power – and how does it differ from authority? According to the 'Encyclopaedia Britannica', and following Max Weber's most influential definition, power is '[…] the capacity to influence, lead, dominate, or otherwise have an impact on the life and actions of others in society. The concept of power encompasses, but is not limited to, the notion of authority. Unlike authority, which implies legitimacy, power can be exercised illegitimately.' (Munro 2024). While power can also be based on the illegitimate and/or illegal use of force, even if only temporarily or for a very short period of time, authority is always associated with legitimacy and permanent structures. Recent sociological and political science, however, has criticised this clear distinction.

The book title 'Castles. Symbols of Power' ('Burgen. Symbole der Macht'), which has become a proverbial reference in German research, might therefore be changed to 'Castles. Symbols of Authority' ('Burgen. Symbole der Herrschaft'), since it is precisely the symbolic demonstration of the power of the nobility, set in stone in the permanence of the castles, that is still associated with these buildings today. Whether power, authority – or power and authority, the castle, and in particular the castle tower as a core element, symbolises the importance of its lord.

Fig. 10 Towers as symbols of power and authority: Notre-Dame in Strasbourg (France) and Münzenberg Castle (Germany) (Photo Münzenberg: Lukas Werther; Photo Strasbourg Cathedral: Jonathan Martz, CC BY-SA 3.0).

The tower (fig. 10) also symbolises power and authority in religious architecture: Gothic church architecture in particular seeks to express the unlimited power of God and his supremacy over all earthly rulers by building ever higher towers. At the same time, these towers also demonstrate the secular power and authority of the Church. In contrast, the Cistercians and later the mendicant orders deliberately avoided towers in their church buildings; only a small, modest bell tower was to express their humble attitude towards God and man.

Power and authority are thus symbolised and manifested in the permanence of towers on castles and churches. If castles and monasteries are viewed from the perspective of **Resource-Complexes** (Froehlich 2023; Krätschmer et al. 2018), this symbolic value is one of several resources used by noble builders or founders to achieve the goal of establishing long-lasting authority in a particular territory. While in the context of castles, resources such as military strength, the architecture of central places and the development and consolidation of territories and material assets play a central role in the formation of aristocratic rule, the monastic foundations of the nobility also focused on the establishment of networks, the acquisition of scholarly knowledge, the creation of identity-forming burial grounds and, above all, the safeguarding of spiritual and thus immaterial resources.

Sigrid Hirbodian / Lukas Werther

settlement and location, for example due to land exhaustion or resource depletion. They include the irregular transport of people, animals and objects, as well as the knowledge and ideas associated with them. Similarly, forced mobility in the form of migration as a result of warlike activities, such as conquest, expulsion, deportation, enslavement, or natural disasters as a pervasive phenomenon in human history (Zeuske 2019; Langer 2021) is a facet of movements to access and develop new places and movements of change. Development movements with strong impacts on socio-cultural systems typically have a transregional character and occur over long distances (Chapter 4.4). In spatial development, so-called spatial pioneers (Neuburger 2017) move to stay in another place, starting from specific causes of out-migration and abandonment of an area. This is combined with the relocation and settlement of people and the consequences of their use and transformation of the new area. This category also includes small-scale intraregional changes of settlement or location for specific purposes. Examples are mining within regions (Chapter 4.3), soil fatigue, resource depletion or other cultural reasons, as well as nomadism, transhumance and alpine farming, where different economic areas are used (Bartelheim et al. 2022), and subsistence-oriented land use systems.

In all the categories of modalities of movement described so far, paths, routes and infrastructure, either existing or to be created play an important role (Chapter 4.3). The same applies to immaterial resources of movement and exchange such as information, knowledge and ideas. The culturally constructed interest in resources thus triggers radial movements towards spatial developments and processes of resource appropriation, which in turn have an impact on social orders and identities. The effects of immaterial resources occur at the interface between a conventional economic definition of resources, a geoscientific perspective of natural space, for example in the course of Bronze Age land use (Henkner et al. 2018; Scherer et al. 2021b), and the creation, preservation and transformation of social relations, for example in the context of integration processes in the Phoenician-Punic West of the 1st millennium BC (Schön/Schäfer 2021). In this context, intangible resources of movement are conceptualised as processes that, like material resources, are subject to spatial and temporal change. This is evident, for example, in the case of roads that initially served to control regions and later served as transport routes for raw materials such as copper and silver (Chapter 4.3). Other examples of strong diachronic changes in resource use include the evolution of Neanderthals and modern humans (Davies 2014; Higham et al. 2014; Kitagawa/Conard 2020) and the role of castles and monasteries in high medieval spatial development (Froehlich 2023).

Irregular but repeated movements lead to change, which includes not only relocation but also adaptation to environmental changes and social developments. With adaptation come knowledge and a learning process, as some mobilities necessarily require skills. These include, for example, not only understanding the geography of a landscape and knowing the routes or weather conditions, but also herding, driving certain vehicles or sailing, among other skills. Thus, movement can be understood and analysed as a dynamic set of intangible as well as tangible resources (Hardenberg et al. 2017).

Even mobility, initially seen as a type of movement, is linked to social processes and interactions that go beyond physical movement in space (Murrieta-Flores 2010). Since ancient times, it is known that human groups interacted in different ways and only exceptionally lived in isolation at certain times, such as on Easter Island (Lipo et al. 2021). In this context, pathways can lead to cooperation and networking, as in the case of Neolithic Mediterranean groups in north-eastern Iberia (Díaz-Zorita Bonilla et al. 2021). This means that sites were already connected in the past, creating roads, crossings, bridges and meeting places. If such physical or material infrastructures are understood as a way of transmitting information and knowledge from one place, person or group to another, information can be considered as another form of communication and, consequently, as a resource that includes all the actors involved in the movement.

4.3 Intraregional Mobility and the Role of Pathways

Intraregional mobility can take many different forms and have many different causes. What they all have in common is that movements are triggered by resources. Conversely, movements can lead to a change in the way resources are handled. Thus, resources can be both causal and dependent on movements. This relationship can be traced in different historical or geographical situations.

Central elements of intraregional mobility are pathways and their associated networks. In this view, pathways can be seen as the main resources for the establishment of movements. Related to this character, pathways play a fundamental role in the establishment of settlement systems. This can be observed in north-eastern Mesopotamia, where pathways, in addition to soil and water, can be identified as primary resources for regional settlement systems (Pfälzner/Sconzo 2015; 2016a; 2016b). In principle, soil, water and pathways are firmly inscribed in the regional geography and are therefore unlikely to be subject to short-term changes. This raises the question of why settlement systems change at all. One possible explanation is that changes in settlement patterns are triggered by changes in path preferences. Except in cases of extreme topography, it is generally possible to choose from a variety of path options within a region. Preference is given to the existence of important movement corridors. Settlements tend to develop along corridors that become major pathways. This applies not only to settlements directly on the main pathways, but also to those in their vicinity. Thus, pathways create settlement corridors. Settlements far away from these corridors lack connections in terms of social ties, exchange opportunities and supply options. This leads to their decline and disappearance. Paths are therefore a significant resource for the region, perhaps its most significant.

A continuous process of defining and redefining pathways can be observed in the history of north-eastern Mesopotamia. This has been driven not only by practical considerations, but also by economic interests or political motives. Economic interests are historically illustrated by the establishment of an important trade route in the Old Assyrian period (2000–1780 BC), connecting Assur on the Tigris with the central Anatolian city of Kaneš (Barjamovic 2008; 2010; 2011; 2018). Its existence stimulated settlement activities in the regions along the route (Pfälzner/Faist 2020). The political motivations for redefining the routes are particularly evident when considering the expansion of powerful lowland states into the Mesopotamian periphery. The expansion of the Akkadian Empire (2350–2200 BC) and the Middle Assyrian Empire (1300–1100 BC) into north-eastern Mesopotamia was largely driven by the intention to control important traffic routes leading from Mesopotamia to Syro-Anatolia. Thus, many of the conquered regions were seen as corridors for the transport of raw materials and products from distant regions. Consequently, the routes of north-eastern Mesopotamia were a basic resource for obtaining other resources, and thus formed a significant factor within the ResourceCultures of Mesopotamia.

Intraregional pathways offer the analytical advantage of being a relatively easily detectable part of the archaeological record. Path reconstruction is possible through the application of geographic tools such as Least Cost Path Analyses (White 2015; White/Surface-Evans 2012), the use of digital terrain models or the analysis of satellite images (Lawrence et al. 2020). Hollow ways around settlements dating back to the Bronze Age are still clearly visible in the modern landscape and can be identified from satellite images, as has been demonstrated for north-eastern Syria and northern Iraq (Wilkinson 1993; De Gruchy/Cunliffe 2020). In very rare cases, pathways can be directly recognised archaeologically. One such case is a Middle Bronze Age (2000–1600 BC) road near the archaeological site of Bassetki in northern Iraq, which was detected by geomagnetic prospection and partially excavated (Pfälzner/Qasim 2017).

Regions with unfavourable agricultural conditions tend to be sparsely populated. They are often occupied by social groups that depend on a high degree of mobility, such as nomadic or transhumant pastoralists. In addition, unfavourable areas often contain important routes linking several densely populated regions. In many cases, the mobile social groups living in the disadvantaged areas act as agents for the movement of goods through their region. Therefore, pathways can be considered important resources even in unfavourable areas where permanent settlements are rare (see excursus 'Favour/Disfavour').

It is argued here that the importance of pathways as a resource is a generally observable phenomenon in different historical and geographical situations, and thus was and is a basic component of the formation of ResourceCultures. This can be concluded from examples drawn from various cultural contexts.

The first example relates to the region of Jiroft in south-eastern Iran. Here, striking discontinuities in settlement systems can be observed over the past 8,000 years (Pfälzner/Alidadi Soleimani 2017; Pfälzner et al. 2019). Phases of dense settlement patterns, e.g. from the Late Chalcolithic to the Early Bronze Age and in the Achaemenid (500–330 BC), Parthian (250 BC – 250 AD), Sassanid (250–630 AD) and Early Islamic periods (630–1200 AD), alternate with almost unpopulated phases in the intervening periods, following a collapse of previous settlement systems. To understand the reasons for this, it is necessary to consider the role of pathways. They played an important, if not decisive, role in these processes. They were the basis for exchange and trade activities during the densely populated periods, as documented for the Early Bronze Age (3000–2000 BC), when goods made of chlorite, calcite and lapis lazuli were exchanged over long distances between Mesopotamia, Iran and Central Asia (Lamberg-Karlovsky/Tosi 1973; Lamberg-Karlovsky 1975; Dales 1977; Kohl 1978; Potts 1982; 1993, Francfort 2020). At that time, the region was connected to supraregional route networks (Kohl 1987; Barjamovic 2018). It can be concluded that settlement densification occurs when there are important pathways and intensive movement along them; settlement disruption and cultural collapse occur when the pathways lose their significance.

This interdependence can also be observed in the modern history of south-eastern Iran. In recent years, the region around Jiroft has experienced an economic and settlement boom. One reason for this is the construction of the Persian Gulf Highway, a road leading from Tehran to the Persian Gulf. Another factor is the resumption of mineral extraction (chromite) in the Jiroft region (Frauen/Klocke-Daffa 2023). This further confirms the strong interrelationship between pathways and movement along them, the settlement system and the use of other resources in a region.

Excursus: Favour/Disfavour

Since the second half of the 19th century, the terms favour and disfavour have been used in geography and archaeology to describe a concept for classifying geographical areas (Miera 2020). A favourable area was defined as one that was suitable for permanent human settlement or use due to climatic factors, altitude and, in particular, the availability of fertile soils (e.g. loess areas). Conversely, low mountain ranges were considered unfavourable areas. This nature-determinist and essentialist way of thinking still exists today, without being reflected upon. The favour/disfavour dichotomy, based solely on agricultural parameters, is not sufficient to adequately explain the historical and current dynamics of land use processes (Knopf et al. 2021). Various new perspectives and approaches have made this clear: modern landscape archaeology (David/Thomas 2012; Bartelheim et al. 2021b), for example, has pointed out that landscapes are generally culturally constructed and the product of specific human needs and choices. The concept of 'marginality' examines spaces in terms of their perception and specific valuation (Neuburger 2022). The concept of 'territoriality' is primarily concerned with aspects of social power and the political control of spaces (Neuburger 2017). The 'liminality' approach focuses on border and threshold spaces, each of which has different socio-political control and religious connotations (Ahlrichs et al. 2018). The concept of complex adaptive systems also challenges traditional views of favourable and unfavourable areas and opens up new perspectives (James et al. 2021).

The key insight is that the assessment of favourability/disfavourability is a matter of perspective. It depends on the time-specific perceptions of the respective people as to which spaces and which resources were considered favourable at which times and for which reasons. The favourability of a space thus becomes a cultural construct. In the same sense, this applies to resources from the perspective of **ResourceCultures**. Here, resources are the means for creating, maintaining and transforming social relations, structures and identities that are created through cultural ideas and practices.

In terms of studying the dynamics of land use in an area or landscape, this means that the perspective based solely on the agricultural suitability of an area must be replaced by a concept that analyses tangible and intangible resources (Bartelheim et al. 2021a). Examining resources and associated networks (seen from the perspective of **ResourceComplexes**) in an area or landscape provides insight into spatio-temporal situations of favour and disfavour based on associated resource uses. In certain historical situations, ores, trade routes or religious connotations have been as important (or even more important) than fertile soils in creating a specific favour of an area (James et al. 2021). Thus, the development of analytical models that capture the links between culturally defined resources, the practices of managing these resources and their impact on groups and identities (seen as **ResourceCultures**) will also be an outcome.

Fig. 11 View of the Baar with modern farmland. The Swabian Alb in the background. Depending on the resource perspective, both areas show different aspects of favour/disfavour at different times, which are not always related to agricultural use (Photo: Thomas Knopf).

Thomas Knopf

What are the driving forces behind the close relationship between movements, pathways and overarching resource concepts? Ethnographic research in the Jiroft region helps to understand the mechanisms at play. The traditional way of life in this region has been characterised by seasonal migrations in the past centuries and partly until today (Stöber 1978; 2002). Regular movements are important for the economic activities of pastoral nomadic groups, consisting of seasonal migrations with animals from a warm climate region in the lowlands to a cold climate region in the mountains and vice versa. This movement gives particular importance to the paths that are traditionally used over long periods of time. They also allowed the group to access the resources of different habitats, such as areas of occasional agricultural activity, which were used as an additional resource (Frauen 2024). In addition to the mobility associated with animal husbandry and the network of pathways along which movement took place, a wide range of other elements were at work to stabilise the social cohesion of the group. Among these, marriage rules and the remembrance of common ancestors are particularly important. These resources are directly related to each other and can be analysed as a specific ResourceComplex. It can be concluded that the ResourceCulture of the region combines paths and movements with an interrelated network of economic, social and ideological practices.

To generalise from the examples discussed above, it can be concluded that the importance of paths for a region is based on their function as a fundamental resource for movements. These movements form the basis of almost all economic, political, religious and social activities in any society. Consequently, the pathways also influence the location and layout of settlements and settlement patterns. They are therefore an important constitutional factor of settlement systems. The movement of people and goods within and between regions takes place along pathways; the extraction and transport of resources is linked to routes; and the social and religious cohesion of a region is also expressed through routes. Therefore, the outstanding importance of mobility in modern society, including the intensified development of route networks (railways, local transport, motorways, etc.) and the search for improved modes of mobility, which is regularly and emphatically demanded by today's stakeholders, is not a new phenomenon. It has shaped the ResourceCultures in many regions for thousands of years.

Looking at different regions and time periods, one can observe various interrelations between paths and cultural practices. They are always deeply rooted in the contingent traditions of a social group or region. Nevertheless, they can be understood as an interplay between different resources of a specific ResourceCulture. For example, regular annual migrations between summer and winter can also be observed in the Himalayan region. Here they are embedded in religious ideas and practices associated with pathways. This is the case with the annual processions for the Hindu gods in the Garhwal Himalayan region of northern India (Polit 2024). The processions bring divine benevolence to the land, livestock and people, thus securing the future of the communities. Of particular importance are the great pilgrimages of individual gods, which take place once a generation. These involve movements of various kinds, each associated with specific practices. The series of ritual acts during the processions can be described as a ResourceComplex. The important relationship between the people and the gods is strengthened during the intense ritual practices of the pilgrimage through a constant exchange of offerings and counter-offerings to the deity, a process understood as the renewal of the deity's power and benevolent influence on the fertility and well-being of the land and the people. During the processions, the pathways shape not only the movements of people, gods and objects, but also human practices in the religious sphere. In this way, paths and the movements that take place on them become a factor in the identity of the social group. They are an important resource for the social organisation of a community. Accordingly, paths and movements (pilgrimages, processions) play an important role in the ResourceCultures of India.

Another aspect of the relationship between pathways and ResourceCultures can be observed in the study of medieval island monasteries in Europe. Some are difficult to access because they are located on very small islands in lakes or seas, such as the Herrenchiemsee Monastery in Germany or the monastery on the British island of Inner Farne near Lindisfarne. *Prima facie*, the island location emphasises the principles of solitude and wilderness, attitudes that are often claimed to have been the motivation for the founding of monasteries. Contrary to their postulated seclusion, however, the early medieval island monasteries on lakes prove to have been traffic hubs for both clergy and secular rulers (Meier 2009, 135–145, Figs. 3–6). European island monasteries served as stopovers in supraregional movements. This shows that, contrary to general expectations, pathways were an important resource even for island monasteries (see excursus 'Islands and Resources, Islands as Resources').

This leads to the question of how pathways are created. For this, it is important to identify the main intentions and the agents of their creation. From a theoretical point of view, it can be argued that paths were formed as manifestations of recurrent movements. These movements gradually imprinted themselves on the landscape in the form of tracks and paths. At the same time, they have to be valorised and maintained through personal networks and social ties. Thus, social factors play an important role in the formation of the landscape in general and the pathways in particular.

In South-west Germany during the Middle Ages, the relationship between pathways and personal networks is clearly illustrated by the construction of castles and their associated monastery foundings. These processes took place in the context of the conquest of space and the formation of leadership in medieval society (Schmid 1853; Krins/Scholkmann 1992; Kienzle 2016; Dendorfer 2017; Krätschmer et al. 2018; Froehlich 2023). It can be shown that personal networks and social ties were important resources for the formation of ResourceComplexes around castles and monasteries. In this period, these networks could only be maintained through routes between the different seats of power. In this way, personal connections were etched into the landscape. These personal ties could have a considerable geographical extension, thus forming large networks of paths.

Can the relationships between personal or social networks and pathways, as demonstrated in medieval South-west Germany, be understood as a general principle in the use of resources for the formation of societies? It can be argued that personal relationships and social ties are generally operative in the establishment of pathways linking communities. It is reasonable to assume that in ancient north-eastern Mesopotamia or south-eastern Iran, for example, the establishment or intensification of pathways associated with changing settlement patterns may have functioned in a similar way. Along with new pathways, new settlements were established and landscapes were restructured. The pathways enabled movements within the territories of groups and polities, including movements of people, ideas, resources and symbols of power. This is how a political landscape is constituted. In summary, pathways and social actors with their personal relations structure space and thus contribute to the formation of characteristic ResourceCultures.

Excursus: Islands and Resources, Islands as Resources

In the 12th century, Abu Abdullah Muhammad al-Idrisi produced a map as part of the 'Nuzhat al-mushtāq fī ikhtirāq al-āfāq' (نزهة المشتاق في اختراق الآفاق). With south at the top, a considerable number of highly visible islands are shown. This is unlike modern maps, where scale can make islands disappear, reduced to a fate of non-existence because of their small size (Dierksmeier 2021, 17). It is important to keep in mind the changing centrality of islands in networks, often linked to their available resources. Sometimes location itself was an important resource for people throughout history (Schön 2020b). Thus, islands in distinct socio-cultural configurations can be approached from the perspective of **ResourceComplexes**.

Al Idrisi's map also introduces an overarching theme of islandness – how islands function between isolation and connection. The thick lines drawn by al-Idrisi emphasise the boundaries with the sea and the mainland, as if to suggest that each island is its own world, isolated from its surroundings. But the number and position of the islands, and the fact that so many were drawn, also suggest how connected the islands were by sea routes. Before air travel, it is hard to overstate the importance of islands as resources for anyone travelling by sea. For provisions of food and water, as trade hubs, navigational reference points, or locations of safety during enemy attacks, storms or shipwrecks, islands were in high demand for economic and geostrategic reasons. People, resources, cultural traditions, religious beliefs, technologies, seeds, diseases, songs, clothing trends and etiquette travelled along island highways on their way to and from the mainland and dispersed across the sea to places as far away as their ships could carry them (Dierksmeier 2021, 18).

Seen from space, the Earth's entire landmass is an island floating in the sea – so islands can be seen as the rule, not the exception (Depraetere 2008). Nevertheless, the physical separation of islands from continental land masses and the clear demarcation of land by the sea has led to the idea in biology and anthropology, for example, that islands can be natural or cultural laboratories for studying evolutionary or cultural developments in the absence of external influences (Dawson et al. 2023). The complex histories of settlement and colonisation, often dating back thousands of years, and the interconnections between islands and the mainland around the world, make the underlying assumption of insular isolation increasingly appear as a topos, influenced by a romanticised continental longing for simpler versions of the world.

Laura Dierksmeier / Frerich Schön / Sophie Hüglin

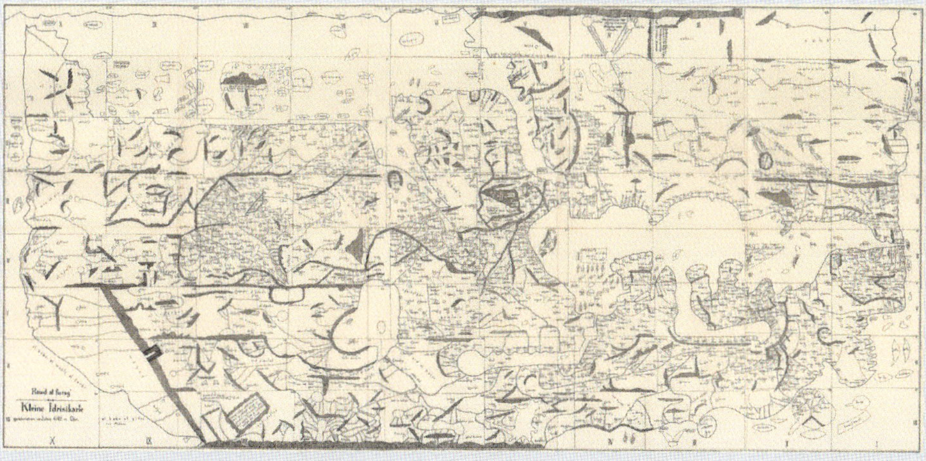

Fig. 12 Smaller reproduction of the Idrisi map of 1192, published by Konrad Miller in his 'Mappae Arabicae' (1926–31) (ULB Halle).

Connectivity is an important prerequisite for movement in space. Therefore, the concept of connectivity needs to be discussed when studying intra- or interregional movements. The mechanisms of connectivity during prehistoric periods have been studied archaeologically in the Iberian Peninsula. In the Guadalquivir region, a high degree of mobility can be observed during the Chalcolithic (3rd millennium BC) and the Bronze Age (2nd millennium BC) (Bartelheim et al. 2021c; 2022; Díaz-Zorita Bonilla et al. 2022; Chala-Aldana 2024). This allows us to identify connectivity as a fundamental resource for the formation of prehistoric societies in the region. In this framework, connectivity refers to the totality of direct relationships and connections between settlements or human groups in the region. Direct evidence of human mobility in the Bronze Age was provided by the study of human skeletal remains from the prehistoric cemetery of Cobre Las Cruces in the province of Seville (Chala-Aldana 2024). Interestingly, the isotopic analyses indicate a mobility of women in particular. In conclusion, both routes and the personal contacts that took place along them played a fundamental role in the movements and associated cultural processes in the Guadalquivir Valley during the Chalcolithic and Bronze Age.

How do ResourceCultures change over time and how are these processes linked to changes in movement and mobility within a region? Another example from Spain provides answers. In the Middle Ages, the common grazing lands (Dehesas) of south-western Spain were the preferred areas for transhumance, i.e. the mobile grazing of herds. Later, transhumance gradually lost its importance (Bartelheim et al. 2021b; 2021c). In the 19th century, the royal protection of transhumance was officially abandoned. This development was influenced by new markets and capitalist economies (Collantes 2009). As a result, more and more land was converted to agriculture. Since the 1990s, a counter-movement has developed, particularly under the influence of non-governmental organisations, which see transhumance as a sustainable way of using land that contributes to the conservation of biodiversity (O'Flanagan et al. 2011). As a result, the number of herds is now increasing again and long-distance migrations have been reintroduced, with more and more herders taking part.

As this example shows, the characteristic ResourceCulture of the medieval Dehesa was based on mobility as a fundamental resource. In this context, the routes and the actors moving along them, including their personal social ties, are of great importance. ResourceCultures change when individual elements of this system are modified, often triggered by external conditions. This also has an impact on movements, pathways and space use as the resource-based landscape system changes.

As we have just seen, movements are associated with a set of cultural practices, all of which are contingent in nature. In general, these practices are structured in a way that creates relationships between different resources. Practices can help to connect pathways to other aspects of movements and other fundamental elements of ResourceCultures. This can be demonstrated through agricultural practices in India (Hardenberg 2016; 2018a; 2018b). In the Odisha region of India, swidden cultivation is practised, which is associated with movement and mobility. Thus, the relationship between agricultural practices, resources, paths, mobility, other types of movement and social order can be illustrated. While the villages are located in the valleys, the fields are on the steep mountain slopes. A network of footpaths has developed to get to the fields. The movements of the farmers represent a form of mobility: not only people move, but also their possessions. Thus, field products are resources that structure people's movements in space and are associated with a specific kind of mobility that forms the rhythm of social life.

Another term for swidden cultivation is 'shifting cultivation' (Riehl 2024), which arises from the fact that fields can only be used for a certain period of time, as the soil loses its fertility. To cope with this problem, farmers have developed a mobile strategy: they keep large uncultivated areas of forest in reserve, where stones mark out places where future villages can be built. When a certain area is no longer arable, the village is abandoned and the people move on. This kind of mobility requires a specific social order of space. For this reason, mountain areas are divided into different territories, which are assigned to specific clans.

In these two examples, the movements of members of a particular group are characterised by two types of mobility: between the village and the mountains, and between old and new villages. The first type of mobility is individual or familial, while the second is collective and village-based, covering longer distances and taking place at intervals of decades.

The various examples from different periods and regions presented above show that movements of people and goods shape many societies in different ways. They can thus be understood as an important aspect of their ResourceCultures. Pathways can often be identified as an essential resource. They form the basis upon which other resources such as food, animal products, exchange goods or dominance are created and valued. Pathways are not only the basis for human economic activities such as agriculture, animal husbandry or trade. These resources, and the ways in which they are used and valorised, have a decisive influence on the nature, extent and dimensions of movements in a given society. Movements in a cultural, resource-based context can be understood in a variety of ways, ranging from transit movements to expansion and regular movements. Repeated movements shape many aspects of people's social lives. This is how movement becomes mobility. Mobility is closely related to a society's religious beliefs, social institutions and political structures. This high degree of interconnectedness is best understood through the concept of ResourceCultures. The processes that make up ResourceCultures do not follow specific mechanisms or rules; rather, the diverse range and contingency of ResourceCultures are clearly evident in the examples described above.

4.4 Supraregional Mobility and Long-Distance Movements

Supraregional mobility is driven by a combination of economic, political, social and cultural factors that have shaped the course of human history and contributed to the interconnectedness of the world. Many of the early migrations were long-distance movements. While these movements accounted for a small proportion of the total – many more movements took place within regions, as noted above – it was the long-distance movements that changed the global maps of societies and settlement distributions. Throughout history, they have been facilitated by various factors, including human evolution, technological advances, trade networks, political alliances and cultural exchanges. Conflicts between migrants and indigenous populations were frequent and led to dramatic events. The socio-cultural context of supraregional mobility depends to a large extent on the processes that took place between potential migrants and the resident population at home, and feeds back into the cultural history of both the migrants and the indigenous population. Key processes of supraregional mobility in different historical periods are migration and expansion, conflict and colonisation.

Migration and expansion of *Homo sapiens* (see excursus 'Migration Past and Present') in the period between about 45,000 and 40,000 years ago (e.g. Higham et al. 2011; Hublin et al. 2020) caused a clear break and profound changes in European human history. Although recent proposals may point to a much earlier presence of modern humans (Slimak et al. 2022), current knowledge supports the idea that the main dispersal of our species occurred primarily during these 5,000 years (e.g. Cortés-Sánchez et al. 2019; Hublin 2015; Mellars 2011). This was accompanied by the gradual decline and extinction of the remaining indigenous Neanderthal populations (e.g. Davies 2014; Higham et al. 2014), which had successfully lived in Europe for the previous 300,000 years (Bolus 2009). The reasons for their disappearance have been controversially discussed (e.g. Banks et al. 2008; Bar-Yosef 1998; Kuhlwilm et al. 2016; Timmermann 2020; Vaesen et al. 2019; Wynn et al. 2016). It appears that this transition has not been uniform. For example, while the migrating *Homo sapiens* found an already depopulated stretch of land in the Swabian Jura around 42,000 years ago (e.g. Conard/Bolus 2008; Higham et al. 2012; Richard et al. 2019), as suggested by sterile boundary layers between the Middle and Early Upper Palaeolithic settlement horizons (e.g. Conard 2002; Conard/Bolus 2006), the presence of the Châtelperronian in its range from eastern France to the Pyrenean region (e.g. Floss 2003; Floss et al. 2017; Soressi/Roussel 2014) marked different transitional scenarios, not least related to encounters and exchanges between these two human forms (e.g. Bar-Yosef 2002; Soressi 2010). One hypothesis is that the late Neanderthals proved less flexible than the immigrating *Homo sapiens* in a period of constant climatic change. As a result, they were forced into last refuges and eventually became extinct.

For Neanderthals and modern humans, the only indirect evidence for social and cultural factors of mobility comes from the study of their tools and production methods. Neanderthals and modern humans clearly differ from each other, and especially with regard to the use of organic materials such as bone, antler or especially ivory, the inventories of the early Upper Palaeolithic show an increase and a more diversified spectrum (e.g. Wolf 2015; Floss/Wolf 2021). However, in their way of life as Ice Age hunter-gatherer groups, these two types of humans are not fundamentally different. On their seasonal migrations and with frequent changes of location, they roamed large territories of several thousand square kilometres (e.g. Bosinski et al. 1986), which requires a high degree of mobility on the one hand, and a good knowledge of the respective habitats on the other, as the availability of essential resources such as food, water, but also lithic raw materials for the necessary production of tools is essential. Lithic raw materials, in particular, can be used as indicators of mobility, and thus of possible movement or transport routes and cycles, due to their site-typical varieties and characteristics (e.g. Cziesla 1990; Wilson 2007). Long-term transport should be interpreted in terms of exchange activities or long-distance movements, for instance during the expansion of *Homo sapiens* into Europe. Here, large river systems such as the Danube or the Rhône seem to stand out as focal lines of human mobility (e.g. Hussain/Floss 2014).

Conflict accompanies the migration and expansion of societies, and many dramatic conflict histories developed from the waves of migration during the Late Bronze Age in the eastern Mediterranean around 1200 BC, although these have sometimes been overemphasised by later generations of historians. For example, the hypothesis of the invasion of the Sea Peoples was based partly on evidence and partly on generalisations of isolated events taken to be typical of larger regions. Millek (2019) assessed the supposed violent movement of the Sea Peoples at the beginning of the 12th century BC, which supposedly caused a horizon of destruction, inhibited interregional exchange and brought about cultural changes in the decades after the beginning of the Iron Age. He was able to show that changes in inter-

Excursus: Migration Past and Present – Or Why We All Have a Migratory Background

There is hardly a more politically charged issue in our societies than that of migration and its real or supposed sociological and demographic consequences. Right-wing extremists abuse this issue for their inhuman political propaganda.

Although today's *Homo sapiens* in Europe owes its existence to various waves of migration – just think of the spread of modern humans 45,000 years ago, the Neolithisation of Central Europe or the early medieval Great Migration, which even owes its name to this phenomenon – sedentarism paradoxically seems to be the measure of all things for humanity, while migration often has negative connotations.

In the Palaeolithic, the oldest and by far the longest period in human history, migration and mobility were among the basic human behaviours, such as reproduction, hunting and sleeping (Floss 2009). Using the **ResourceAssemblages** perspective, we can distinguish between different levels of migration over time, but also at the same time. While at a meta-level we are talking about whole population shifts, as the examples above show, there are also migrations at a regional and daily level.

Ice Age populations migrated in a seasonal cycle from one place of residence to another, adapting to environmental conditions and resource availability, with the rhythm and frequency of these migrations also depending on climatic conditions. In a cold, dry environment, Ice Age hunter-gatherers maintained a more structured lifestyle, which could include longer periods of sedentarism. In more temperate climates, with greater vegetation cover and the decline of large herd animal populations, movements were more frequent.

In principle, a distinction can be made here between so-called residential mobility between centralised campsites and logistical mobility undertaken from a central residential site. In a cold, dry environment, groups of Ice Age hunters could easily cover hundreds of kilometres per year in this way.

In modern archaeology, the areas they travelled through and where they once lived can be determined, for example, by analysing the origin of objects and lithic raw materials they carried, by studying stable isotopes or by genetic analysis.

However, the world of the Ice Age and that of today are very different, so we do not want to trivialise the migrations that can be observed today. The loss of homeland is always painful and is all too often caused by climate change and war, or a combination of factors. Nevertheless, on a global scale, societies are always and everywhere subject to dynamic change, not just in exceptional cases. Despite all the challenges, especially for those who have lost their homes, we should also see the consequences of migration, at least for the receiving communities, as an enrichment of diversity.

Harald Floss

regional exchange did not correspond to a dramatic transition from the LBA to the Iron Age. Furthermore, there is little evidence that the trading ports on the southern Levantine coast were destroyed around 1200 BC. Nor could destruction have been an obstacle to interregional exchange (Millek 2017; 2019, 147–188; see also Millek 2018; 2023).

For example, trade patterns were already changing before 1200 BC, while other flows of trade goods continued afterwards. Trade in both Cypriot and Mycenaean pottery largely ceased between 1300 and 1250 BC, well before the end of the Late Bronze Age, while trade in certain goods from Egypt and trade in non-local metals continued and even increased after 1200 BC (Millek 2022). Thus, while interregional exchange acted as a cultural resource, bringing new goods and ideas that were incorporated into the cultures of the region, changes in this exchange did not overtly affect the well-being of the southern Levant. The perspective of a ResourceAssemblage helped to catalyse an examination of how, over time, dynamics in interregional exchange may have acted as a key factor in the transition to the Iron Age.

Overall, migration, expansion and conflict have been deeply intertwined throughout history, with movements of people often being both a consequence of and a catalyst for various forms of conflict. In ancient times, conflicts between rival empires and nomadic tribes often led to population displacements and forced migrations. The migration of nomadic peoples and 'barbarian' tribes also played a significant role in the transformation of ancient empires, such as the Roman Empire. In the late Roman period, waves of 'Germanic', 'Hunnic' and other tribal migrations into Roman territories contributed to political instability, economic disruption and social upheaval. The Visigoths' sack of Rome in 410 and the migration of the Huns into Europe are notable examples of how migratory movements could trigger conflict and contribute to the collapse of established political orders. The medieval Crusades, a series of religious wars waged by European Christians against Muslims in the 'Holy Land', led to significant population movements and migrations during the Middle Ages. The Crusades also stimulated cultural exchange and trade between East and West, but often resulted in violence, plunder and the forced conversion or expulsion of non-Christian populations.

Colonisation refers to the process, driven by mobility, of establishing control over foreign territories or peoples for the purposes of exploitation and settlement. Diachronic discontinuities in settlement systems over the last 8,000 years have been described, for example for the Jiroft region (Pfälzner/Alidadi Soleimani 2017; Pfälzner et al. 2019), which was used as an example of regional mobility in the previous chapter. The settlement systems of the region were particularly dense from the Late Chalcolithic to the Early Bronze Age, but collapsed and disappeared completely at the end of the Early Bronze Age. After almost 1,500 years of abandonment, the region was recolonised during the Achaemenid period (500–330 BC). Dense settlement patterns, similar to those of earlier periods, reappeared. After 1200 AD there was another collapse. Once again, the region was almost completely abandoned. It was not until modern times that a third period of intensive settlement developed in the region. The repeated recolonisation of the region can be seen in the context of supraregional movements over long distances. During the Early Bronze Age from 3000 to 2000 BC and the later Achaemenid and Parthian periods, Jiroft was connected to supraregional road networks (Kohl 1987; Barjamovic 2018; Wiesehöfer 2013). Stones, especially chlorite and diorite, from the Jiroft region (Potts 1989; Pfälzner/Alidadi Soleimani 2017), but also from Oman, were exchanged over long distances between Mesopotamia, Iran and Central Asia (Lamberg-Karlovsky 1975; Dales 1977; Kohl 1978; Potts 1993), as evidenced by decorated chlorite vessels of the 3rd millennium BC (de Miroschedji 1973; Kohl 1975;

Lamberg-Karlovsky 1988). Lapis lazuli from Afghanistan reached Mesopotamia either via India or via south-eastern Iran through the corridors of the Jiroft region (Lamberg-Karlovsky/Tosi 1973; Potts 1982; Francfort 2020). Other goods, such as textiles and grain, were probably also traded and transported over long distances with the stones. But it was not only material resources that were transported along these routes. People also moved along with them transmitting knowledge and ideas (Vidale 2015). The strengthening of relationships and personal contacts as well as the gain in prestige of the actors through the exchange of gifts and knowledge is another part of the ResourceCultures of the Jiroft area, as evidenced for example by finds of prestige objects in tombs or temples, such as imported chlorite vessels in Mesopotamia as valuable votive offerings (Kohl 1975). In addition to economic motivations, socio-cultural reasons certainly also initiated supraregional movements. This example shows that the valorisation of rocks alone cannot explain the phases of abandonment and recolonisation of the Jiroft area, but ResourceAssemblages of material and immaterial resources around long-distance transport allow a coherent and comprehensive understanding of the abandonment and recolonisation of a region. In this context, permanent networks of routes within and between regions, which can be reactivated again and again, are an important stimulus for colonisation. This effect can also be observed in the modern history of south-eastern Iran. In recent years, the region around Jiroft has experienced an economic and settlement boom. Some possible reasons are the construction of the aforementioned 'Persian Gulf Highway' (see Chapter 4.3), as well as the start of chromite mining in the Jiroft region (Frauen/Klocke-Daffa 2023).

The Iron Age also saw exceptionally relevant 'colonial' movements, for example when 'Phoenicians' moved to the western Mediterranean in the 9th and 8th centuries BC and founded Carthage and many other places, and the 'Great Greek Colonisation' between the 8th and 6th centuries BC (Schön/Töpfer 2016). Evidence of the socio-cultural history of the Phoenicians and Carthaginians is provided by the origin of the trade goods found in their main destinations. Interestingly, they did not trade mainly with their own pottery, but also with many Sardinian-Nuraghian products and smaller quantities of products from Cyprus and Greece.

Concerning the Greek colonisation phase of the Iron Age (Bernstein 2004; Tsetskhladze 2006), a number of theories had emphasised pull factors such as the amount of mining opportunities in the destination areas as well as grain production possibilities. However, these theories did not seem to be able to explain the whole process of 'Greek colonisation'. Coin minting and production sometimes occurred earlier in colonies, such as in Akanthos on the Chalkidiki Peninsula, compared to its metropolis on the island of Andros, from which Akanthos was settled. If silver mining had been the motivation for settlement, one would have expected silver to be used in the metropolis first. Similarly, the focus on grain can only explain part of the colonisation process. Grain does not appear on many of the coin images of the Black Sea colonies, but other items such as sturgeon do. Finally, push factors such as the Persian Empire threatening the Ionian cities, and overpopulation in central Greece could explain parts of Greek colonisation. Overall, similar to intraregional mobility, the supraregional movement that led to Greek colonisation can be interpreted as a ResourceComplex consisting of many different resources, including immaterial components such as the identification of former migrant families – and also many families of indigenous ethnic origin – with the Greek metropolis. This suggests that colonisation was a complex process with very dynamic and multifunctional identities and motivations.

In terms of supraregional mobility, the connectivity of regional networks of pathways is particularly important for long-distance movements. For example, long-distance connectivity has been studied archaeologically for prehistoric times in the Guadalquivir region of the Iberian Peninsula. Least-cost path analyses intersecting with historic livestock pathways in the present day area (Chala-Aldana 2024) showed a high degree of connectivity between regions via livestock routes. The coincidence of archaeological settlements with historically known transhumance routes suggests a high degree of transhumance in the region. For the Bronze Age, supraregional exchange is best evidenced by typological similarities of ceramic forms throughout the Guadalquivir Valley (Bartelheim et al. 2021c; 2022; Díaz-Zorita Bonilla et al. 2022), suggesting close cultural, economic and probably also personal contacts.

These examples show that connectivity is a significant element in understanding social and cultural processes within and between regions. During the Chalcolithic and the Bronze Age on the Iberian Peninsula, actors and their personal relationships are considered as relevant resources, which, in addition to raw materials and agricultural products, led to the formation of characteristic ResourceCultures of that time. Long-distance movements can be seen as a decisive motor for the formation of these ResourceCultures, both in terms of pathways and their connectivity.

In addition to connectivity, food is a crucial element of mobility. For example, in Viking Age societies in the Early and the beginning of the High Middle Ages, trade, migration and the famous plundering activities implied a higher mobility in parts of society related to the distribution of food (Maravall/Baten 2019; Baten et al. 2019). In Scandinavia, the distribution of food resources was quite different compared to other societies in Europe at the same time. This difference was influenced by the mobility component of the Viking-era ResourceCulture. Nutritional inequality between social groups and between men and women was relatively modest, i.e. smaller than in other European regions of the time. One mechanism may be that trade, migration and plundering require the active participation and motivation of each individual seafarer and warrior. The Vikings could not carry out these activities with unmotivated individuals. In contrast, farming and craftsmanship activities in other regions of Europe do not depend on such extraordinary movement. Some simple tasks in agriculture can even be performed by slaves, and the ruling elite do not depend so much on the individual motivation of each participant. A second mechanism that could explain the relatively strong position of women is that trade, shipping and plundering meant that men were absent for large parts of the year, and women had to take over agricultural and craft work, along with a specialisation in cattle farming and dairy production. The role of women was stronger than in cereal farming because animal husbandry relied less on the upper body strength typical of the male body. This greater role in production strengthened their position within the household.

In Central Europe during the migration period of the 5th and 6th centuries, the inequality of human stature, which indicates differences in food consumption, was significantly lower than in the 7th and 8th centuries (Meinzer et al. 2019). Mobility played a smaller role than during the turbulence of the *Völkerwanderungszeit*, and a highly structured society developed with only modest geographical mobility. The ResourceCulture differed markedly between partially mobile and immobile societies. It shifts from mobility to food at its centre and includes agricultural knowledge, tools and especially land for cultivation.

4.5 Synthesis

The last thousand years have witnessed remarkable developments in mobility and movement, reflecting the evolution of societies, technologies and global connections. In ancient times, mobility was mainly limited to walking, horse riding and sailing. Overland travel was arduous and time-consuming, and people relied on footpaths and primitive roads. Intraregional mobility within a region included daily and regular movements for subsistence, socio-cultural and religious gatherings and transit movements. Horse riding offered a faster alternative for those who could afford it, but was still limited by the endurance of the animals and the quality of the terrain. Meanwhile, maritime trade routes linked distant regions, facilitating the exchange of goods and ideas across continents. Supraregional mobility and early migrations became a driving force of socio-cultural development, helping to connect regions and further shape the course of human history. The Middle Ages saw the emergence of more structured transport systems, particularly in Europe and Asia. The development of road networks and the use of pack animals facilitated the growth of trade and the spread of cultural influences. Over time, almost all kinds of material and immaterial resources, from minerals to ideas, were mobile and shaped the socio-cultural fabric of the societies involved, both within a region and between regions. Resources were both push and pull factors, acting along pathways and their connectivity or networks, which were a central element of movements.

In order to understand the drivers and consequences of social movements and their impact on landscapes and natural environments, it is necessary to analyse a specific set of resources in the context of a ResourceComplex. Intraregional values change over time due to mobility, and a similar phenomenon occurs in neighbouring regions connected by mobility. ResourceAssemblages facilitate an analytical understanding of this phenomenon. Both approaches help to show that it is not individual resources such as gold, knowledge or inventions that cause mobility. Depending on how they are valued and how they are embedded socially, politically, symbolically or religiously, they can act as triggers, although they are not generally considered as such. Rather, it is always different combinations of material and immaterial values and their resources, seen as ResourceComplexes and ResourceAssemblages, which can explain the motivation for movements as well as their modalities, course and accompanying socio-cultural developments. The same can be said for questions about the targeting, planning and control of movements, the involvement of specific social groups and the nature of communication and contact established. The answer can be found in the concept of ResourceCultures, which emerge, develop and disappear as representatives of a certain space-time structure when their resources change in value or state.

The theoretical concept of ResourceCultures and illustrative examples show that movements represent spatial cross sections in historical developments that influence both social identities and cultural symbolism, and thus interact with developments (see Chapter 3) and valuations (see Chapter 5). Mobility can be triggered by resources, but it can also entail a change in the way resources are handled or lead to a reassessment of resources. Resources can thus be causal for movement or dependent on movement. This interplay can be observed in a variety of historical and geographical contexts. Pathways, connectivity and food are key elements in this interplay between resources and movement, serving multiple functions for mobility as a mechanism for settlement, the conservation of power, land management and belief.

The field of mobility is expected to undergo further significant changes in the future. These changes will be driven by innovations such as autonomous vehicles, hyperloop transport systems and urban air mobility based on renewable energy. These developments may lead to the emergence of ResourceCultures in societies that prioritise sustainability and environmental protection. Such changes could have a profound impact on the mobility landscape and contribute to the long-term stability of democratic structures.

5 Values, Valuation and Value Creation of Resources

Roland Hardenberg, Irmgard Männlein, Thomas Thiemeyer

In the context of resources, terms such as 'value', 'valuation' or 'value creation' are very often defined in economic terms. As Jochen Henning (2007, 84–87) points out in an overview, the term value is understood in economics either in the sense of results or positions that people have achieved through their efforts, or as a designation for a standard in the exchange of goods. A distinction is made between (subjective) use values and (objective) exchange values. Depending on the approach, value is then expressed in the desire to buy, in the price or in the perceived services, functions or attributes (Henning 2007, 87). Value creation is understood from the point of view of companies as a 'process of creating added value through processing' ('[…] wird als *Wertschöpfung der Prozess des Schaffens von Mehrwert durch Bearbeitung* bezeichnet' [Müller-Stewens/Lechner 2005 as cited in Henning 2007, 84; italics in original]) and from the point of view of the buyer as a 'welfare gain [...] through the consumption of products and services' ('Der Wohlfahrtsgewinn […] durch den Konsum von Produkten und Dienstleistungen' [Henning 2007, 87–88]; translations by the authors). Resources, such as material or immaterial goods, are what make this creation of value possible. This definition in economics is tailored to specific economic fields of action and differs in part significantly from the use of the term in other disciplines, as can be seen from Krobath's (2009) overview of concepts and discourses on values in philosophy, psychology and sociology. Two approaches can be distinguished that have been highly influential in these fields, focusing more on the symbolic and social dimensions of values. One goes back to Clyde Kluckhohn, who understood values as the 'desirable' (Kluckhohn 1951, 395) and thus as an orientation for social action. This definition also influenced Rokeach (1973), who, however, put more emphasis on the meaningfulness of values and defined values as 'an enduring belief that a specific mode of conduct or end-state of existence is personally or socially preferable to an opposite or converse mode of conduct or end-state of existence' (Rokeach 1973, 5). This approach also underpinned the psychological work of Schwartz and Bilsky (1987) and was the starting point for the attempt to develop a model of quasi-universally valid values (Schwartz 1994). In sociology and psychology, this approach has been used to discuss questions about the origin of values, their relationship to the self and personality, the motivation of actions by values and changes in values (Hitlin/Piliavin 2004). In the case of resources, this raises the question of the extent to which certain values – perhaps cross-culturally – have motivated the search for and use of resources, and whether changes in the use of resources also reflect changes in the values of a society. The implicit universalism of this approach was the subject of intense debate in the 1970s and 1980s, sparked, for example, by questions of rationality (Wilson 1979; Hollis/Lukes 1982; Geertz 1984). The cultural relativist position in this debate offers a second approach to values. It has its origins in American cultural anthropology, particularly the work of Franz Boas, Alfred Kroeber and Ralph Linton. While universalists argue that there is a global core of values that differentiate under different environmental conditions and historical processes, cultural relativists postulate the uniqueness of cultural value systems (Heintz 2009, 5). According to this approach, a value is understood as something supra-individual, learned, which integrates cultural elements into a meaningful whole (e.g. Kroeber 1952). Value in the sense of meaning creates relationships or systemic connections. This idea is also found in the value theory of Louis Dumont, for whom values in non-modern societies 'express the order of relations that permeate the local ideas of the cosmos, which includes the society' (Berger et al. 2010, xvii). Based on this understanding of values, symbolic or structural anthropology (e.g. Platenkamp 2003) or archaeology (e.g. Hodder 1982) examines holistic systems of relations and meanings that are evident, for example, in ritual actions, architecture, settlements or the arrangement of grave goods. Conflicting values or conflicts arising from the loss of existing values or the impact of new values receive little attention in these approaches. According to this understanding, resources are symbolic, i.e. meaningful means by which values are expressed in terms of social hierarchies or valued orders.

More recent comparative perspectives seek to overcome the oppositions articulated in earlier approaches between universalism and particularism, holistic integration and conflict and between society and the individual. Robbins (2007), for example, argues that holistic value systems are most relevant to societies in which a central value is dominant, while theories of individuals' decision-making processes take hold when there is conflict between different values and people consciously reflect on their values and their meanings. This approach can be fruitful in the context of resource development issues in foreign spaces, for example, where different value systems or resource valuations clash as a result of colonisation and foreign domination. Robbins also combines psychological and cultural perspectives on values, because for him values have a subjective and an objective dimension: they are subjective because they are relevant to individual action, and they are objective because they exist beyond the individual as an aspect of culture (Robbins 2015, 220). For him and his co-author Sommerschuh, values are social in the sense of 'the good', that is, 'something people want or should want to attain' (Robbins/Sommerschuh 2016, 2). In relation to resources, this approach draws attention to the question of which concepts of the 'good' underlie the use of resources in a society.

Other approaches focus more on the social and symbolic dimensions of valuing things and material resources. As early as the 1970s, Michael Thompson (1979) developed the so-called 'garbage theory', according to which things often go through a phase of complete devaluation in the context of revaluation processes. Igor Kopytoff (1986, 73-77) has also made an important contribution to this question, arguing that through discourses and practices, such as processes of production or exchange, people make things unique, give them a 'biography'. The ethnologist Karl-Heinz Kohl (2003) also shows how the value of things depends on their cultural and historical context and changes. An example of this is provided by Kollewe (2007), who examines how prehistoric, anthropomorphic artefacts change their value and the role that social change plays in this process. In relation to resources, this means paying attention to what subjective and communally shared values are attached to a resource, how these values depend on time and context and how they may change. Following Bender and Taves, one could therefore understand the concept of value as 'the claims that people make regarding the importance of something (anything), regardless of whether the importance carries a positive or negative valence' (Bender/Taves 2012, 10).

The concept of 'value creation' is not yet well established in cultural studies. Studies of cultural heritage use the term to describe the economic, emotional and political dimensions of these resources. Thus, Bendix et al. (2007, 10–11) understand 'value creation'

('Wertschöpfung') as both the nobilisation of cultural heritage, i.e. the symbolic elevation through the award, and the associated, usually veiled, economic gain that results from the appeal of cultural heritage to the masses (see excursus 'Cultural Heritage'). 'Value creation' can also be used to describe the processes by which concrete values turn something into a resource for social and political relations, entities and identities. The term then refers to the emergence of practices or forms of organisation through which people appropriate something of value, make it understandable to others and use it. In particular, this also brings to the fore various forms of representation and symbolisation, such as myths, legends, historiography or translations, as well as narratives through which political claims are legitimised, knowledge is preserved or remembered and change is made possible. These forms of value creation can thus themselves become a resource for the community. Graeber (2001), who sees value as the 'importance of actions' i.e. the recognition of the capability of human beings to act and not just to produce, shifts the focus from the capacity to generate value through exchange to an analytical horizon that includes the symbolic and material perspectives of actions.

In summary, taking into account these different cultural approaches to values, three dimensions of resources can be identified to structure this section: first, the symbolic dimension (Chapter 5.1), i.e. the meaning or significance of resources. This perspective draws attention to the ideas and conceptions that subjectively and objectively define the values of resources, to the ways in which these resources are represented in different media and to the practices and processes by which resources are valued and marked as particularly desirable. Second, the social dimension (Chapter 5.2). From this perspective, we look at what social values resources express, how social relations, orders and identities are created through the use of resources and, finally, how societies change in the long term as a result of the use of resources. Third, the political dimension (Chapter 5.3). This focuses on the often conflictual processes of resource appropriation and also examines the options available to individuals. This perspective draws attention to crises and forms of exploitation, for example in the context of colonialism and nationalism, and to the political significance of knowledge about resources, which can both legitimise and destabilise rule, open up or restrict potential for action, and be both accumulated and exchanged. In the following three sections, these three dimensions of resources, which are often intertwined and therefore cannot be treated as clearly delimited perspectives, are discussed in detail and illustrated with examples.

5.1 The Symbolic Dimension of Resources

'In all its dimensions, including the social and the material, human existence is symbolically constituted, which is to say, culturally ordered' (Sahlins 1999, 400). When we speak of the 'symbolic' dimension, we are referring to objects, actions, events, relationships that serve as vehicles for ideas and conceptions and are to be understood as 'models for' and 'models of' the world (Craik 1943; Geertz 1973, 93–94). The symbolic dimension of resources thus describes a meaning or meaning-making of resources that is not *a priori* assigned to them: it is the subjective as well as the objective, culture-specific as well as cross-cultural ideas of human actors that are decisive in turning something into a resource with a very specific meaning or symbolic value. The symbolisation of resources is of course not a simple matter: it is based on ideas and concepts, it arises through habituation, rupture or transformation, it arises historically or is constructed.

Excursus: Collective Memory

The idea of a 'collective memory' ('*kollektives Gedächtnis*') was popularised in the 1990s by the Egyptologist Jan Assmann (2011 [German original 1992]). He states: 'It is fundamental to the thesis that the past only comes into being insofar as we refer to it.' (p.17). This means that the past is only constructed in memory. Assmann calls this 'memory' ('*Gedächtnis*') because, with reference to Maurice Halbwachs's theory of memory (1992), he assumes that individual memory can only emerge in the community. It is acquired through socialisation, i.e. it is learned. Memory is therefore 'collective' when it refers to groups rather than individuals. It is thus closely linked to collective and individual identities (Thiemeyer 2021; Nora 2009).

Assmann divides 'collective memory' into two areas: 'communicative memory' and 'cultural memory'. The former is intergenerational. It is based on shared 'biographical memories' (Assmann 2011, 34–35) of contemporary witnesses and is tied to their lifetimes because it is passed on orally. It appears in many different forms because it tolerates the coexistence of numerous, sometimes contradictory stories about the past. 'Cultural memory', on the other hand, is artificial. It is supported by institutions of remembrance such as museums, memorials or archives and is given a fixed place in people's everyday lives through monuments, rituals, festivals or holidays so that they can remember certain people or events permanently (Nora 1984–1994). 'Cultural memory' does not spread by itself (like 'communicative memory'), but must be specifically organised. It is subject to controlling bodies (institutions, researchers) that authoritatively regulate access. The transition from private 'communicative' to publicly negotiated 'cultural memory' is the moment when history is made. It marks the transition 'from the memory struggle [*Erinnerungskampf*] to the culture of memory [*Erinnerungskultur*]' (Frei 2005, 26).

The concept of 'collective memory' has been widely accepted in cultural studies because it makes clear that our view of the past (transmitted as history) (Munslow 2007) is always culturally shaped and learned. However, it has also been criticised for its tendency to essentialise: the concept of 'collective memory' suggests that there is such a thing as a naturally evolved (national) culture of memory that determines people's identity. One must submit to it and cannot change it. Since people *de facto* actively contribute to the formation of memories and constantly change them, 'collective memory' can be understood as a central element of a **ResourceAssemblage**, which would expand the concept of 'collective memory' to include contingency. Second, the idea of 'collective memory' is based on homogeneous societies with a uniform socialisation of all their members, which is binding for everyone. In societies characterised by migration and the fluctuation of their members, however, this premise is increasingly questionable.

Thomas Thiemeyer

Fig. 13 The memorial to the murdered Jews of Europe in Berlin (Photo: Wolfgang Staudt from Saarbrücken, Germany, CC BY 2.0).

Moreover, the origin, ownership and transmission of resources are represented and legitimised through multiple symbolic forms, through 'local narratives' (Gudeman 2009, 66–67), such as myths or other forms of memory. In this context, symbolic orders in the form of texts, standardised actions (rituals), architecture (such as palaces or sacred complexes/temples) or sites of memory (see excursus 'Collective Memory') can be resources for specific social units, as the symbolic value of resources plays an important role in the identity of communities. However, this is not always the case, i.e. the symbolic encoding of resources and value conversions interact with each other. In this context, symbolisation (Firth 1973, 15) and conversion contribute to the contextual change of meanings of resources and thus make possible different value creations of one and the same resource (e.g. a once valuable resource can become valueless and vice versa, see Chapter 5.1.2). Not only are the phenomena of symbolisation of resources, their processes and mechanisms, complex, resources themselves also have multiple symbolic meanings. In principle, symbolic resources are always supported by social and, in many cases, also political meanings and are co-conditioned by them, so that the multidimensionality of cultural systems of meaning (e.g. Kertzer 1988) becomes very tangible: this includes condensation, when resources have different meanings at the same time and an interaction between them takes place, but it also includes the multivocality (Turner 1967, 45–52) of resources, when they are interpreted differently within a community, for example, and it also includes the resulting ambiguity, which can easily be functionalised politically – especially in the area of the symbolic dimension of resources. Resources can also have a metaphorical meaning, which is then closely related to symbols within cultures as part of systems of meaning. These meanings, 'dominant', 'key' or 'core symbols', and the actions associated with them are to be interpreted as representations of a social system, its value systems and religious ideas (Firth 1973, 77).

Some examples of symbolic resources in the narrower sense are presented below. These can be identified in the field of written literature and poetry as well as in the visual arts (Chapter 5.1.1) in the form of mythically conceived realities from people's lives, with regard to the change in the attribution of meaning of elementary resources such as soil and water in a settlement in north-eastern Mesopotamia (Chapter 5.1.2), or in the field of religion in the form of certain practices and the resulting infrastructures as well as revaluations (Chapter 5.1.3).

5.1.1 Representations of Symbolic Resources

For the symbolic charging of resources, the modes of representation in different media play as important a role as the practices and processes by which resources as such are valued or devalued. Resources can become recognisable as such and acquire meaning in oral traditions as well as through written texts, images and material objects. Symbolic resources can thus be both immaterial and material. Interesting dynamics and interactions can be seen here: for example, texts with certain aetiological myths, such as Hesiod's Greek poetic texts from the early 6th century BC ('Theogony' and 'Erga' with the myth of Prometheus or 'Erga' with the myth of the ages), which also shed light on the meaning of certain rituals practised by humans (e.g. sacrifice) (Männlein-Robert 2021; Herren 2021), themselves acquire the value and meaning of a resource relevant to their own culture. This can be seen in the equally intense and lasting reception of Hesiod in ancient literature, the intertextual meaning and poetic tradition-making of the symbolic values of the resources he both handed down and constructed, as can be seen, for example, in Hellenistic times on the basis of Aratus's

'Phainomena' (Clausing-Lage 2019). Not only does Aratus use Hesiod's 'Erga' as a pretext (in the sense of intertextual theories), but he also gives this archaic text a new value for his contemporaries by using its astronomical descriptions as a model and thus demonstrating his appreciation. In this way, the Hellenistic author Aratus creates a new symbolic value for Hesiod's text as a significant literary and scientific resource.

A particularly interesting symbolic complexity can be identified in Hesiod's 'Erga', for example, in relation to grain. Grain is central to Hesiod as the basis of human nutrition and thus also as a material resource: the archaic poet places agriculture at the centre of the farmer's activity. Grain, together with olives and wine, is the basis of nutrition in ancient Greece, which is why the sowing, harvesting, storage and cultivation of grain are extensively depicted (Herren 2021, 208–218). At the same time, however, the resource of grain is revealed as a complex, multiple symbolic resource, as religious, social and poetological dimensions of it can be identified in various narrative patterns and myths in which Hesiod focuses on the foundations and basic conditions of the farming community from different perspectives. The religious dimension comes into play at the very beginning of the poem, since the relationship between man and the gods is implied here and needs to be explained. As Hesiod recounts in mythical form in the famous world age myth (Erga, verses 109–201), from the Golden Age through the Silver, Heroic and Iron Ages to the current Iron Age, the situation of humanity has gradually deteriorated as a result of increasing ethical and moral cruelty. In the context of the cultural descent model, the focus is on agriculture, including grain. The resource grain, which is difficult for humans to obtain, is thus essentially dependent on the favour of the supreme god, so that grain clearly has a symbolic meaning with religious connotations. The closely related social dimension can be grasped, for example, when Hesiod explicitly denies the use of grain to the violent, antisocially acting brazen age, thus suggesting a connection between dietary practices and the ability to form a community (Herren 2021, 33–34, 101). The archaic poet Hesiod thus makes clear in the poetic textual medium his realisation that the individual is fundamentally dependent on good social as well as religious relationships, a realisation that crystallises in the construct of the mythic narrative in the corresponding symbolic connotations of the central resource of grain.

The same resource – grain – is also represented together with wine as a key resource in Euripides's 'Bacchae' (cf. Schweizer 2021), an Attic tragedy from the late 5th century BC. The fact that grain is understood as a gift of the goddess Demeter – together with the alternation of growth-favourable phases in the course of the year represented by her daughter Persephone – and wine as a gift of the god Dionysus, clearly demonstrates a sacral connotation of these two elementary resources in the tracing back to gods (Eur. Ba. 274–279). Grain 'is' Demeter, wine 'is' Dionysus – according to the religious understanding of the Greeks at this time – so that these resources are not only reflected in a sacred context through identification, but are also firmly embedded in the respective cult and represent central elements of the ritual acts (Obbink 1993, 85). In addition to the literary medium, there are also figurative representations of the deities mentioned by Euripides or figures with a similar reference to grain and vine, for example on early classical so-called Locrian votive reliefs. Both the Attic tragedy and the images thus attest to mythical and ritual forms of reflection on grain and wine as basic agricultural resources and their contextual valuation. In both representations, Demeter and Dionysus, i.e. bread and wine, function both as basic foodstuffs and as a means of forgetting hardship and sorrow, as *pharmakon*, Demeter specifically also as a figure of the invention and dissemination of agriculture, Dionysus specifically as the inventor and disseminator of wine with all its complex connotations. Here, an extended conceptualisation of resources as a means of constructing, maintaining and changing identities that depend on cultural and social appropriation and valuation emerges (cf. Hardenberg et al. 2017, 14–15), in particular a conceptualisation in which resources are seen as part of networks or bundles of tangible and intangible elements of social and material spaces that are inconceivable without each other (Latour 1996, 369; 2005; Hodder 2014, 19–20). Figurative representations of, for example, Demeter, Persephone or Dionysus were seen as figurative embodiments of all these ideas in sacred spaces. In this sense, Euripides's descriptions of Dionysus and the *Bacchae* (his cult crowd) can be seen as paradigms for the image and cult of Dionysus that inspired visual artists (Isler-Kerényi 2015, 213; Porres Caballero 2013, 162). Admittedly, dancing women are already depicted on Late Archaic vessels used for symposia with gestures and objects that are described almost a hundred years later in Dionysian processions of initiates or cult participants in Euripides's 'Bacchae'. The viewer of the early 5th century BC, who was himself part of a symposium and thus of a kind of Dionysian ritual, thus saw figures of Dionysus, satyrs and maenads acting in cultic situations. Questions about the nature of the religious experience reflected in these pictorial media are therefore particularly relevant today (Osborne 1997, 208). In the case of the uninhibited satyrs, for example, who are visual symbols of the power and influence of the wine god Dionysus, no real or cultic correspondences can be demonstrated. As imaginary figures, they serve to embody cultural metaphors such as the suspension of morality or the dissolution of established social norms of behaviour (Stähli 1999, 167, 171). The dancing women, the *Bacchae*, were characterised in groups or individually in a kind of trance (cf. *Enthousiasmos*) with specifically ecstatic dance movements and abnormal postures (cf. Eur. Ba. 862–864). Satyrs and maenads are depicted on some vessels together with Dionysus in a quasi-mythical context (cf. Stehle 2012, 194). The aforementioned modes of representation and practices of visualising the key resources of grain and wine, which are represented in literature and images by Demeter and Dionysus (along with accompanying figures), are constitutive elements here, as are practices of consumption and their valuation, which can therefore also be regarded as symbolic resources in their own right (cf. Rüpke 2018, 25).

5.1.2 Valuation, Revaluation and Devaluation of (Symbolic) Resources

In the long period from 7000 BC to 2000 AD, fundamental changes in regional settlement patterns with regard to the distribution, number and size of settlements can be observed time and again in the settlement region of predominantly agrarian north-eastern Mesopotamia (in present-day northern Iraq) (Pfälzner 2010; Pfälzner et al. 2018; Wissing 2017). This shows that the settlement development of the region does not follow a linear or evolutionary development, but is characterised by breaks and shifts. But why do settlement locations change constantly, and why does a new settlement pattern emerge in each period? Water (in an arid landscape) and soil (to provide food for an agrarian community) were initially the most important resources for settlement. Due to the constant climatic and geomorphological conditions of the region, the resources of water and soil did not change significantly during the period under study. Rather, they were constantly available resources throughout the periods. Consequently, settlements should have always been located in the same places. This is not the case, however, and demonstrates that even resources such as soil and water can be subject to changing cultural valuation (see also excursus 'Freshwater Scarcity on Islands').

In fact, these basic resources were repeatedly reassessed over time (see also Chapter 4.3). Paths as resources – both factual and symbolic – also played a crucial role. They determine the connections between settlements, and between settlements and the outside world and other regions (e.g. Wilkinson 2014; Wilkinson et al. 2010). If such contacts are restructured – for a variety of reasons – a re-evaluation of the resource path takes place, which can also lead to a re-semantisation and new semantisation of the resources water and soil. The resources water, soil and paths form a thoroughly changeable network of valuation. For example, the contacts between settlements via roads are determined by numerous economic, political, social and cultural factors, which are determined by the demography and population density of a region, its social structure and its main economic activities (trade, agriculture, livestock, production) or political framework conditions (administration, defence) (Wilkinson 2014, 23–32). Thus, the external conditions of a historical period in the mentioned settlement area shape the requirements for settlements, their contacts and routes. These variables stimulate the re-evaluation of the basic spatial resources of water, land and roads in a region. In the course of these processes, resources can of course also be devalued, e.g. by making certain locations unattractive, although the resources water and soil are still available and can in principle be regarded as favourable. In this way, settlements become resource aggregates because they are the result of a comparative evaluation of a variety of resources and external factors. For the community in question, however, the settlement itself is ultimately the most important resource because it combines the valuations of all the resources that are important to the community.

5.1.3 Sacralisation of Resources

In cultural and religious studies, different aspects of the sacred are emphasised and studied (e.g. Hardenberg 2017): the sacred can refer to that which is protected from the profane (e.g. Durkheim [1912] 1976; Eliade 1959), it can be associated with extraordinary emotions of fear and awe (e.g. Otto 1917), or it can be experienced as something ambivalent, forbidden and dangerous (e.g. Freud 1950). The sacred is used to denote the liminal and transitory (e.g. Douglas 1966). It can denote people, objects, places or practices associated with gods, spirits, divine or magical forces (e.g. Marett 1909), be applied to dangerous, violent or destructive actions and forces (e.g. Girard 1972) and refer to rare, external objects of immense value (e.g. Sahlins 2013).

Combining these different approaches leads to a 'polythetic' understanding of the sacred that can be applied to resources, understood as those means that are essential for the formation, continuation and transformation of social groups or social identities (Hardenberg et al. 2017, 14). Resources, according to this understanding, can take many forms and be used in different ways depending on cultural ideas and values. This concept of resources has some similarities with Durkheim's theory of the relationship between religion and society. For Durkheim ([1912] 1976), sacral powers are in fact those forces that create social bonds and hold society together. Religion, understood by Durkheim as the separation between the sacred and the profane, imposes those collective constraints that individuals also experience when they act as members of a community. If resources are seen as values and forces that also shape social relations and identities, then it is to be expected that such resources will evoke feelings of respect or awe. People then attribute to these resources – or their representations – a certain power on which they feel dependent. When this happens, the resource itself becomes something sacred in the sense of the definitions above: it becomes forbidden, ambivalent, dangerous, rare and highly valued.

These reflections on the sacred contribute to a better understanding of an elaborate ritual that is performed approximately every 19 years in the town of Puri in eastern India until today (Hardenberg 2008; 2011; Gäbel 2017). The ritual is called *nabakalebara*, which literally means 'new bodies' because the main purpose of the ritual is to renew the wooden statues of the four main deities of the Jagannatha temple in Puri. For this purpose, a special group of priests goes in search of four trees that can be identified by certain signs. These trees are felled, brought to Puri and then used to make the new statues. This period of renewal combines elements of the above-mentioned 'polythetic' definition of the sacred, as it is a time of seclusion, ambivalence and danger. Thus, in the temple, the deities are only accessible to a very specific group of priests, everyone else is excluded. These priests are given the difficult task of 'burying' the old statues, 'creating' the new ones and transferring a special 'life substance' (*brahmapadartha*) from the old to the new. During this liminal period, the deities are separated from the 'profane' and the usual ritual routines are suspended. In a very Durkheimian sense, the deities represent the community: if they are sick and suffering, so are the people. Conversely, their renewal also represents a regeneration of society. The entire temple community and the millions of pilgrims who visit Puri every year are seeking the same resource when they worship the deities of the Jagannatha temple: divine power (cf. Hardenberg 2017). The prerequisite for participating in this divine power is the existence of physical images of the deities that the devotee can see and behold. It is said that the power emanates from the eyes of the deities and can be received by the worshipper when he looks (*darshan*) at the images (Eck 1998). Another way to gain divine power is to eat temple food (*prasad*). Dozens of temple cooks work daily in the kitchen to prepare food for the gods six times a day. Once prepared, the food is offered to the gods in a complex ritual by special priests. Some of this food is given as payment to the temple priests and the rest is sold to pilgrims and their representatives in a special place called the 'market of bliss' (*ananda bazar*). By eating this food, devotees can materially participate in the divine power of the gods (Züfle 2017).

The divine power transmitted through sight (*darshan*) and food (*prasad*) is thus a sacred resource that creates and sustains a complex system of worship involving – among others – the temple congregation, the city administration and even various state institutions responsible for transport, housing, sanitation and so on. In other words, the renewal of the deities is literally a renewal of society and the state as a whole.

5.2 The Social Dimension of Resources

Dealing with resources presupposes human interaction, and this means that resources are on the one hand embedded in existing social practices and orders, but on the other hand also have a formative influence on forms of human relations and actions. These social dimensions are explained in more detail and illustrated with examples in the following sections. First, it is discussed how resources lead to the emergence of resource communities (Chapter 5.2.1), which may also include non-human actors. The next step focuses on the link between resources and identities of social groups (Chapter 5.2.2), especially in the context of the creation of cultural heritage through processes of authentication and musealisation of 'old' things. The third part then focuses on how the use of resources can transform social practices and orders (Chapter 5.2.3).

5.2.1 The Emergence of Resource Communities: Planning and Contingency

The concrete use of resources is always shaped by specific spatial conditions, changes over shorter or longer periods of time and depends on specific social orders. Beyond all differences in time and space, however, there are also recurring characteristics in the handling of resources, because their use always presupposes a network of certain components. These include, of course, people with their specific knowledge, experience, practical skills and social relationships. This world of relationships includes not only other people, but usually also elements of the environment, such as animals or plants, and not infrequently also relationships with visible as well as invisible actors or forces of cosmological worlds. Such a network also consists of material and immaterial resources that are available in specific environments or techniques that people have developed to use resources (Bartelheim et al. 2017, 15).

Teuber and Schweizer (2020) give the example of cod, which is a fundamental resource for fishing communities in the Atlantic. A resource community formed around the use of this resource would be shaped by a number of variables – for example, the quality and taste of the cod, but also certain natural conditions such as the temperature and quality of the waters. It would require intangible resources, in particular knowledge of the techniques used to catch cod and knowledge of the habitat and behaviour of the fish. With this knowledge, such a resource community would be able to use certain material tools, such as boats or nets, which, by their very nature, would have an impact on the concrete practices of fishermen. Over time, different fishing techniques could develop, requiring the existence of infrastructures such as ports, transport vessels or cooling techniques. On a symbolic level, myths or stories around cod could emerge, preserving knowledge, expressing values and contributing to the creation of identities. Teuber and Schweizer call this possible network around the resource cod a ResourceComplex which would consist of equipment, infrastructure, knowledge, practices and symbols or myths or stories (Teuber/Schweizer 2020, 13).

Such networks, which emerge as people use a particular resource, can be conceptualised more generally using Gudeman's idea of a 'base' defined by him as 'a community's shared interest, which include lasting resources (such as land and water), produced things, and ideational constructs such as knowledge, technology, laws, practices, skills, and customs' (Gudeman 2001, 7). Elsewhere he elaborates his ideas about the base with further examples: 'Consisting of entities that people appropriate, make, allocate and use in relation to one another, the base is locally and historically formed. In the Latin American countryside, a farmer considers as base his house, land and crops; a university's base includes its library, laboratories, offices, communication systems and concepts linking researchers [...]' (Gudeman 2005, 97). At the centre of such a 'base' in the sense of Gudeman (2001; 2005) or a ResourceComplex in the sense of Teuber and Schweizer (2020) (see also Chapter 2) are thus material and immaterial resources in which a group has common interests and for whose use a historically grown network of specific things, knowledge and practices is necessary. These do not necessarily have to be 'lasting resources' as Gudeman suggests, i.e. natural raw materials ('land and water'), but can include all elements of community life that are considered essential for the desired form of existence.

These networks are partly the product of human planning, i.e. people use the individual elements in a targeted way, taking into account their characteristics and changing them if necessary. Such planned networks are always subject to processes of change, also due to contingencies or developments that have little or no relation to the use of these specific resources. For example, new technologies and knowledge, political crises, economic turmoil, climatic changes, new pathogens and many other events, large and small, originating in other places and times, can change the networks around Atlantic cod in quite unforeseen ways. This results in the formation of heterogeneous, dynamic networks of components, the composition of which is more contingent and can be better conceptualised as ResourceAssemblages (see Chapter 2). From the perspective of assemblage theory, networks are characterised by the fact that they establish links between elements of resource use that were previously unconnected, for example because they originate from distant times or spaces. Over time, through appropriation, adaptation and planning, these components can be integrated into the use of resources and thus become part of the existing network. Any network that develops in the use of resources is thus always both planned and accidental, has structure and is nevertheless characterised by innovation and change. In this respect, the terms ResourceComplex and ResourceAssemblage do not designate different types of networks, but rather complementary perspectives on any structure that develops through the use of resources.

This is illustrated by the example of the contemporary use of millet by the Dongria Kond, an indigenous community living in the highlands of Odisha, India (Hardenberg 2018b; 2021a; 2021b). The Dongria Kond cultivate a variety of crops, but millets are particularly significant as a resource because of their importance to their entire way of life. Millets can be used to realise multiple values, such as a healthy diet adapted to work, payment for labour services, the establishment of marriage relationships, communion through commensality and the cultivation of relationships with the cosmic forces on which one's life depends. Applying the concept of ResourceComplex to this situation, it becomes clear that this use of millet depends on a network (or 'base') made up of many items, such as the knives, winnowing fans and baskets needed for harvesting, transport and storage, the cookers, pots and cutlery needed for food preparation, but also the extensive knowledge of the weather, environmental conditions, the characteristics of the different varieties of millet, the control of pests, etc. necessary for a successful harvest. In addition to family members, this network includes village helpers and the myriad spiritual beings that the Dongria Kond think influence their lives. The use of millet has thus created a special resource community.

Changes in this network and resource community are partly due to dynamics that originate far outside the sphere of influence of the Dongria Kond habitat (Hardenberg 2021b). In recent decades, for example, rice consumption has gradually become established in the highlands, while millet as a food has been pushed out of the public and especially the religious sphere. This marginalisation of millets and the entire millet-related network could also be observed in the mountains of Odisha, until a new, countervailing development began. It had its origins in debates among national and international economic and political actors, who had realised that millets had a significant role to play in achieving the United Nations Sustainable Development Goals (SDG). Various organisations began to promote millets as a 'miracle grain' in Indian cities, and new shops and restaurants sprang up offering a range of millet-based products. The increased demand for millet in turn had an impact on Indian farmers, who now had economic incentives to grow more millet. Using the theory of ResourceAssemblage, it is possible to identify the development of new, heterogeneous resource communities. Whereas millet was previously consumed only in very remote areas by a population classified as 'poor' and 'backward', it is now also consumed by the urban population and by many types of intermediaries who connect urban and rural areas.

5.2.2 Valorisation of Resources: Social Identities and Contested Objects

As we have just seen, resource use requires coordinated action and interaction between human and non-human actors. Resource communities emerge in which certain values, ideas and practices are shared to a certain extent in order to generate or appropriate resources and use them within the framework of culturally shaped interests. This close association between certain resources and the groups that use them often means that the resources, especially in their materiality, become a representation of the community. In this sense, resources stand for or symbolise the identity of certain groups within the framework of self-attributions and attributions by others (see excursus 'Identity'). Demarcation processes play an important role here, i.e. the attempt to mark social boundaries between groups on the basis of certain resources. The perceived differences between resources are often transferred to the claimed differences between social groups.

A particular role is played by resources that are identified with the past of specific communities, or that one or more groups claim as an expression of their historical heritage. Such resources are usually identified as historical or archaeological relics through the knowledge and authority of experts. Because of their visual, material and functional 'otherness' – caused, among other things, by a temporal distance from contemporary expectations and viewing habits – such 'old' things are often perceived as 'special'. As a result, they can sometimes become museum objects or cultural heritage through processes and practices of valorisation. This happens, for example, through the attribution of authenticity, which is constructed through references to their alleged uniqueness, materiality or age, or on the basis of 'scientific objectivity'. However, these processes of attribution can be highly conflictual, especially when the divergent interests of different groups collide (Theodossopoulos 2013, 337–360).

Such processes of resource valorisation can be illustrated using examples of historical watercraft now exhibited in museums (Schade 2020; 2021; 2022). The 'Bremen Cog' (Schade 2020), for example, dates from the 14th century and is now on display at the German Maritime Museum (DSM) in Bremerhaven. When it was recovered in the 1960s, the wreck was already identified as a 'cog' of the 'Hanseatic era'. This classification meant that the find was considered to be the first archaeological evidence of a 'cog' previously known only from historical sources. It also meant that the find could be linked to the 'Hanseatic League'. At the same time, however, the cog also became the subject of controversy. For example, it was found in Bremen but exhibited in Bremerhaven, although the latter is a more recent city and has no direct connection with the 'Hanseatic era'. Supporters of this solution, however, pointed to Bremerhaven's maritime tradition, particularly in shipbuilding and seafaring. The fact that the 'Bremen ship' was nevertheless able to become a 'Bremerhaven ship' is largely due to the institution of the DSM. As a founding object, visitor magnet, unique selling point and research object, the 'cog' was and still is an identity-forming element for the museum. Conversely, the museum also has an identity-forming effect on the city of Bremerhaven, which as a former 'shipyard city' has undergone a major structural change. Thus, over time, the 'cog' could be linked to the local identity via the museum (DSM) (Schade 2020). This process of identity formation was reinforced, among other things, by the practice of rebuild-

Excursus: Identity

The Latin term *idem*, from which the modern term identity is derived, was originally used in mathematics and philosophy (Gingrich 2011, 143) and describes something that is and remains the same. As summarised by Andre Gingrich, the term identity was first discovered by psychology and politics at the turn of the 19th to the 20th century. While the former used it to describe the characteristics of persons or groups, the latter (mis-)used it to mobilise masses for war and mass murder in the name of nationalism. Since the second half of the last century, the term has undergone an inflationary expansion that can be understood as a reaction to a globalisation that began to affect all spheres of life and uproot people. People were therefore looking for something they could rely on, something that would remain the same. In the age of a fleeting modernity, everything (and therefore nothing) became a subject of identity. The resulting 'whateverism' of the term has led some scholars to question its usefulness for science. But I argue with Gingrich that it is nevertheless needed, because otherwise the crucial question 'Who am I?' would be ignored (Gingrich 2011, 143).

However, a reasonable use of the term avoids confusing identity with an idea of individuality, since the latter is a local product of European history and only one way, among others, of experiencing personhood.

Identity always presupposes alterity, since, for instance, someone who identifies as a female homosexual also claims not to be male and heterosexual. If it were not for these alternating concepts of 'being male' and 'being heterosexual', it would hardly make sense to identify as female and homosexual. The example shows that identity always has collective dimensions; there is no identity in the singular. These dimensions must be carefully examined in order to distinguish elementary forms of identity from superficial displays of individuality.

This leads to two other basic characteristics of identity: plurality and relationality (Baumann/Gingrich 2004, 6). Identity is plural, because no one is just an academic, for instance, and relational, because in a football stadium it is surely more important for an academic to know which team s/he supports (or plays for). Identities may be contingent, but they are not arbitrary. They must satisfy plausibility criteria, the most important of which are continuity, coherence and consistency (Thiemeyer 2021). These criteria are realised through the prime quality of humans: narrative. Contingencies are constantly woven into a meaningful plot (Straub 2004). Only when this fails do people or groups suffer an identity crisis. These criteria apply to single persons as well as to communities. Communities are also characterised by the fact that their members identify in a similar way with things that are important to them. They therefore attach similar values to them.

This is where the concept of resources comes in: it addresses the question of how tangible and intangible elements become the basis of personal and collective identities, in order to identify the different ways in which resources can shape identities. Schweizer (2022), for example, describes ancient *heroa*, the tombs of mythical heroes, as foci of collective identities. He shows that the concrete semantics inscribed on a *heroon* can change drastically over time, depending on the demands of the times. Nevertheless, they have served the citizens of a *polis* for centuries as resources of a shared identity.

But resources are identity constitutive in any case, because they reflect the way a people lives and is in the world. In my contribution in the same volume, for example, I show how livestock has an important symbolic dimension for a group of former pastoral nomads in Iran, symbolising a shared past and ideas of freedom and belonging in the world – and thus expressing a shared identity.

Wulf Frauen

ing and preserving the ship over a very long period of time, partly in public. As a result, part of the local population was able to 'grow up' with the 'cog' (and the museum) (Schade 2020).

Another example of such a controversial valorisation of resources in the museum is the 'Nydam Boat', dated to the 4th century (Schade 2021). Exhibited in the Archaeological Museum at Gottorf Castle in Schleswig, it has a longer history of research and reception than the 'cog'. It was found by a Danish archaeologist in 1863 and assembled, completed and exhibited in Flensburg, which was Danish at the time. After the end of the Second Schleswig War (1864), the former Danish territory around Flensburg became German and with it the boat. The boat then went to a museum in Kiel. The Nydam Boat quickly became a bone of contention between Germany and Denmark, primarily over the question of who owned it, but also over whose past it represented. During the Second World War, the boat was stored outside the city to protect it from bombing, and after the war it came to the city of Schleswig. In the ensuing discourse about its 'new' location, Schleswig was increasingly constructed as the 'home' of the 'Anglic' boat. For a long time, however, the ownership remained disputed. The situation calmed down when the exhibit was loaned to the Danish National Museum in 2003–2004 and the boat subsequently returned to Germany (Schade 2021). Today, the boat is seen as evidence of a shared history and cooperation in the German-Danish border region. Especially for the museums in Flensburg, Kiel and Schleswig, the Nydam Boat was and is a particularly valuable object. This is due to the archaeological uniqueness of the find, as it is the only ship of its type that has been recovered, reassembled and exhibited. It is also a special 'highlight' for the Schleswig Museum because it is an important research object and the museum's identity was linked to the exhibit at an early stage (Schade 2021).

Another example of the valorisation of museum resources is the 'Kon-Tiki', a 20th century raft now in the Kon-Tiki Museum in Oslo (Schade 2022). Unlike the Bremen cog and the Nydam Boat, it is not an archaeological object, but a recent watercraft that has been museumised not because of its actual age, but because of its integration into the narratives created by and about Thor Heyerdahl. Over the decades, however, this replica, built by Thor Heyerdahl in the 1940s from historically imagined models, has itself become 'old' and something of an 'original'. The Kon-Tiki and the Kon-Tiki Museum are located on the Bygdø peninsula, where the Viking ships, the Fram and the Norwegian Maritime Museum are also located, touching on central aspects of Norway's past and identity. In addition, the Kon-Tiki has an identity-forming effect on the museum and the story of Thor Heyerdahl and his legacy (Schade 2022).

In summary, all three objects are historic vessels that differ in terms of date, development and location, but also have clear similarities. First, all three objects are now recognised as central and valuable exhibits in museums. The value of all three objects is based on attributions of uniqueness and originality as well as materiality and visuality. Second, all three exhibits have been constructed as assemblages through certain processes and practices, both in material terms, such as salvaging, assembling, adding, restoring, conserving or exhibiting and in their immaterial dimension, such as narratives or scholarly publications. Third, processes of identity formation are linked to these objects. These identities are often contested and may change over time. In all cases, however, the display of objects in museums plays an important role in shaping local, regional or transregional identities (see also excursus 'Cultural Heritage').

Excursus: Cultural Heritage

The Oseberg ship was hidden under a burial mound for more than 1,000 years. After it was excavated by archaeologists in 1904/1905, a museum was built in Oslo to house it permanently. This building became the Viking Ship Museum (in the future: Museum of the Viking Age), where three large, well-preserved ships from the Viking Age and many other finds from the period were on display.

The Oseberg ship is exemplary for an object that has been elevated to the status of cultural heritage: from a functional object, a ship, it became a ritual object for a burial, then an object of archaeological curiosity and now an object of cultural heritage. Its 'authenticity' increases its effectiveness and value, and the emphasis on the originality of the material creates uniqueness. Material objects are also intangible resources of knowledge – they may contain technical, chemical, craft or artistic knowledge that was necessary for their production. Authenticity in cultural heritage is also achieved by giving objects a story that makes them special and extraordinary, in this case the two women buried in the Oseberg ship opened up speculation about why women were given such a high status burial (Holck 2006). Enthusiasm for the Viking Age has grown since the 1970s, not only in Scandinavia but also in Europe (Hannam/Halewood 2006), and is actively expressed in a participatory culture of remembrance through activities such as festivals, concerts, craft workshops, fairs or tourist pilgrimages to Viking Age sites and museums (Staecker 2005a; 2005b). The network of tangible and intangible resources, their development and the contingencies surrounding cultural heritage shape the **ResourceCulture** for these nations and groups. Resources can also provide narratives of common history and origin, enabling the formation of community and identity.

At the same time, the destruction or alteration of cultural heritage is a way of influencing the memory culture of a group or even a nation, and of promoting ideological ideas. In 2001, the Islamic Emirate of Afghanistan passed a resolution to destroy all statues of non-Islamic origin (Manhart 2001). Even though UNESCO intervened, several important sites, such as the Buddhas of Bamiyan, were dynamited (Manhart 2015). Another example of such a shift in understanding and attitudes towards cultural heritage in the context of violent conflict can be seen in the Russo-Ukrainian war, where the Russian destruction of Ukrainian cultural and symbolic urban spaces and heritage has led to techniques and strategies for its preservation, while at the same time the Ukrainian population and government have developed a critical attitude towards the display and maintenance of Soviet and Russian cultural heritage, with the intention of clearing public spaces of markers that symbolise Russian imperial claims to Ukrainian lands (Rachkov 2022, 34). These examples help to illustrate how cultural heritage is part of a **ResourceAssemblage** that responds to changes in the social and political order and becomes an expression and representation of (changing) collective identities.

Gabriele Alex

Fig. 14 Heritage (© Museum of Cultural History, University of Oslo, Norway/Olaf Væring. CC BY-SA 4.0).

5.2.3 Social Change and Resources

Resources are based on cultural interests (Hardenberg et al. 2017, 14). Therefore, the answer to the question 'What is a resource?' depends strongly on the cultural context. This context is determined by the specific practices, valuations and interpretations of the social actors who have an interest in particular resources. It changes as ideas and values about resources change and as new knowledge, materials, technologies, infrastructures, practices or social contacts influence the use of resources. Interest in resources is culturally variable and always influenced by historical change (see Chapters 2 and 3). On the one hand, interest in resources, which is always culturally constructed, leads to processes of social change, such as the emergence of new resource-related practices, orders and entities. On the other hand, such resource communities change under the influence of environmental and climatic changes, demographic developments, globalisation, political or economic crises or conflicts and many other factors. This, in turn, can affect the interest in and management of resources. The following archaeological and ethnological examples from different times and places illustrate these interdependent dynamics.

In the predominantly agricultural north-east of Mesopotamia (in present-day northern Iraq), fundamental changes in regional settlement patterns in terms of distribution, number and size of settlements can be observed over the long period from 7000 BC to 2000 AD (Pfälzner/Sconzo 2015; 2016a; 2016b). This appears to be related to changes in a particularly important resource, roads. In Mesopotamia, these determined the connections between settlements and the connections of settlements to the outside, to other regions. When contacts between settlements were restructured – for a variety of reasons – there was always a reassessment of the resource of routes (see Chapter 4.3). This usually involved a reassessment of other resources such as water and land. The contacts between settlements along the paths were determined by many economic, political, social and cultural factors. These could be the demography and population density of a region, the social structure, the main economic activities (trade, agriculture, pastoralism, manufacturing) or the political framework (administration, defence). The external conditions of the respective period shaped the demands on the settlements, their contacts and thus the routes as a decisive resource (see Chapter 4.3).

Another archaeological example of the change of and through the culturally determined interest in and use of resources is the use of flint in north-western Bavaria during the Early and Middle Neolithic (Scharl 2017). The distribution and use of flint as a knife presupposed practices for the extraction of flint, the production of knives from the raw material, the improvement or recycling of tools and the distribution or exchange of raw and processed material. From the Early to Middle Neolithic, a specific type of flint, the so-called Plattenhornstein from Abensberg-Arnhofen, became the preferred resource (Scharl 2017, 117), while other materials lost importance. This was accompanied by a re-evaluation of the resource, which went from being an occasional and mostly subsistence product to a commodity (Scharl 2017, 119). According to Scharl, following Roth, new forms of mobility seem to have facilitated this change (Scharl 2017, 119). Thus, it seems possible that people from distant settlements were sent to the Franconian Alps to acquire raw silex in exchange. It is possible that these people took more raw material with them than they needed to live on, in order to have a barter commodity to trade at their place of origin (Scharl 2017, 119).

These new forms of mobility seem to have been linked to other social developments of the time. For example, settlements became larger and denser, and villages became more important as units of social solidarity. As a growing population needed to be supplied with flint, trade became the predominant form of exchange (Scharl 2017, 119). The rapidly changing fashions of increasingly complex pottery can be taken as an indication of increasing communication and, given the construction of circular earthworks throughout Central Europe, one might even speak of denser communication networks (Scharl 2017, 120). Such sites presumably facilitated the exchange of experience, knowledge and material goods such as flint tools, and thus had a decisive influence on social orders, the valuation and handling of resources.

This kind of dynamic, in which resources are rediscovered or re-evaluated, setting in motion social developments that in turn have an impact on resources and their use, can also be observed in the present. In recent years, for example, Islamic religious knowledge has become an increasingly important resource in the Central Asian country of Kyrgyzstan (Hölzchen 2017; 2023). This is due to a new interest that has spread among the Muslim population since the collapse of the Soviet Union and the country's independence in 1991. Islam arrived relatively late in Kyrgyzstan, especially in the northern mountainous regions, and was severely repressed during the 70 years of Soviet rule. With independence in 1991, the country opened up and attracted religious institutions from various countries, based in Turkey and Saudi Arabia. They spread various forms of religious knowledge in Kyrgyzstan and have successfully stimulated discourses on the Quran, the good and bad life and the right faith. According to Hölzchen (2017; 2023), the networks and infrastructures for access to Islamic knowledge are expanding massively. Throughout the country, religious schools (*madrassas*), places of worship (mosques) and pilgrimage networks (*darvat*) provide local people with access to religious knowledge.

A comparison of these examples shows the significance of places and infrastructure for studying resource change and resource-induced social change. In most cases, communication with and exchange of resources is concentrated in central places such as sanctuaries, pilgrimage sites, temples and mosques or ritual sites. It is at these places that interaction, and therefore the management of resources, is concentrated which is seen by a community as central to its way of life. However, these places are only points of concentration in larger networks of communication, which are primarily created by infrastructures – from paths to digital systems. On the one hand, these infrastructures shape the way resources are handled; on the other hand, the material and symbolic aspects of resources create the concrete manifestations of the ever-changing infrastructures.

5.3 The Political Dimension of Resources

Politics, as defined by the 'Politiklexikon', 'refers to any kind of exertion of influence and shaping as well as the enforcement of demands and goals, whether in private or public spheres' (Schubert/Klein 2020). Those who engage in politics are guided by interests and seek to shape communities – societies, cultures, milieus, etc. – in a certain sense. It is therefore about the agency of actors, which is usually not so much an individual merit, but to a large extent structurally embedded in a political system and institutionally supported (Greenwood/Hinings 1996). Power is only partly acquired, but always conferred.

Power is understood as threatened and contested, as a matter of negotiation that can look different depending on the organisational form of a community. Of particular interest are the knowledge orders on which power positions are based. Knowledge orders define what communities know about themselves and others, and shape the ways in which they can perceive and talk about the world. These orders define what kind of knowledge is considered relevant and what can legitimise or delegitimise it (Reckwitz 2006; Knoblauch 2020).

5.3.1 Power and Agency

The analysis of power relations looks at who has what status in a group and who can and cannot represent their interests. Max Weber's distinction between power and domination is classic. Weber defines domination as 'the probability that a command with a given specific content will be obeyed by a given group of persons' (Weber 1978, 53). He sees power as 'the probability that one actor within a social relationship will be in a position to carry out his own will despite resistance, regardless of the basis on which this probability rests' (Weber 1978, 53).

Thus, while domination tends to refer to direct command exercised by individuals (the king) or institutions (the parliament), power refers to relationships between individuals and groups (see excursus 'Power'). Power is often organised in a decentralised way because it can be exercised by many and is not necessarily linked to a clearly identifiable centre. Power can be exercised in a community by anyone who disciplines, patronises or supervises others, thus influencing what others do (cf. relevantly Foucault 1995; Gramsci 2010). And power can also be exercised by small, marginal groups who rebel against, actively fight or subtly undermine supposedly hegemonic institutions.

An example of this is the attempted appropriation of the Obersalzberg near Berchtesgaden by right-wing extremists or Nazi nostalgics that has been investigated by anthropologist Julia Gilfert (2023). The site of Adolf Hitler's representative residence – the Berghof – is still an attraction, destination and place of remembrance for right-wing extremist visitors. Located not far from the Obersalzberg Documentation Centre, the now forested site became a site of hegemonic interpretations of history immediately after the end of the Second World War (Dahm 2016). From the outset, the educational work of the documentation centre, which has existed since 1999, competed with other forms of appropriation, such as the taking of souvenirs or the creation of photographic then and now collages.

As Gilfert puts it, the staff of the documentation centre counter what they see as undesirable appropriation practices by claiming the site for themselves, for example by offering more tours of the site. The fact that the appropriation of historical sites is not a one-off act but a dynamic process became particularly clear in the context of the redesign of the Obersalzberg Documentation Centre. Since practically no guided tours could take place between 2020 and 2023 due to extensive remodelling and expansion work, 'undesirable' uses of the Berghof site again became increasingly visible.

5.3.2 Conflicts

Such conflicts of interest, manifested in the practices by which groups appropriate places, have always been of interest to cultural studies analyses of politics (Koster 2021; Adam/Vonderau 2014). The institutional settings in which the conflicts take place are also always significant (Douglas 1986).

Fundamentally, approaches that tend to assume a consensual coexistence in societies that reaches some kind of social consensus through compromise (which many authors would attest to in most democracies) differ from those that assume that there are fundamental clashes of interests in societies that cannot be bridged. Marxist theory would attest to this for fundamentally capitalist societies. Following Antonio Gramsci, political scientist Chantal Mouffe (2013) has called this 'agonistic' politics. Mouffe argues that in politics there are always conflicting interests (antagonisms) that can be pacified, but which can always be expected to become irreconcilable opposites. Those who think in this way no longer assume that there is some kind of consensus or compromise where, if found, all antagonisms can be resolved and reconciled. Their understanding of society is thus based on the assumption of permanent power struggles – latent or overt – that can only be contained by institutions. The central concept in this kind of theorising is that of hegemony. For Mouffe, any order given to societies by themselves is associated with hegemony, that is, the domination of a particular group over the interests of all others. 'We [she and Laclau] call "hegemonic practices" the practices of articulation through which a given order is created and the meaning of social institutions is fixed. According to this approach, every order is the temporary and precarious articulation of contingent practices.' (Mouffe 2013, 2).

Hegemonic discourses are characterised by the fact that they eventually become invisible and taken for granted. 'What is at a given moment accepted as the "natural" order, jointly with the common sense that accompanies it, is the result of sedimented hegemonic practices.' (Mouffe 2013, 2).

Such 'sedimentary hegemonic practices' describes anthropologist Manuel Respondek (2025) in his research about Riotinto, Andalusia. Here, in the mid-19th century, the British mining company Riotinto Mining Company Ltd (RTMC) opened what was then the world's largest copper mine. Until then, the Cuenca Minera region had been characterised by extensive agriculture and large land holdings (Delgado Ramos 2008, 185–186). Mining then brought about a radical transformation of the agrarian society into an industrialised one in which stable employment relationships created new values.

In this sense, the mine as the main resource and centre of society and life in Cuenca Minera is the starting point for a change in values (cf. Respondek 2025). More distant villages remained unaffected. Work in the mine became a universal value of a bourgeois industrialised society, associated with ideas of security, individual perspective and progress, but also with class thinking, specialisation of labour and alienation from the landscape and agriculture (Ruiz Ballesteros 1998, 195).

According to Respondek, the process of resource appropriation by the RTMC was fraught with conflict. In addition to the mining rights, the mining company acquired the entire territory of the community of Riotinto. It built a segregated and gated Victorian-style neighbourhood for the British management, with its own church, English club and sports facilities. The workforce, which migrated to Cuenca Minera from all over the Iberian Peninsula, was housed in newly built, low-cost neighbourhoods. The lack of housing and the omnipresent control and paternalism of the company divided society. Poor working conditions and the emission of sulphur gases led to protests by workers and landowners in the surrounding villages, culminating in a massacre of around 200 demonstrators in 1888 when the military regiment called in by the mining company fired into the crowd. In the following decades, Cuenca Minera became a centre of trade union and socialist movements (Ruiz Ballesteros 1998, 118–119).

The massive environmental destruction in Cuenca Minera has led to processes of revaluation in which resources that are central elsewhere, such as clean drinking water, forms of extensive agriculture and the landscape itself, have been devalued or revalued. These revaluations can be seen today in a range of narratives and myths. For example, there is a popular consensus that the deep red river contaminated with heavy metals is natural (Olías/ Nieto 2015). The mining landscape is also charged with meaning and associated with the planet Mars, rather than addressing environmental issues. Change in the sense of renaturalisation projects is rejected because the landscape is part of the mining identity and testifies to people's individual experiences of mining activities. Other actors point to a living British cultural heritage that continues in British achievements and practices (football, tea, language) in everyday life (cf. Ruiz Ballesteros 2011, 21). Underlying these narratives to this day is the continued 'sedimented hegemonic' discursive power of mining companies, expressed primarily through the cultural institutions they operate, such as the Mining Museum or the company's own cultural foundation.

As Respondek (2025) puts it, value creation in Riotinto is an ongoing process, as can be seen in the heritagisation of mining culture and its marketing to tourists. The way in which cultural heritage is created as a resource depends strongly on the actors who proclaim, use and legitimise it. In Cuenca Minera, this reveals (discursive) power structures that persist to this day, defining certain aspects (technological progress, British heritage) as cultural heritage. They manage it as such, while ignoring other aspects of life (labour struggles, human-nature relations). On the one hand, the mine generates income through tourism as a revalued resource. On the other hand, heritage also serves to legitimise mining in the past, present and future. It generates knowledge about the history of a region, which in this case is closely linked to a specific industry.

5.3.3 Knowledge Orders

Knowledge can be understood as based on a three-step process of experience, perception and interpretation: people do not perceive their world without presuppositions (Strawson 1950; Stalnaker 1998); rather, they select and assign cultural meaning to what they perceive. The orders that essentially configure this sense-making are based on previous experiences and forms of memory, what is called 'situated knowledge' (Haraway 1988). What is perceived and remembered can be consciously reinterpreted. Knowledge is thus the result of a multilayered process of receiving, storing and mentally processing information about the world, which we call knowledge orders (Oliver Martin 1957).

The preservation and transmission, but also the continuous updating of knowledge is of central importance in all cultures. This is done through the creation of objects such as images, texts, monuments or institutions such as guilds and schools, which play an active role in the creation and transmission of knowledge through practices such as narratives and rituals. Knowledge thus involves three dimensions: a corpus of propositions, forms of representation and the social practices and institutions that accompany them (Foucault 1995; 2002; Barth 2002). Such a perspective always asks about shared knowledge orders that lead to specific shared ways of acting or identities. The focus on the knowledge orders that underlie the collective knowledge of groups is directly related to the foundations of identity (Thiemeyer

2021) and is strongly correlated with power relations. Following Foucault, the relationship between knowledge and power is often described as reciprocal: not only does knowledge imply power, but conversely, knowledge orders are established and changed through power mechanisms (Foucault 1995; 2002; Weiler 2005).

Such interactions between bodies of knowledge and political power relations demonstrates historian Nadja Mozdzen (2024) in an early example from ancient Greece. After the defeat of Athens in the Peloponnesian War, an intensive discourse on the appropriate form of social organisation developed in the early 4th century BC. The democratic order in Athens was blamed for the defeat by parts of the population, and several voices were raised in favour of an oligarchic alternative. In 403 the democratic side prevailed, but the pressure to justify itself remained. In the end, Attic democracy emerged strengthened from the so-called crisis of the 4th century. In these debates, knowledge of Athens' (supposed) past plays a special role, as it is supposed to strengthen one's position as a political resource. But how does such knowledge become a resource in political discourse, and on what levels is it expressed?

Next to philosophical literature, the Attic orations are the most significant written sources on political disputes in 4th century Athens. It is in the speeches that political argumentation can be directly grasped. Scholars such as Demosthenes refer in their speeches to the (supposed) history of democracy, thus turning constructed knowledge of the past into a political argument and consequently into a political resource. An essential point of reference is the *patrios politeia*, the order of the ancestors. In this way, central figures of Attic history or mythology are associated with democracy: Cleisthenes, Solon, Drakon and Theseus. As the pressure on democracy increased, from about the middle of the 4th century onwards, the alleged origins of democracy were pushed further and further back into the past, until finally Theseus was regarded as one of its founding fathers. At the same time, the debate about democracy is intertwined with the question of how to deal with the rising Macedonians. Here, too, the constructed knowledge of democracy's past becomes a resource in political discourse (cf. Mozdzen 2024).

Parallel to these intellectual debates, the public space was fundamentally transformed by a flurry of building activity. In the course of the 4th century, some 60,000 m^2 of space was built; for comparison, in the 5th century it was around 20,000 m^2. Several large-scale projects, such as the Theatre of Dionysus, the third complex of the *Ekklesia* at the Pnyx, the Panathenaic Stadium, the *Gymnasion* at the Lykeion or the successive extensions of the *Agora*, brought about a higher degree of participation and thus perpetuated the ideas of Attic democracy in the built environment.

Intellectual discourse about the political order in ancient Athens is closely related to constructed knowledge about the past. Knowledge becomes a decidedly political resource, especially in speech. The building measures and administrative reforms also show a high degree of creative will – especially in the question of social organisation. The increased building activity can be interpreted as the materialisation of intellectual discourse and thus also as a political resource.

5.4 Synthesis

The examples used to illustrate the symbolic, social and political dimensions of the valuation paradigm highlight three aspects: 1) Resources are constituted through situated practices of valuation and valorisation. Depending on space and time, different things or phenomena are seen as resources or lose this status. 2) These processes of valorisation and devaluation always take place in networks of things, ideas, institutions, knowledge orders, etc., which we can address and analyse as ResourceComplexes or as ResourceAssemblages. 3) These processes have analogous structures, whether they take place in the distant past or in the present in Europe, Asia or Latin America. In this sense, interpreting resources as practices of valorisation is a perspective that can be used regardless of the context.

6 Case Studies – Defining and Analysing ResourceCultures

Raffaella Da Vela, Wulf Frauen, Beat Schweizer

The following case studies are constructed independently of each other, but together they rest on two common pillars: 1) the resource concept as presented in Chapter 2 and 2) transdisciplinary discussions on key issues such as 'Insularities', 'Landscapes' and 'Cultural Memories and Identities'. These discussions in turn contributed to the applied 'resource perspective' and further to the identification and description of ResourceCultures.

The concept of ResourceCultures, with the perspectives of ResourceComplex and ResourceAssemblage, was not understood as a restrictive corset, but as a very flexible approach to the very different regional and historical contexts on which the case studies from distinct disciplines focus. The contributions cover a time frame from the 6th millennium BC to the present day, with different spatial and temporal scales of observation, from a single village to macro-regions and even global interactions. Societies in Latin America, Europe and Asia are examined and socio-cultural dynamics related to resource use are considered in short-term interaction practices as well as in long-term developments in mobility, housing, worship and cultural identities.

The case studies clearly demonstrate that applications of the concept of ResourceCultures can vary depending on the nature of the data, the sources available and disciplinary traditions and methodologies. The high degree of adaptability of the concept (see Chapter 2) to different scientific settings proves to be an asset. The following case studies illustrate the benefits of this multifaceted framework and how it can be used (and adapted) in various ways to gain knowledge and new perspectives for specific disciplinary research. The resulting diversity of the case studies represents the variety of ways in which ResourceCultures can be discussed inter- and transdisciplinarily.

Nevertheless, all the case studies focus on the same key cultural and social aspects of the 'becoming or emergence of resources', such as materiality, symbolic valuations, encounters, transformation, knowledge and power. Some of the contributions also share certain characteristics and have therefore been sorted according to topics: the first part (Chapter 6.1–3) serves as a kind of introduction, as it both illustrates our basic understanding of resources and gives examples of what a specific ResourceCulture looks like in practice. Wulf Frauen's contribution on *Dām-Dārī* (animal husbandry) shows how resources, social structure and collective identity form a contingent but remarkably stable compound for a small community of mountain dwellers in the Iranian province of Kerman, which is described as a ResourceCulture. Elena Revert Francés's contribution also focuses on animal husbandry, but in the case of cattle as a resource in the Late Bronze Age and Early Iron Age on the Iberian Peninsula. She shows how people's lives were shaped in various ways by cattle and how cattle as a resource became a crucial part of their society, far beyond being a mere supplier of milk and meat. Marta Díaz-Zorita Bonilla, Edwin Mauricio Marciales Daza and Monice Timm add an important aspect to our understanding of resources by showing how, during the Bronze and Iron Ages in Menorca (Balearic Islands, Spain), certain raw materials were sometimes used as important resources and at other times ignored. This is a good illustration of how a community's perception of resources is not only subject to environmental affordances, but is also shaped by cultural ideas.

In the following part (Chapter 6.4–6) the resource concept is applied by reflecting on the work carried out around specific key issues. From the perspective of the ResourceComplex, Laura Dierksmeier, Sophie Hüglin and Frerich Schön show in their diachronic contribution on 'Insularities', how water on islands is affected by the specific geography, exploitation and use techniques, governance, history and religion of the respective islands in the Mediterranean and in the Atlantic. Two contributions on landscapes deal not only with different time periods and settings, but also with contrasting but equally valid understandings of how the resource concept can be linked to notions of landscape. The case study by Mohammad Karami and Benjamin Glissmann shows, through detailed and careful archaeological description, how Chalcolithic settlements in south-eastern Iran in one case and routes and pathways in northern Mesopotamia of the Bronze Age in another formed key resources for the local population at the interface of human-environment interactions. Attila Dézsi's contribution on a 'utopian' (or rather dystopian) settlement in northern Paraguay illustrates how landscapes shape communities just as they are shaped and reshaped by humans. The late 19th century colony of Nueva Germania is interpreted as a ResourceCulture around trails or yerba mate in which power and hierarchy are projected into and realised through the landscape.

The case studies that follow (Chapter 6.7–11) address the crucial link between resources, resource use and collective identities. Christian Kübler's contribution asks why the medieval Palatines of Tübingen have hardly left any traces in the collective cultural memory of the region and benefits from the resource concept by conceptualising specific aristocracies of Baden-Württemberg as ResourceCultures, for which he uses the term 'AristocraticCulture'. Wulf Frauen invites the reader back to the mountains of present-day Kerman (Iran) by describing in detail an important aspect of the ResourceCulture presented in the opening case study. The recourse to a mythical ancestor is understood here both as a *lieu de mémoire* in the sense of Pierre Nora and as a ResourceComplex. The ResourceComplex perspective is used here to show how a site of memory establishes social cohesion for a kinship group. Beat Schweizer takes a similar stance in his contribution on ancient *Heroa*. Schweizer's *Heroa* are similar to Frauen's ethnographic example, in that the founders of the respective ancient cities became mythical figures through ritual acts at the funerary monuments. These examples show that the ideas surrounding the *Heroa* were highly flexible and constantly changed according to the needs of the time, but despite all this changeability they always provided orientation and identity for the people of the Poleis. Accordingly, Schweizer sees them as an expression of a specific ResourceCulture. A different kind of ResourceCulture can be found in the Indian state of Uttarakhand. The people who live here experience their landscapes as shared with deities. In their contribution, Lokesh Ohri and Karin Polit show how, in this context, cultural heritage becomes a resource to fight against the exploitative tendencies promoted by the government. This heritage is seen by the authors as a ResourceComplex for environmental activism, which can be used to construct identities and ultimately lead to a ResourceCulture based on respect for all human and non-human agents living in the region. The final contribution draws a clear line from resources through the perspectives of ResourceComplexes and ResourceAssemblages to the description of ResourceCultures. Raffaella Da Vela's liquid stones are funerary stones that were able to shape social identities and power relations over generations in the Northern Apennines during the Iron Age. This ResourceCulture of 'stones of memory' invites us to rethink conventional explanatory models that use problematic terms such as 'colonialism', which are questionable in this context.

In summary, the case studies presented here show a wide variety of applications of the resource concept in different periods, areas and disciplines. This common framework allows researchers to understand and relate to each other's work across divides that might otherwise have seemed insurmountable. It is the focus on a 'resource perspective' as well as the flexibility and adaptability of the concept of ResourceCultures that has brought the authors of this contributions together in this joint perspective.

6.1 Some Thoughts on Resource Management and its Socio-cultural Implications – A Case Study from the Mountains of Kermān (South-east Iran)

Wulf Frauen

Introduction and Research Question

This case study deals with a hitherto controversial issue in social and cultural anthropology: the relationship between people's livelihoods and their culture. Cultural materialists such as Marvin Harris, for instance, would explain cultures in terms of their respective natural environments (including the available resources), while others, following Clifford Geertz, would argue that even the recognition and interpretation of an environment surrounding humans could never take place without cultural beliefs. The only thing everyone can easily agree on is the simple fact that there must be a connection. This contribution attempts to shed light on this question by starting from a seemingly simple insight: every livelihood has certain resources at its core. To illustrate this, I will use an example from the mountains of Kermān,[1] a province in south-eastern Iran. Here I have studied the community of a remote mountain village through extensive fieldwork. In this case study I will describe in detail one of the livelihoods of this community using the ResourceComplex perspective. In this perspective, resources do not exist in isolation, but only in clusters with other elements that are needed to make the resource usable and thus constitute it in their interaction. It is a useful tool for two purposes: to identify the central resource and to show why it is a resource in the first place. Both together can show how a particular resource can become a crucial livelihood for a community under study.

The aim of this contribution therefore goes beyond a more descriptive endeavour. The guiding research question is: how does the use of a resource by a group shape the social life and culture of that group and, conversely, cause that resource to become the basis of its way of life? There are two difficulties with this research question because it uses the terms 'resource' and 'culture'. While 'culture' seems illusive in a world where, at least in everyday discourse, the term is used for almost everything, 'resources' are so far mostly seen from an

1 All names of places, persons and events are transcribed according to the transliteration guidelines of the DMG the first time they are mentioned and subsequently referred to in a Latinised form for better readability. Exceptions to this rule are specific terms for which no adequate Latinised form seems possible without causing potential misunderstanding.

economic perspective, in the sense of raw materials or means of production. Neither of these everyday definitions is very helpful for the purposes of this case study. I therefore understand culture broadly, following Clifford Geertz's notion of 'webs of significance' (Geertz 1973, 5), as 'contested systems of meaning' (Hardenberg et al. 2017, 17) and resources as 'means to create, sustain and alter social relations, units and identities within the framework of cultural ideas and practices' (Hardenberg et al. 2017, 16). These definitions already indicate that I see resources and cultures as interdependent. This interdependence is manifested in another key concept: ResourceCultures. The concept of ResourceCultures is not as easy to summarise in a short, understandable sentence as the previous concept of ResourceComplex, because it shows the subtle interdependences between resources and cultures. One of the main advantages of this approach is that it can show both that culture cannot be seen merely as an adaptive system that reacts to environmental conditions and available resources, as cultural materialists[2] argue, and that it would be equally wrong to understand a particular way of managing resources as being solely determined by culture, without taking environmental aspects into account.[3] Rather, it is the interplay of the two that shapes both. The concept of ResourceCultures can best be understood when it is applied in practice and illustrated by a specific example. The main objective of this case study is therefore to show how a specific community in the mountains of Kerman lives and how their livelihood is linked to their culture and social structure.[4] In doing so, the case study will also demonstrate that the key concept of ResourceCultures provides a useful tool for understanding the interdependencies of resources and cultures in concrete cases.

Theoretical Framework – ResourceComplexes and ResourceCultures

The theoretical framework has already been outlined in this book (see Chapter 2). It is therefore not intended to repeat it here in its entirety. However, I would like to remind the reader that I understand resources as 'means to create, sustain and alter social relations, units and identities within the framework of cultural ideas and practices' (Hardenberg et al. 2017, 16). This quote nicely emphasises that resources are constitutive of social structures and collective identities of certain groups. Resources are always cultural, but they cannot be seen in their genesis independently of natural affordances. The basic aim of this case study is to show this by describing a ResourceCulture through a concrete ethnographic example. This ResourceCulture integrates social and cultural ideas with values and nature, forming a nexus. This nexus became a web of significance for the group under study, and it remains at the core of their identity.

Kermān and the Village of Bāġ-e Borǧ

The following description is based on my previous work on the same region (Frauen/Klocke-Daffa 2023). It provides the reader with an understanding of the geographical context and environmental background that will be relevant to the discussion that follows.

2 Such as Marvin Harris (1979) or Jared Diamond (2000).

3 As some scholars understand Geertz, although I doubt that he ever took such a radical stance.

4 The same example has been described in less detail and without the notion of ResourceCultures in a previous article by me for the journal Anthropos (Frauen 2024).

The Province of Kerman

The Kerman Province is located in the south-east of Iran, bordering Yazd Province to the north/north-west and South Khorasan Province to the north-east. It is bordered to the east by Sistan and Baluchestan, Iran's easternmost province, and to the south by Hormozgan, the Gulf province with the important port of Bandar-e Abbās. The eastern provinces are separated from Kerman by a natural barrier, the 'Plain of Emptiness' (Dašt-e Lut), a salt desert reputed to be the hottest and driest area on the planet. Although not comparable to the Dašt-e Lut, the province is also separated from Fars to the west by another natural boundary in the form of patches of wasteland (notably the Kavir-e Namak-e Sirjān, which locals refer to as Kafa-ye Qaṭru) (Abbott 1855, 66; Borjian 2017). These environmental frameworks make it highly likely that the corridor to the south, to the Gulf, has always been of vital importance to the area, and this is indeed assumed by archaeologists (Pfälzner/Alidadi Soleimani 2017, 106).

Kerman is Iran's largest province, covering approximately 11 % of the country's territory (Borjian 2017) (fig. 15). However, it is not densely populated, with a population of less than three million (2011 census, SCI 2014), and the urbanisation rate is significantly lower than the national average (58 % to 68 %, Zanjani/Nejatian 2017). The topography of Kerman is dominated by the typical interplay of mountain ranges and plains of different altitudes of the Iranian Plateau, which forms a specific environment between the deserts of Arabia and the steppes of Central Asia. Like most of the Iranian Plateau, Kerman suffers from water scarcity and consists of either arid or semi-arid zones (Borjian 2017).

As Borjian argues in the 'Encyclopædia Iranica', the typical binary division of the Iranian Plateau into zones of warm and cold climate (referred to as *Garmsīr* and *Sardsīr* respectively) is central to understanding Kerman and the way people live there. The north around the cities of Sirjān, Rafsanjān and Kerman is dominated by the highland climate of *Sardsīr*, while the south around the cities of Kahnuj, Manujān and Ǧīroft is dominated by the lowland climate of *Garmsīr* (Borjian 2017). *Sardsīr* has warm to moderate summers and cold winters, while *Garmsīr* has hot summers and moderate to warm winters. The two distinct micro-climates correspond to two distinct flora and faunas, and indeed the most common definitions are based not on altitude or average temperatures, but on vegetation and agriculture: *Sardsīr* is defined as the area where date palms can no longer be grown (Bobeck 1952, 75). Instead, trees and shrubs such as pistachio and almond dominate. Grain is occasionally grown in the highlands (Bobeck 1952, 75). For the sake of completeness, it should be mentioned that two additional terms are sometimes used: a temperate region between *Garmsīr* and *Sardsīr*, which is simply called 'moderate' (*moʿtadel*), and a highland region at altitudes beyond *Sardsīr*, where no agriculture can be practised, only pastoralism, which is called *sarḥadd* (de Planhol 2000; Bobeck 1952, 75). In Kerman, on the other hand, I rarely heard anyone talk about the moderate transition zone,[5] and the term *sarḥadd* was mostly used as a synonym for *Sardsīr*.

The entire Iranian Plateau is characterised by mountain ranges that run parallel to (and can be seen as a continuation of) the Zagros massif, and by highland basins between them. The city of Kerman, the capital of the province, lies in one of these basins, and the two chains that make up the Kerman Basin come closer together south of the city, colliding into a single massif and, at least apparently, splitting again further south, so that the whole structure

5 And if so, it was referred to as *band* and not as *moʿtadel*.

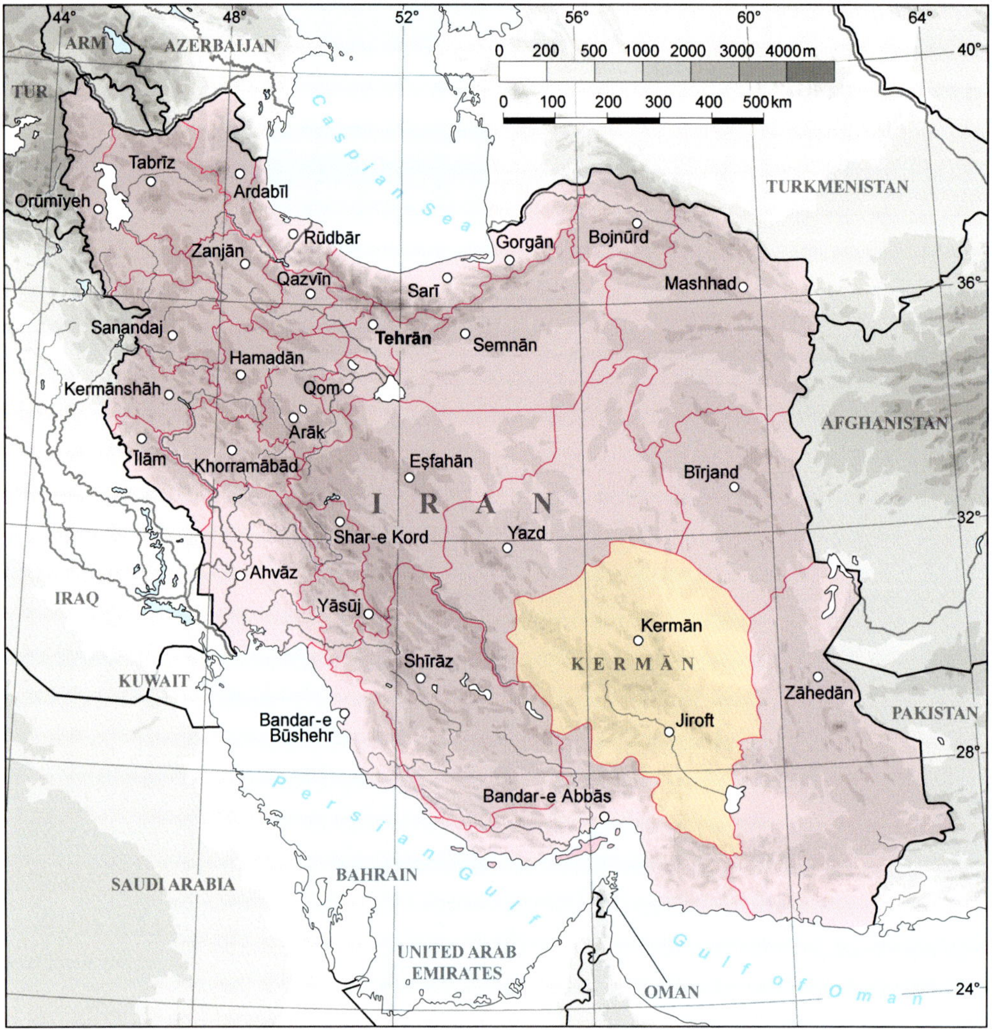

Fig. 15 Map of the Kerman Province within Iran (Graphic design: Richard Szydlak).

looks like an odd X on a map. They form the lowland basin of Jiroft, which, like Kerman, is surrounded by two mountain massifs, one to the east (Jabāl Bārez) and one to the west. The western massif is often referred to in the literature as Sardouiyeh. However, the locals prefer to call it Kuh-e Baġ-e Borǧ – the mountains of Bagh-e Borj.

Bagh-e Borj

A rare local account called 'Ǧoġrāfīā-ye ṭabīʿī-e Esfandaqeh' mentions that a high peak in the aforementioned massif used to be called 'Tower', in Persian *Borǧ* (Ibrāhīmī/Ibrāhīmī-Pūr 2017, 95). This explains why the small village at the foot of the summit is called 'The Garden of the Tower', in Persian *Bāġ-e Borǧ*. Like many villages in Iran, Bagh-e Borj is divided into two parts: Bagh-e Borj 'up' and Bagh-e Borj 'down'. The two parts are about three kilometres apart at an altitude of 2,300 m, with the upper part (*bālā*), as the name suggests, a little higher than the lower part (*pāyīn*), although the difference is hardly worth mentioning. In the 2006 census conducted by the Islamic Republic, the two parts were treated as independent villages with populations of 118 and 48 respectively (SCI 2006). However, as argued elsewhere, the villagers themselves would reject the notion of two independent villages, as the two form one social unit: all villagers belong to an extended family called Eskandery. The Eskanderies tend to refer to themselves as *ṭāyefa*, and people in relatively nearby villages would also describe them as such.[6] *ṭāyefa* is an indigenous term that could be translated into English as 'tribe', although this term is of course somewhat problematic (Gingrich 2015). The Eskanderies are endogamous and believe they share a common ancestor, two points that will be important below. On the rare occasions when the Eskanderies are found in literature, they are usually referred to as *ʿašīra*/*ʿašāyer* (Ibrāhīmī/Ibrāhīmī-Pūr 2017, 96), a term usually translated into English as 'nomads', which in Persian describes people who follow a seasonal migration pattern and whose livelihood is based on animal husbandry. Although they no longer follow a seasonal migration pattern, animal husbandry remains one of the primary livelihoods in the village and is described in detail below. More important is the mining of stones, which began in 1950 with the discovery of chromite in the area and has contributed significantly to the sedentarisation of the group.[7] Nowadays, in addition to chromite, certain gemstones are also mined and sold, notably garnet. The complex processes of changing valuations and perceptions of mineral resources and their impact on the village community are described elsewhere (Frauen/Klocke-Daffa 2023) and cannot be elaborated in the context of this case study, which focuses on livestock. Accordingly, the previous descriptions have focused on providing the reader with the background necessary for the further discussion of animal husbandry. However, it should at least be noted that the area is also rich in minerals and thus offers more than one natural affordance.

The village has some basic infrastructure (a primary school, electricity and, of course, a mosque), while other aspects that could generally be considered part of fundamental infrastructure are completely lacking (such as paved roads and any kind of shopping facilities, which also applies to most basic things like bread or water). The village has some of the characteristics of settlements that have come into being more or less 'suddenly' over a short period of time rather than successively over a longer period, such as the absence of a clearly visible centre and a seemingly random arrangement of houses. The impression of having been 'put there overnight' rather than 'grown organically over time' is, in my experience, not atypical of villages in Iran where former nomads have settled (fig. 16).

Dām-Dārī – Animal Husbandry in the Village of Bāǧ-e Borǧ

The Persian term *Dām-Dārī*, like its English equivalent animal husbandry, refers to all forms of systematic animal husbandry. Accordingly, in the case of Bagh-e Borj, it seems sensible to mention first which animals are meant when referring to animal husbandry: *Dām-Dārī* in this case refers to the raising of sheep (*gūsfand*) and goats (*boz*). This can be explained by

6 Actually, there is hardly any village 'close by' when Middle European circumstances serve as a measure.

7 Although this was probably not the only reason why the seasonal migration was abandoned. The eventual sedentarisation of the entire community must be seen as a process that began in 1950 and ended in 1970, when the Eskanderies swapped their tents for solid houses. The effects of the White Revolution, which reduced the amount of pasture available, and a severe drought in the region, which has affected livestock farming but not mining, must also have played a role. However, it is clear that the emergence of the mining sector in the area played a crucial role, as this is the reason given by the villagers themselves when asked.

Fig. 16 The village of Bagh-e Borj (Photo: Wulf Frauen).

environmental factors: as Bagh-e Borj is located in a cold mountainous region, animals are needed that can cope with the aridity, sparse vegetation and low temperatures in winter. In particular, the goat, whose main characteristic is that it is undemanding and easily satisfied, has long been considered the classic mountain animal – the 'Babylonian Tree', an epic poem of Middle Persian literature, recounts the contest between a goat and a date palm (Digard 1989). The poem reflects not only the dichotomy between animal husbandry and agriculture (in the sense of crop raising), but also between the highlands and the lowlands. The goat can be seen as a symbol of the mountains, just as the date palm is a symbol of the lowlands.[8] The sheep is not as undemanding as the goat, but in its Kermani variety (Kermani sheep) it is still undemanding enough to find sufficient food at an altitude of between 2,000 and 2,500 m above sea level, although supplementary feeding is necessary in winter. In addition to sheep and goats, almost every household has some smaller animals, such as chickens (*morġ*), which are used to supplement the diet but are not otherwise of great importance to the village. Larger animals such as donkeys (*ḫar*) or camels (*šotor*) are no longer kept in the village.

From the ResourceComplex perspective, the basis of the resource network around sheep and goats is grazing in combination with work. In this respect, the village benefits from its relatively remote location in the mountains to the far west of the Esfandaqeh plain: as the area is not densely populated, there are no disputes over the use of the pastures. Another fact contributes to this: the area is not suitable for extensive farming, so very little of the land is used for agriculture. However, small-scale agriculture is practised by almost every household and is naturally combined with animal husbandry: small gardens provide supplementary feed for the animals, and the animals in turn provide manure that is used in the gardens. When temperatures drop below zero in winter, the animals need shelter. Because of the small number of animals in most households, shelter is not provided by large barns (*tawile*), but by small structures made of clay, mud and stones called *āġāl*, which look like small pyramids in the shape of an inverted V. The final element in the ResourceComplex

8 These symbols date back to ancient times. The date palm, for instance, is already a common theme in the iconography of the chlorite artefacts of the so-called 'Jiroft style' (Perrot 2008).

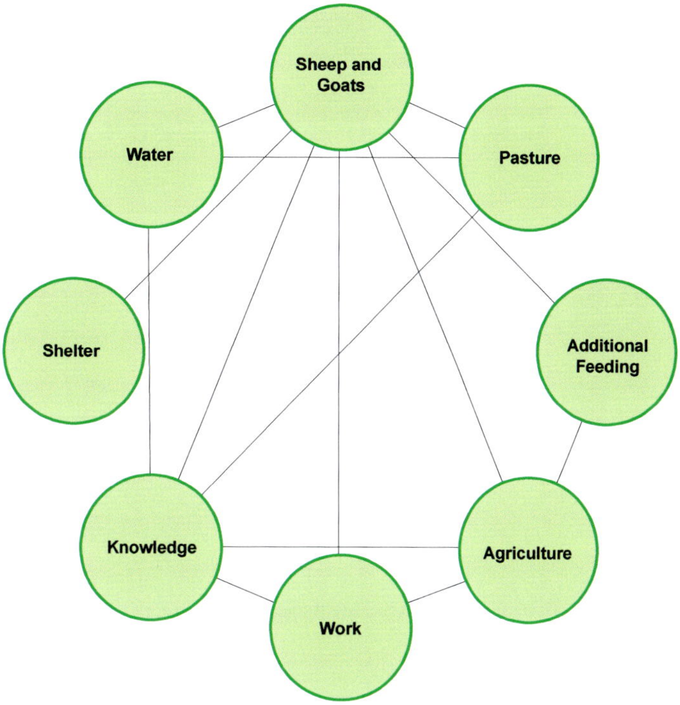

Fig. 17 Diagram of the Resource-Complex around animal husbandry in Bagh-e Borj (Data: Wulf Frauen; Graphic design: Richard Szydlak).

perspective on animal husbandry is perhaps the most crucial: knowledge. Knowledge is required not only to produce most of the products derived from animal husbandry, as will be shown below, but also to keep the animals alive, since it includes the knowledge of where to find water in the vast mountains and other crucial aspects. This specific knowledge can and should be understood as cultural capital in the sense of Bourdieu. In summary, an illustration of the crucial elements that can be seen as a ResourceComplex around animal husbandry is depicted above (see fig. 17).

A wide range of products can be derived from sheep and goats. It may come as a surprise that the most obvious product, meat (*gūšt*), is by no means the preferred use of the animal. The reason is quite simple: meat requires the slaughter of the animal, and no other products can be obtained from it. In fact, only older male animals are slaughtered, if at all. In this respect, Bagh-e Borj is no exception, as this quotation from the 'Encyclopædia Iranica' shows:

'The preferred uses for the goat, however, are those that permit extraction of products from the living animal' (Digard 1989).

I will therefore concentrate in my account on the non-meat products derived from sheep and goats. The most important of these, at least in the case of Bagh-e Borj, are dairy products. A brief but complete overview is given below. The lactation period of both sheep and goats lasts about five months after the birth of the lambs. During the first two months, the milk is usually kept for the lambs and then used for dairy products, of which there are many. A complete picture of the production cycle is depicted in fig. 18. *Tarf* is a liquid substance produced during the production of *kašk* by pressing the unfinished *kašk*. It is then boiled and dried (in the shade, not in the sun). It is used to sour soups. If there is an English translation, I am not aware of it.

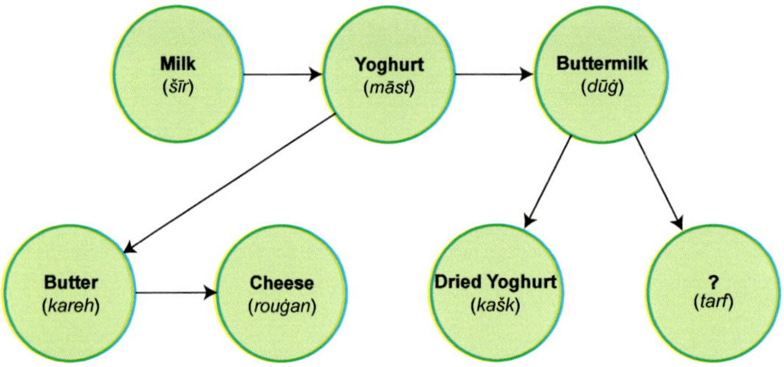

Fig. 18 Milk processing (1) in the village (Data: Wulf Frauen; Graphic design: Richard Szydlak).

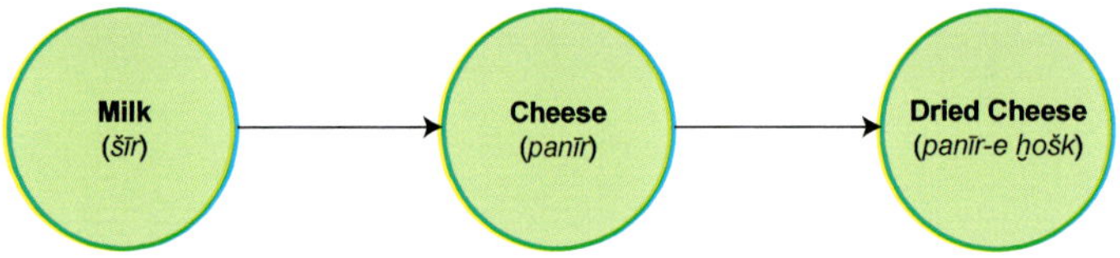

Fig. 19 Milk processing (2) in the village (Data: Wulf Frauen; Graphic design: Richard Szydlak).

Alternatively, milk can be processed into cheese (*panīr*), although this is much less common (fig. 19). It should be noted that dried cheese is usually referred to as *panīr* in everyday conversation as well, rather than *panīr-e ḫošk*.

One of the reasons why the first of the two production chains shown is usually preferred is that the second requires rennet. Rennet is obtained from the stomach of a lamb, which cannot be more than a few weeks old. Since rennet cannot be stored for long, the period during which cheese can be made is limited and a lamb has to be slaughtered. Nevertheless, if a household has taken the trouble to make cheese, it will certainly be proudly presented to visiting guests (*mehmān*). Cheese, like most of the dairy products described above, is not economically important to the village and is almost exclusively produced for self-sufficiency. The only exception may be *kašk*, which is sometimes produced in relatively large quantities. *Kašk* is made by salting drained *dūġ*. The salted mass is formed into small balls which are dried in the sun for about five days. They are a protein-rich food that can be stored for a long time. When relatively large quantities of *kašk* are produced, some of it is sold to Esfandaqeh, although my interlocutors emphasised that their main aim is also self-sufficiency (fig. 20).

Apart from dairy products, goats and sheep also provide hair (*mū*) and wool (*pašm*). The economic dependence on wool reported by Ansari-Renani et al. (2013) for certain communities in the nearby district of Baft has no equivalent in Bagh-e Borj.

In general, it can be said that from a purely economic point of view, animal husbandry has lost the dominance it enjoyed for centuries before the sedentarisation of the group. This can be seen from a simple fact: at the time of the study (2018 and 2019), only one household reached the mark of 120 animals needed to fully cover its expenses.[9] Nevertheless, *Dām-Dārī* is omnipresent in the village and there are few households that do not have at least some goats. Each day begins with a cup of fresh milk, which is milked before the animals are taken to their pastures around the village. The milking and subsequent drinking of the milk, in its

9 According to Daniel Bradburd (1980, 609). It should be mentioned that other authors give different estimates.

Fig. 20 *Kašk* in relative high quantity (Photo: Wulf Frauen).

uniformity, almost resembles a ritual that welcomes the dawn. The fact that *Dām-Dārī* has lost much of its economic significance can also be explained by the fact that it is no longer practised in its most efficient form as it has been for centuries. It therefore seems fruitful to take a look at the community's past. As will be shown, this broadening of perspective will also help to understand the role that animal husbandry plays in the village today.

The Past of the Community: Pastoral Nomadism

Talking about the past of the village community means talking about pastoral nomadism. Until seventy years ago, the villagers lived as pastoral nomads, following a seasonal migration pattern. Although there is a wide variety of nomadic lifestyles in Iran, some basic common characteristics can be identified. The first is that this way of life is always associated with animal husbandry, which of course applies by definition to nomadism in general.[10] Nevertheless, in the Iranian case this basic criterion seems to be of particular importance. In his entry on nomadism in the 'Encyclopædia Iranica', Eckart Ehlers states that the first basic characteristic of pastoral nomadism (in Iran) is 'dependence on domesticated animal husbandry' (Ehlers 2011). The second basic characteristic, according to Ehlers, is 'migration along established routes between focal grazing areas' (Ehlers 2011). This migration pattern has a common feature that can be found throughout the country: it can be described as 'mountain' or 'vertical nomadism' (Ehlers 2011), as it is always based on seasonal migration from a warm climate area in the lowlands (*qešlāq/Garmsīr*) to a cold climate area in the mountains (*yeylāq/Sardsīr*) sometime between March and May, and vice versa from August to November (Ehlers 2011[11]). Iran's distinctive mountainous features, together with its arid climate, allow for the combined use of winter and summer pastures, combining two types of habitat for livestock in a system based on mobility. Ehlers rightly points out that it is '[...] the big achievement of the nomads [...] to have successfully combined the use of these high and low habitats into an economically fully viable form of animal husbandry' (Ehlers 2002, 41).

10 The term nomadism derives from the ancient Greek term *nomádos* ('being on pastures') (Fischer 2011, 304). It should be noted, however, that while animal husbandry is constitutive of the term, nomadism is now often used to refer to forms of mobility that have nothing to do with animal husbandry. Thus, in her entry on nomadism for Andre Gingrich's 'Lexikon der Globalisierung', Fischer distinguishes between 'reale NomadInnen' and 'metaphorische NomadInnen' (Fischer 2011, 304).

11 Astonishingly, Ehlers, one of the most respected experts on nomadism in Iran, confuses the terms *qešlāq* and *yeylāq* in his entry on nomadism in the prestigious 'Encyclopædia Iranica'. The way it is given in the text above is correct.

This form of mobility does not mean that the nomads are on the move the entire year. In most cases, only two treks are undertaken each year. Apart from these two regular annual movements in spring and autumn, the nomads remain in their *qešlāq/Garmsīr* and *yeylāq/Sardsīr*. Georg Stöber, in his entry on 'The Nomads of Kermān' in Tapper's 'The nomadic Peoples of Iran', the most comprehensive work on nomadism in Iran, states that

'[T]he nomads return to the same pastures and camp sites year after year. In their winter quarters (*garmsir, qeshlaq*) they have stone walls over which they pitch their tents, and in their summer quarters (*Sarhad, yaylaq*) some families have trees, fields, and even simple houses' (Stöber 2002, 253, italics in original).[12]

The distance between winter and summer quarters varies considerably. In Kerman, where the binary climatic division between cold and warm zones is even more pronounced than in the rest of the country (Borjian 2017), they can be up to 200 km apart, but it can also '[…] be a matter of [just] several kilometres' (Stöber 2002, 253). In any case, this way of life is a form of pastoral nomadism, as Stöber rightly points out (Stöber 2002, 253).[13] Generally speaking, the nomads of Kerman are organised in 'numerous tribes', the majority of which are 'very small, consisting of perhaps ten families' (Stöber 2002, 253). With only one documented exception,[14] these tribes are not and have not been organised into larger political entities such as confederacies, which is a major difference from the Zagros nomads. I therefore share Richard Tapper's view that it is not very useful to take the major Zagros confederacies (Qašqā'ī, Baḫtiyārī) as some kind of obscure model or, as Tapper puts it, 'archetypes' for a 'uniform tribal structure in Iran' (Tapper 1997, 13). The fact that the nomads of Kerman tend to be organised in smaller social units characterised by rather short-distance migrations does not mean that they are doing anything 'wrong' just because they do not fit into established (western) schemata. On the contrary, the fact that the hitherto almost unstudied tribes of Kerman show significant differences from the Zagros tribes, which have been the focus of anthropological research since Fredrik Barth's influential 'Nomads of South Persia' (Barth 1961), shows how fruitful this area is for further research. Stöber, in his work on nomadism in Kerman, the most comprehensive account of Kerman Province published to date, attempts to provide a register of all the tribes in the region (Stöber 1978, 277–286). Stöber himself makes no claim to completeness, but the Eskanderies of Bagh-e Borj can be found in the register (Stöber 1978, 283), although by the time of Stöber's fieldwork (1973–1975) most of them must have abandoned their seasonal migration.

Prior to the process of sedentarisation that began in 1950, the Eskanderies used to make the seasonal migration from *qešlāq/Garmsīr* to *yeylāq/Sardsīr* and vice versa on an annual basis. The village of Bagh-e Borj was their *yeylāq/Sardsīr*, the summer pasture in a region with a

---

12 Stöber uses the term *sarḥadd* instead of *Sardsīr* because the two words are used synonymously in the Kerman Province. In most areas of Iran, however, *sarḥadd* refers to elevations where agriculture is no longer possible (de Planhol 2000). Therefore, I decided to use the term *Sardsīr* instead to avoid misunderstandings.

13 This statement seems self-evident, but it is not. Some authors have a very poor understanding of pastoral nomadism, even if they otherwise produce quite insightful and sophisticated works. Abbas Alizadeh, for instance, writes that Iranian pastoral groups should not be labelled nomads since they 'differ fundamentally from steppe nomads' and 'spend only a fraction of the year on the move' (Alizadeh 2010, 354). It seems that Alizadeh considers 'nomads' to be synonymous with 'hunters and gatherers', which is fundamentally wrong.

14 The Afšār have been studied in detail by Stöber (1978). It should be mentioned that during the Pahlavi period, the political structures of the tribes were dismantled throughout the country, so that the confederacies *de facto* no longer exist (Ehlers 2011). In general, therefore, it seems increasingly questionable to explain Iranian nomadism in terms of its political sphere at the higher levels and not in terms of its kinship-based social sphere at the still intact lower level.

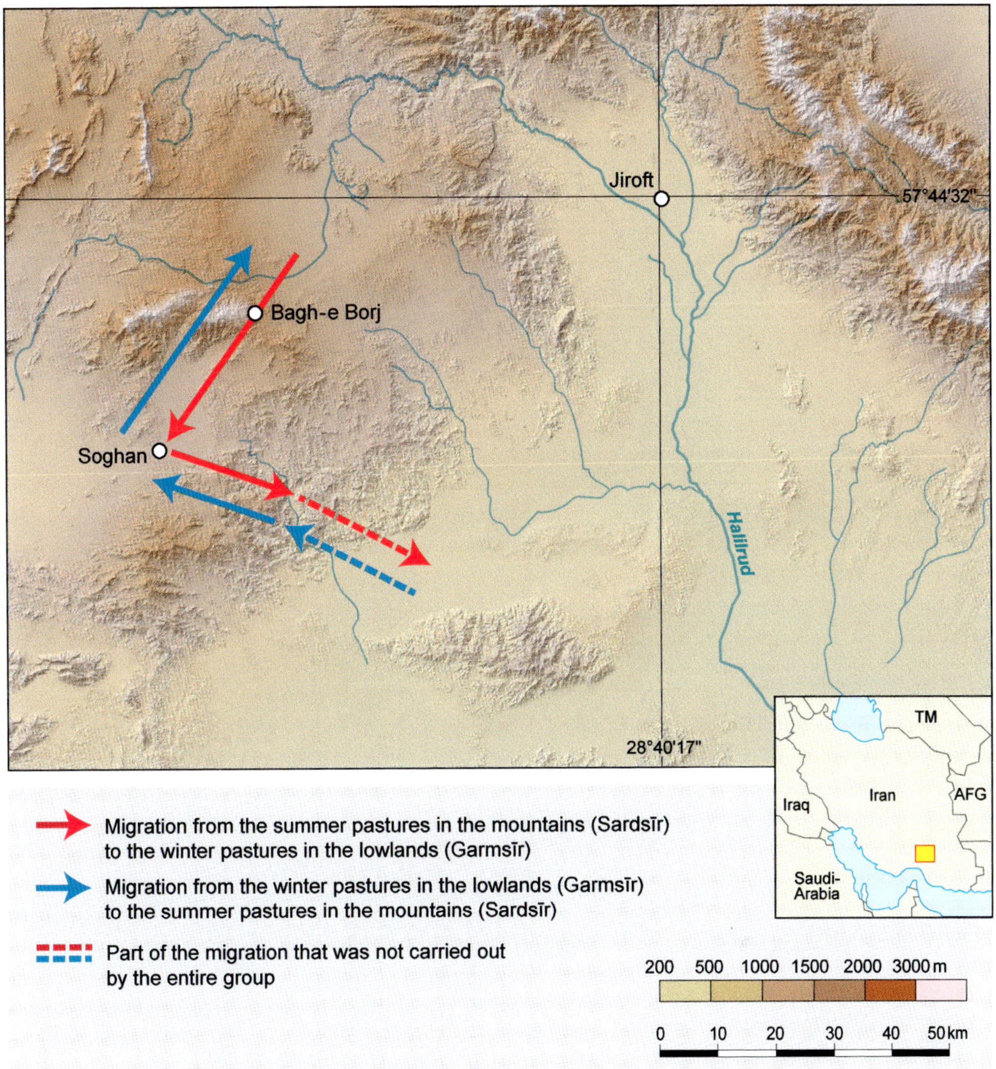

Fig. 21 Route of the former seasonal migration (Data sources: ESA COP30; Map: Wulf Frauen; Graphic design: Richard Szydlak).

cold climate, and it would probably be more accurate to say that it was not a village, since the Eskanderies lived in tents, although some permanent and stable structures already existed. As discussed above, this is by no means atypical for pastoral nomads in the Kerman region. As mentioned, Bagh-e Borj is located at the outer end of the highland plain of Esfandaqeh at an altitude of 2,300 m. South of the mountains of Bagh-e Borj lies the plain of ṣūġān. In the little more than 30 km from Bagh-e Borj to Soghan, the altitude drops from over 2,300 m to 1,500 m.[15] At no point along the way does the altitude exceed 2,426 m. The climate and flora of the Soghan plain can already be considered part of *Garmsīr*, and several tribes have

15 All information concerning distance, elevation, etc. is obtained from internet sources such as Google Maps and Google Earth.

their winter pastures here (Stöber 1978, 127). The Eskanderies, however, did not have their *qešlāq* here, but continued on to the plain of Fāryāb, which lies another 50 km south-east of Soghan at an altitude of 650 m above sea level. While some Eskanderies settled in the Faryab plain for the winter, most of them had their *qešlāq/Garmsīr* in a place called Zahmakān. Zahmakān is located in the centre of Tang-e Mordan, a gorge formed by the riverbed of the Mordan River, which connects the plains of Soghan and Faryab (Pfälzner/Alidadi Soleimani 2017, 111) (fig. 21).[16]

The annual migration from Bagh-e Borj to Zahmakan and vice versa can serve as a classic example of the pattern described above, combining two habitats into a way of life perfectly adapted to animal husbandry. While the pastures of the highlands were exploited for half the year, they had the other half of the year to recover while the livestock grazed in the lowlands. The same was true for the lowland pastures, which were also used for half the year and left to recover for the other half. In addition, seasonal migration gave the group access to both lowland resources (date palms, citrus fruits, etc.) and highland resources (walnuts, fruit trees, etc.). Ehlers notes that the '[...] nomadic way of life always tried to (and had to) maintain an equilibrium between the resources of the natural environment and the needs of the people' (Ehlers 2011). In this way of life, '[... a]nimals and herds thus became an essential link between man and his natural environment' (Ehlers 2011). Contrary to what one might assume, however, such a resource-based *modus vivendi* cannot be reduced to herding sheep and goats and other most obvious features. The following section illustrates the deeper dimensions of its socio-cultural implications.

Socio-cultural Implications of Animal Husbandry

In 1961, Barth published an account of the Basseri, a tribe of pastoral nomads in the Fārs Province around Šīrāz, based on fieldwork in the region in 1958/59. The book, entitled 'Nomads of South Persia', was the first comprehensive work on an Iranian nomadic tribe. Barth was also the first to identify a fundamental problem that required explanation. He noted that the nucleus of Basseri society consisted of individual tents (which he called *khune*[17], which literally means 'house') and that these basic units were characterised by a high degree of egalitarianism in the sense that they 'are units of production and consumption; represented by their male head they hold rights over all movable property including flocks; and they can even on occasion act as independent units for political purposes' (Barth 1961, 11).[18] The

16 Two aspects of this seasonal migration are striking: first, the Eskanderies did not use the nearest *qešlāq/Garmsīr* for their winter pastures, which would have been the plain of Soghan. This could be explained by the fact that the rights to use the land may not have been with them, but with the other tribes who had settled there. Second, most of them did not follow the Tang-e Mordan all the way to the Faryab plain but stayed in Zahmakan. The area of Zahmakan is fertile, but offers far less grazing than the plains of Soghan and Faryab. One explanation is that the tribes have always supplemented animal husbandry with other sources of income, two of which were helping to transport goods along the caravan routes that passed through their territories or, alternatively, raiding the caravans. Stöber speaks of 'positive' and 'negative' influences (Stöber 1978, 249). Whether the Tang-e Mordan was once part of a major caravan route remains to be proven, but if it was not, it would certainly have provided an excellent hiding place after raids into the nearby plains of Faryab and also Boluk. This is, of course, pure speculation. When I asked the villagers about their means of subsistence in the days of seasonal migration, they always only referred to animal husbandry (combined with small-scale agriculture).

17 DMG: *ḫāne*.

18 There is one major exception, the *čupān*: young men who have not yet established their own households and therefore work for a time as shepherds for wealthy households until they are able to establish their own (Barth 1961,

problem that Barth became aware of is that, unlike in a sedentary community, in a nomadic society the individual households are first highly egalitarian and therefore independent, and second highly mobile, since they live in tents and can, by definition, transport all their belongings from A to B. This becomes particularly important in times of seasonal migration, when '[... e]veryday the members of the camp must agree [...] on whether to move on, or to stay camped, and if they move, by which route and how far they should move' (Barth 1961, 25). But even at times when they were not migrating, individual households could potentially (for example, in the event of a dispute) simply pack up and leave. At the same time, however, Barth notes that the erosion of the group would be harmful in the long run, since a number of animal husbandry practices can only be carried out most efficiently when more than one household is involved. Barth therefore concludes that a '[...] community of nomads can only persist through continuous re-affirmation by all its members' (Barth 1961, 25), which constitutes '[...] the difference between growth and prosperity of the herds, or loss and poverty' (Barth 1961, 26). It could be added that staying in a group is also much more desirable for safety reasons when moving through remote and sparsely populated areas, although this is not mentioned by Barth. The question of what exactly is the glue that holds the group together has been discussed and answered differently ever since. Barth himself understood the Basseri primarily as a political unit and therefore emphasised the subordination of all Basseris to their chief (Barth 1961, 75). However, he already realised that this could not be the whole picture and therefore looked for an additional explanation. What Barth suggested can be confirmed by the study in Bagh-e Borj: nomadic societies in Iran are held together by kinship ties.

The strong tendency towards endogamy that Barth found among the Basseri (Barth 1961, 35) is still reflected in the Eskanderies of Bagh-e Borj, even though they abandoned their seasonal migration 70 years ago. This principle is reinforced by the strong belief in a common ancestry: all Eskanderies are descendants of a common ancestor known among them by the name of Karbalā' Bārānī. Karbalā' is said to have come to the mountains of Bagh-e Borj some 350 years ago, and most Eskanderies can trace their own lineage back to this mythological figure, as the following anonymised example from an elderly villager illustrates (see fig. 22).

The reference to a common ancestry and to Karbalā' is still very active in the village and is reinforced by various practices. These usually include a pilgrimage to the tomb of Karbalā', which is believed to be located on a small hill overlooking the only notable well in the vicinity of the village. The villagers literally call going there *ziyarāt kardan*, which means 'to make a pilgrimage'. In the past, the site was even associated with social practices that are no longer practised. For instance, newly married couples used to go there three days after their wedding and sacrifice a goat, the meat of which was distributed among the villagers. This practice can be interpreted as both seeking the blessing of the ancestors and establishing a new social position within the group. In addition, all practices that referred to a common ancestry served to strengthen the group and provided social cohesion. This is precisely what Barth referred to in the quote above as 'continuous re-affirmation' (Barth 1961, 25).

Barth understood the tendency towards endogamy among the Basseri from a strictly functional perspective. The preference for remaining with agnatic and matrilateral relatives is thus described by him as a means by which the heads of households 'establish as many strate-

13–14). The *čupān* is directly attached to another household, but given this exception, the households are theoretically completely self-sufficient and therefore seemingly independent from each other.

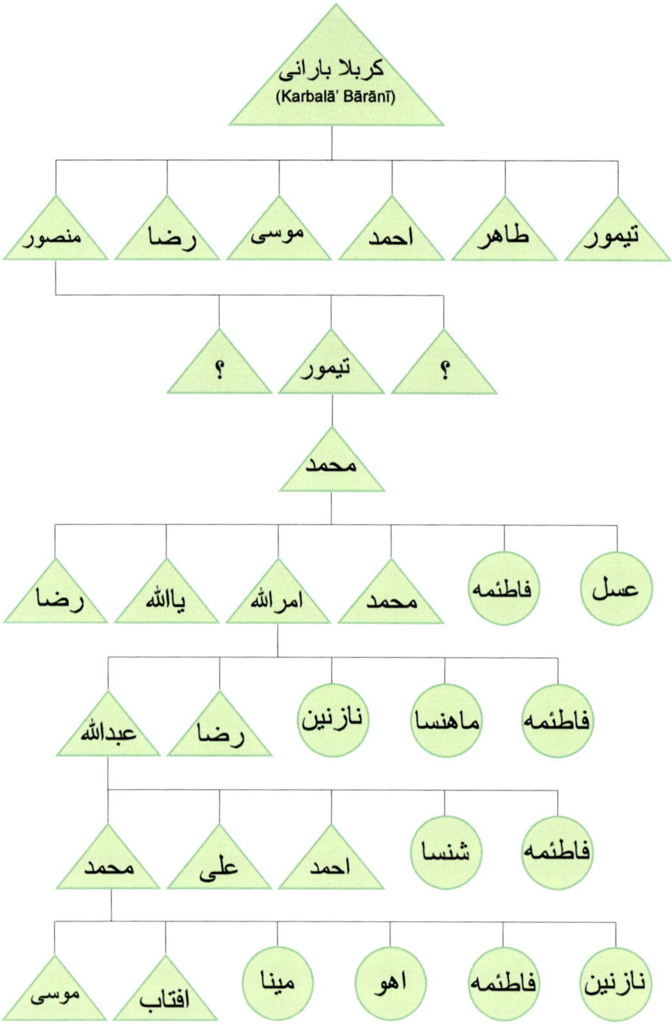

Fig. 22 A genealogy of a villager that leads back to the mythical common ancestor Karbalā' Bārānī. The illustration is not complete: occasionally the names of ancestors were unknown (marked with '؟' in the diagram) and female ancestors occur 'only' from the fourth generation after Karbalā' on. Note also that it is anonymised by changing most names (Data: Wulf Frauen; Graphic design: Richard Szydlak).

gically placed affinal bonds [...] as possible' (Barth 1961, 35) in order to extend 'his influence over other members' (Barth 1961, 35). In the case of Bagh-e Borj, it would be very difficult to understand the persistent endogamy from such a perspective, since due to the now settled lifestyle and the (economic) dominance of other resources over animal husbandry, the more important ties of individual households are not within the village but outwards. Nevertheless, the principle that a descendant of Karbalā' Bārānī may only marry another descendant of Karbalā' Bārānī is still practised without exception. Therefore, at least for Bagh-e Borj Barth's functional approach needs to be broadened. Although the social structure of the community is indeed linked to a way of life based on animal husbandry, its continuity cannot be understood solely in terms of efficient resource management. Its basic function, at least in Bagh-e Borj and today, is identity formation. If, as in the example above, the villagers can trace their individual descent back to the mythical ancestor Karbalā' Bārānī, this is of primary importance to them, as it is fundamental to their own identity and understanding of their place in the world. Kinship and marriage patterns are also an expression of this process of identity formation. The following concluding section will bring all these loose ties together and describe animal husbandry, *Dām-Dārī*, as a ResourceCulture.

ResourceCulture Animal Husbandry

As has already become clear, in the case of Bagh-e Borj, sheep and goats are more than just a means of providing milk, dairy products, occasional meat and, in some cases, economic income. All this is true for the villagers: of course, the animals are milked, the milk is drunk for breakfast and processed into *Kašk* and cheese, and these products are consumed and, depending on the household's production, sold, just as the animals are sold in certain cases. All these practical uses can be found in the village and at least some households still depend on their livestock for both subsistence and economic income. But there is more to the story. The animals and the practice of animal husbandry link the villagers to their past as pastoral nomads, to their origins represented by a mythical ancestor and to a particular way of life that feels 'right' to them. It is crucial to their individual and collective identity and is a major source of not only their economic but also their social existence. This makes it a resource.

As is well known, for *ʿĪdu l-Aḍḥā*, the Islamic Feast of Sacrifice, a sheep or a goat is usually sacrificed. As throughout the Muslim world, this is done in Iran, where the feast is called *ʿEid-e qurbān*. On this occasion, every household in Bagh-e Borj sacrifices one of its precious sheep or goats. It seems obvious that sheep and goats are an appropriate gift to God, not only in the eyes of the Eskanderies but also throughout the Muslim world. However, it should be emphasised that goats and sheep were also crucial to other practices, such as the above-mentioned custom of sacrificing a sheep at the tomb of Karbalāʾ Bārānī three days after the wedding. It should also be remembered that the meat was then distributed among the villagers. It has already been mentioned that such an offering strengthens the social bonds between donor and recipient, and hence the social cohesion of the group as a whole. However, looking at the bigger picture, these practices are multi-dimensional and create a bond not only between the villagers but also between different spheres: the distribution among the villagers reaffirms their connection as a social group (the point that was so crucial for Barth); the fact that the sheep was sacrificed at the tomb of Karbalāʾ Bārānī further establishes a connection to the ancestors and, through the ancestors, to the land, with which the Abrahamic figure of Bārānī in particular is inextricably linked; and the *ʿEid-e qurbān* establishes a connection to the religious sphere of Shia Islam. All this forms a continuum in which sheep and goats serve as mediators. As an aside, it should be remembered that *ʿEid-e qurbān*, the Islamic Feast of Sacrifice, refers to the story of the Qur'anic prophets Ibrāhīm and Ismāʿīl. Abraham sacrifices an animal instead of his son, which is accepted by God. Like the Eskanderies until 70 years ago, Abraham was a pastoral nomad whose livelihood was based on animal husbandry (Nippa 2011, 8).

In the days of seasonal migrations, animal husbandry structured not only the days of the Eskanderies (milking in the morning, taking the animals to pasture, etc.) but also their annual cycle (through migrations from the highlands to the lowlands and vice versa). Their social structure was adapted to this way of life through endogamous marriage patterns, which only allowed marriages between members of the same lineage. In a broader sense, the recourse to a common ancestor and even the idea of being a group that must stay together can also be interpreted in this context. Sons who left their parents to start their own households were given part of the herd, so that the reproduction of the herd was inextricably linked to the reproduction of the social group.

Animal husbandry as a ResourceCulture thus shapes the social structure of the group and forms the basis of its individual and collective identity. Accordingly, the central resource 'sheep and goats' cannot be easily replaced, even if seasonal migration is abandoned and animal husbandry loses its former economically dominant position. Its basic characteristics are social equality, cooperation and the maintenance of a relative balance between natural affordances and human needs, as Ehlers rightly points out in the 'Encyclopædia Iranica' (Ehlers 2011). This is not a romantic fiction, but the logical response to the demands of this way of life. It should also be mentioned that 'social equality' should not be confused with 'economic equality'. It is not the case that if household A has 21 sheep and household B has only 19, one sheep is given from A to B so that the two are equal. This was never the case, and in fact no one ever argued that it was: not Barth, not Ehlers and certainly not me. Egalitarianism and social equality refer to the self-sustaining character of the individual households, which were theoretically independent of each other. Cooperation and social cohesion were therefore a response to this theoretically atomising tendency of the group and were guaranteed by various culturised practices that were crucial for the survival of the group – and subsequently became the basis of its identity. This is an important point because it explains why these patterns can still be seen in the village today, although the days of seasonal migration are over. Nevertheless, the identity-constituting aspect of the ResourceCulture animal husbandry remains crucial to the group. In accordance with these dominant principles of the ResourceCulture, it was not spatially expansive, but rather circular.

Concluding Remarks: Recent Changes

The descriptions in the previous paragraphs and the discussion in this last one should have provided a comprehensive understanding of how the livelihood of a social group, in this case animal husbandry, can be linked to the socio-cultural existence of the group. It should also have demonstrated that a cultural constructivist understanding of resources and the concept of ResourceCultures can help to make these connections visible. This was the main aim of this case study and it is therefore comprehensive in terms of its research question. However, in relation to the village of Bagh-e Borj, it only tells part of the story, albeit an important part.

As I have mentioned before, from an economic point of view, animal husbandry is no longer the most important livelihood in the village, although it is not entirely unimportant. However, if I were to mention just one resource, it would be stone quarrying, which has become more important since the early 1950s when chromite was discovered and then mined near the village. As a result, more and more villagers gave up their seasonal migration and went into mining. Animal husbandry became a sideline activity for most households, but not the mainstay of their existence, at least not economically. The exact reasons that led the Eskanderies to abandon their traditional way of life, which they had practised for centuries and which was fully adapted to the needs of animal husbandry, are diverse. It should be noted that stone quarrying can also be understood and described as a ResourceCulture in its own right, which is another advantage of using the concept of ResourceCultures: they can be compared and contrasted. For the case of Bagh-e Borj, however, this would go far beyond the scope of this case study. For the sake of completeness, it will suffice to say that animal husbandry is merely a supplementary source of income for most households, while the economic base is constituted by something else. Nevertheless, I would refrain from calling it subordinate to other resources. For most villagers, it is still quite the opposite. I would like to conclude with an anecdote that I think is very telling: when I was conducting fieldwork in the village with an Iranian colleague, we returned from work to our host's house in the evening. While we were away, a sheep had given birth to a lamb, and on our return we found our host taking care of the newborn lamb. Later we were talking about the situation and my colleague, a Tehrani himself, asked me with surprise, if I had noticed how happy our host was holding the newborn sheep in his arms. I had indeed noticed it, and although our host's main occupation was stone quarrying rather than animal husbandry, I doubt that chromite or any other stone could have brought the same smile to his face.

6.2 Cattle as a Resource During the Transition from the Bronze Age to the Iron Age in the South-west of the Iberian Peninsula

Elena Revert Francés

Introduction: Cattle as a Resource

Cattle were, and still are, a valuable resource in the Iberian Peninsula (fig. 23). In all archaeological excavations where soil conditions allow, a large number of bovine bones have been preserved, testifying to this interaction with humans. The most recent example is the industry built around the fighting bull, which is bred mainly in Andalusia.

Cattle are very versatile in comparison with other animals that are often also bred for human consumption, such as pigs, goats or sheep. The degree of utilisation of its different components, such as meat, milk, skin, bones and horns, but also its labour, is very high.

Fig. 23 Bull in foreground and cow behind him in the Sierra Morena Mountains (Photo: Elena Revert Francés).

Bovids are gregarious animals, i.e. they are grouped in herds that can vary in size. They are flexible and highly adaptable to environmental and landscape requirements (Hesse 1997, 442). Nevertheless, they are capable of shaping the landscape in which they live and can have a significant impact on the environment. Since their main food is grass, the best places to raise cattle are those where there are large areas of grassland with abundant, varied and fresh grass (Pérez Ripoll 1999, 97). Therefore, deep land, ideally with a neutral pH, where this grass can grow well, will be the most suitable. By returning much of the fertility of the pasture they consume to the soil through their excrement, cattle help to maintain the ecological balance of the places where they are kept, as long as there is no overgrazing. Ideally, herds should be kept outdoors for as long as possible, leaving the stall barn for less favourable periods of the year. They also need quite a lot of water per day. Depending on the feed they have eaten, cows can drink up to 200 l of water a day when in full milk production. If the farmer cannot provide this water artificially, the cows have to find it naturally by grazing near springs or wells with constant water, which can lead to overgrazing around these natural springs. An advantage of cattle farming is that it can be combined with crop farming, occupying land that is less suitable for agriculture and thus increasing the economic value of cattle. Furthermore, as mentioned above, the excrement could be used as fertiliser in the fields, which supports the hypothesis of a combination of crop and livestock farming in the Lower Guadalquivir Valley in the Early Iron Age. Transhumance could have taken place between the Sierra Morena Mountains, still an ideal place for livestock farming, and the fertile fields of the Lower Guadalquivir Valley, where these fields would have been fertilised (fig. 24). However, the arm of the sea known as the *Lacus Ligustinus*, which reached what is now Alcalá del Río, could have limited the size of the agricultural and grazing areas compared to the present landscape (fig. 25 + 27). Recent studies of prehistoric cattle on the peninsula suggest that the simple topography of the Guadalquivir Valley and the high levels of solar radiation would have provided very good conditions for cattle rearing (Nieto-Espinet et al. 2021).

The productive and reproductive cycles of cattle are influenced by the breed and its characteristics, the type of feed, climate, soil conditions in the breeding area, herd management, hygiene and health. Failure to adapt to the environment of cattle transported to new locations would result in loss of weight and lustre of the coat, reduced milk production in

Fig. 24 The Lower Guadalquivir Valley with the town of Villaverde del Río (Seville) seen from the foothills of the Sierra Morena Mountains (Photo: Elena Revert Francés).

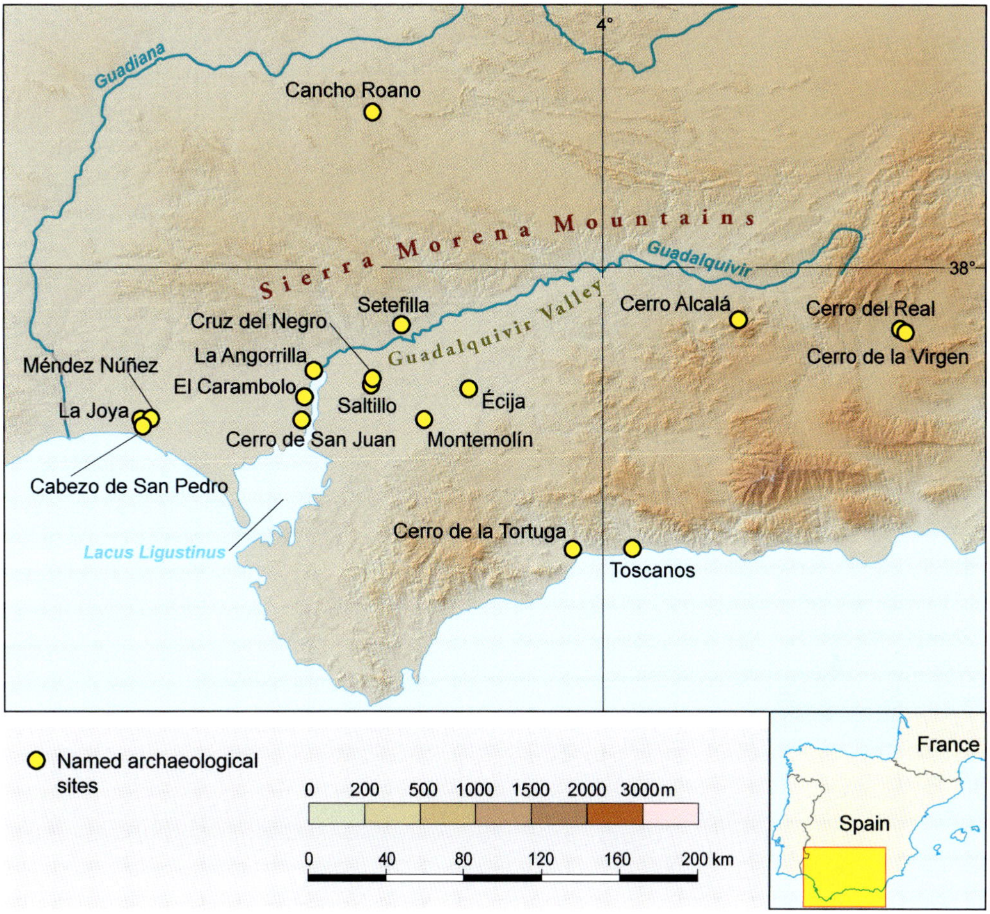

Fig. 25 Map of the region of Andalusia with all the archaeological sites named in the text, the Sierra Morena Mountains, the Guadalquivir Valley and the *Lacus Ligustinus*. The drawing of the *Lacus Ligustinus* is based on the research of Arteaga (Arteaga/Roos 1992; Arteaga et al. 2016) (Data sources: CGIAR-CSI SRTM; Map: Elena Revert Francés; Graphic design: Richard Szydlak).

cows, susceptibility to disease and production difficulties (Spanish Ministry of Agriculture, Fisheries and Food 2023).

The castration of bulls seems to have been a common practice since Neolithic times. It has also been documented in the Iberian Peninsula. The aim was to make castrated bulls docile, to reduce sexual competition with other bulls in the herd and to obtain individuals with greater working power, as well as them being considered a meat reserve in times of crisis (Pérez Ripoll 1999, 101).

A priori, cattle seem to be a multifaceted and economically interesting resource. But what is a resource and why have cattle been chosen as a resource since time immemorial? Resources are the driving force of societies, what creates, sustains and transforms them. They are something that is valuable to people and worth caring for and preserving. Cultural and social aspects are interwoven with the choice of these resources (Bartelheim et al. 2021a, 11). The traditional economic meaning of resources, understood primarily as 'natural resources' or raw materials, has evolved from the 18th century to the present. However, the nature of resources unfolds in a range of possibilities: in addition to resources derived from nature, there are economic, social, cultural and thus symbolic resources (Bourdieu/Waquant 1992).

But the most important thing to be clear about is that, in our understanding, there is no such thing as a resource *per se*, in a natural sense. Valuing something as a resource is the end of a process in which several steps have to be taken: first, humans perceive the potential of a given element; second, we perform actions that lead us to have an experience with that element; finally, we interpret that perception and experience and judge whether the element in question can be considered a resource. In this case, bovids went through this initial process of interaction with humans and were, and still are, a valued and appreciated resource (Bartelheim et al. 2021a, 9–21).

The Materiality and Immateriality of Cattle as a Resource: The ResourceComplex

Resources never exist in isolation, but are part of networks that bring together other components, both material and immaterial. The use of any resource requires the interaction of different elements integrated in cultural practices, such as people and other living beings, things, norms, forms of knowledge, places, infrastructures. Thus, an open and changing network is configured around resources, which has come to be called ResourceComplex (Bartelheim et al. 2021a, 13–15). The concepts of materiality and immateriality are key categories of analysis when examining the different elements that make up these resource networks.

As shown above, the materiality of cattle unfolds in a variety of ways. Due to the size of these animals, which in some cases, especially with current specimens, reach up to 1.5 m at the withers and weigh up to a tonne, depending on the breed and possible castration in the case of males, the range of products is much richer than for other farm animals. Cattle are a major source of meat, which can be consumed in a variety of ways, including raw, roasted, boiled, air-dried or preserved in salt or vegetable oil. The bones of cattle can also be used not only for food (marrow) but also as raw material for tools, decorative elements, buttons or even musical instruments. Their horns were also a valuable raw material. Milk has also been one of the most commonly consumed bovine products since prehistoric times, as well as derived products such as cheese and yoghurt (Sherratt 1981; Dudd/Evershed 1998; Spangenberg et al. 2008). Bovine hides, after tanning, were also a widely used raw material and very necessary in the daily life of prehistoric societies.

By analysing nitrogen isotopes in animal and human bones, it is possible to detect an artificial contribution of fertiliser of animal origin to the prehistoric fields where the plants that these animals and humans ate grew. Undoubtedly, cattle dung would have been one of the most important sources of fertiliser, as mentioned above.

In the Iberian Peninsula, cattle were usually draught animals used for agricultural work. On the other hand, the use of bovids as mounts has not been demonstrated, neither today nor in archaeological remains, as has been the case in other areas and, in some cases, with non-Iberian breeds, such as in various regions of Africa and Asia, or in parts of Europe, such as Greece (Brentjes 1972, 12–23).

But it is not only these material aspects of cattle's contribution to society that are important. The role of bovids as carriers of a strong symbolism in pre- and protohistoric ideology throughout the Mediterranean also held great importance, the meaning of which is very difficult to grasp with traditional archaeological methods such as excavation or work on the finds. These immaterial dimensions were represented materially in the Early Iron Age necropolises of the Guadalquivir Valley, such as that of La Angorrilla, which shows us that the skeletal remains of these mammals were charged with a symbolism that is currently beyond our grasp. Of the 69 tombs in the necropolis (54 inhumations and 15 cremations), 20 contained animal offerings (19 inhumations and one cremation). This symbolism was expressed by the fact that in 80 % of the inhumations with animal offerings, these consisted almost exclusively of *Bos taurus* pelvises. In two other inhumations, tombs 37 and 38, offerings of mammalian animal bones were found in a very fragmented and poor state of preservation, which prevented their classification, and in inhumations 18 and 41 the offerings consisted of ostrich and chicken eggs respectively. The only cremation with animal offerings contained goat and sheep bones (Pajuelo Pando/López Aldana 2014). The symbolic importance of the species *Bos taurus* is evident in this necropolis, in this case clearly related to the inhumations.

Another example of the importance of *Bos taurus* in the imagination of these Early Iron Age populations in the south of the Iberian Peninsula is the necropolis of Cruz del Negro, where both inhumation and cremation rites coexist, the latter being the most numerous. A total of 179 structures have been documented here, of which 112 are burials (of which 10 are inhumations) and 67 *ustrina* (Amores Carredano/Fernández Cantos 2000, 158–159). Animal sacrifices were found in 22 structures. Although goats are the most common species offered in these burials, whose bones were cremated before being deposited, there are seven cattle whose bones were deposited after the cremation of the deceased (Bernáldez Sánchez et al. 2013, 331).

The settlement and necropolis of Setefilla are located in the foothills of the Sierra Morena. Both are only a short distance apart. This necropolis contains 10 burial mounds for both cremation and inhumation (fig. 26). Each mound contains a different number of burials. The

Fig. 26 A view of the foothills of the Sierra Morena Mountains with the sanctuary (white building) and the castle (tower behind the sanctuary) of Setefilla. The prehistoric settlement of Setefilla is situated under the castle (Photo: Elena Revert Francés).

Nr.	Site	Location	Province
1	Cancho Roano	Zalamea de la Serena	Badajoz
2	Cabezo de San Pedro	Huelva	Huelva
3	La Joya	Huelva	Huelva
4	Méndez Núñez n° 7–13	Huelva	Huelva
5	La Angorrilla	Alcalá del Río	Sevilla
6	El Carambolo	Camas	Sevilla
7	Cerro de San Juan	Coria del Río	Sevilla
8	Palace house of the Marquis of Saltillo	Carmona	Sevilla
9	Cruz del Negro	Carmona	Sevilla
10	Setefilla	Setefilla	Sevilla
11	Montemolín	Marchena	Sevilla
12	Écija	Écija	Sevilla
13	Cerro de la Tortuga	Málaga	Málaga
14	Toscanos	Vélez-Málaga	Málaga
15	Cerro Alcalá	Torres	Jaén
16	Cerro del Real	Galera	Granada
17	Cerro de la Virgen	Orce	Granada

Fig. 27 Archaeological sites named in the text (Data: Elena Revert Francés; Graphic design: Richard Szydlak).

mound known as Tumulus A, 29 m in diameter, is the largest in size, but also contains the largest number of cremations, 65 in total, in addition to the central chamber. Animal bones were deposited in the burial structures, the most numerous being cattle bones, but they were not directly related to the burials. Nevertheless, cattle bones were deposited as burial goods in grave 11 (Aubet Semmler 1975, 82, 157–158).

This deposition of bovine bones in burials, following a certain pattern within each necropolis, was not a new custom in the south of the Iberian Peninsula at the beginning of the Iron Age; it does not seem to have been introduced from outside by the Mediterranean populations that frequented the peninsular coasts from the 1st millennium BC onwards. The Early Bronze Age populations of the south-eastern Iberian Peninsula have also established patterns of animal offerings in burials, with bovids and ovicaprids being by far the most represented species. Bones from the appendicular skeleton of bovids (femur, humerus and tibia) were deposited in these graves according to a pattern whose cultural and social norms remain unknown. A distinction between the biological sex of the deceased and the type of animal sacrifice has not yet been established (Aranda/Esquivel 2007).

It makes sense to think that what is contained in a tomb is directly related to the deceased buried there. But what exactly does this mean? It is not plausible that this question can be answered unanimously for all regions, times and cultural attributions on the planet. Sometimes interpretations assume that everything in a tomb belonged to the deceased in some way, and that his or her most precious possessions must therefore accompany him or her into the afterlife. In other words, the richer the grave goods, the greater the material wealth of the deceased in life. This dogmatic explanation is too simple and certainly does not apply in all cases. Leaving aside the rest of the grave goods, such as ceramic vessels, metal objects or jewellery, why were certain bovine bones deposited in some burials and not in others? What was the purpose of this? Was it the expression of an ancestral collective cultural memory in the sense of Aleida and Jan Assmann (1988; 1992; 1999; 2006), going back centuries or even millennia? One of the most widely accepted hypotheses today is the practice of commensality, with a ritual banquet attended not only by those present, but also by the deceased. By depositing part of the animal sacrificed for the occasion in the tomb, the deceased would participate in this banquet (Aranda/Esquivel 2007, 115; Delgado Hervás 2008). This deposit of part of the animal in the tomb would therefore be a central aspect of the ritual. A very plausible hypothesis, in line with our own studies of organic residues in ceramics from Bronze Age burials in the southern peninsula (still in progress), is that of a banquet, or rather the offering of cooked food in ceramic vessels deposited inside the grave. In this case, the offering itself would be the content – the food – and not the container – the ceramic vessel. Are the cattle bones deposited in the burials related to the cooked food offered in the ceramic vessels?

But it was not only in burial sites that the symbolism of these hoofed animals was present. In various Early Iron Age sanctuaries in the Lower Guadalquivir region, such as El Carambolo, the palace of the Marquis of Saltillo, Cerro de San Juan, Montemolín or the sanctuary excavated in the plot number 7–13 of the Méndez Núñez street in Huelva city, the sacrifice of cattle has also been documented (Ramos Soldado 2021).

The immateriality of cattle is not only represented by these offerings deposited in necropolises and sanctuaries. An element present throughout the Mediterranean in the 1st millennium BC is the so-called oxhide altar, which is said to represent the stretched hide of a sacrificed bull or ox, as seen in the sanctuaries of Cancho Roano, El Carambolo or Cerro de San Juan (Celestino Pérez 2001, 28; Fernández Flores/Rodríguez Azogue 2010, 227 fig. 22; Escacena Carrasco 2001, 88 fig. 6). The peculiar shape of this altar is also found in other objects, such as the gold pectoral from the Tesoro de El Carambolo, the ritual plate from tomb 16 of the necropolis of La Joya or the Cypriot copper ingots (Kukahn/Blanco 1959, figs. 1, 6, 8; Garrido/Orta 1978, 53 fig. 26; Buchholz/Karageorghis 1973, 279). The appearance of an oxhide in the myth of the founding of Carthage by Dido/Elyssa is also of interest (Scheid/Svenbro 1985, 330–331).

The mysticism of cattle, especially bulls, is present throughout the Mediterranean, especially in Greek mythology, as evidenced, for example, by the legend of the Cretan Minotaur or two of the mythical labours of Hercules. The seventh of these was the capture of the Cretan bull, who was the biological father of the Minotaur. In turn, King Minos – the husband of Pasiphae, the mother of the Minotaur – was the son of the Phoenician princess Europa and the bull-shaped Zeus. There are even classical authors who place the tenth of these labours of Hercules in the study area: the theft of the oxen of King Geryon, the mythological king of the equally mythological Tartessos. The oldest known mention of the story is in the 'Theogony' of Hesiod, a Boeotian poet who lived at the turn of the 8th and 7th centuries BC. He places the action on the island of Erytheia, without specifying its location. Strabo, Avienus, Virgil, Pliny and Diodorus Siculus place it in Iberia. Hecataeus of Miletus and Arrian, on the other hand, place the events in Greece, where the myth of the labours of Hercules originated. In any case, this myth became popular from the 6th century BC onwards on Greek vessels decorated with representations of Hercules fighting Geryon, but it would have had little or no connection with the Phoenicians and the indigenous inhabitants of the south of the Iberian Peninsula, judging by the lack of echo of this type of iconographic representation in the area under study (Blázquez Martínez 1983).

The Greek – or more specifically the Proto-Corinthian and Corinthian – influence in the representation of bulls seems to have played a relevant role once again in the use of Greek vessels as models for numerous ceramic vessels with painted zoomorphic decoration found in the study area and dated between the 7th and 6th centuries BC. The best preserved examples are those from Carmona, Montemolín, Cerro Alcalá and Écija, with a total of more than 60 vessels (Carrasco/Fernández 2020; Schattner 2022).

From the perspective of the ResourceComplex, indigenous actors from the Iberian Peninsula, as well as Greeks and Phoenicians, were an active part of the resource network around cattle in the south of the Iberian Peninsula during the 1st millennium BC. It should be emphasised that this would have been the case particularly in the first half to mid-millennium, before the Roman conquest, as the Romans modified any pre-existing paradigm around livestock and crop cultivation on the peninsula.

The materiality of bovines would have been linked to their immateriality, given on the one hand by the symbolism of cattle in rituals through sacrifices in sanctuaries and necropolises, and on the other by figurative representations in ceramics and in the so-called oxhide form. The network that these elements bring together is very extensive and complex, starting with the people who raise the cattle, their distribution networks, the people by whom and the places where the material obtained from the cattle (meat, milk, hide, bones, etc.) is processed, the people in charge of carrying out the sacrificial acts in which the cattle are offered, those who witness them, the potter's workshop where the ceramic vessels with these representations of the bulls are made, the intention behind them, the customs, rules and objects that concern all these activities and a long etcetera. In this way, it has been shown how this elaborate network was created around the resource of cattle, which is here referred to as a ResourceComplex.

The Emergence of a ResourceCulture

Over time, this complex network of resources can change under the influence of certain factors, external or internal, which act as catalysts or inhibitors of certain changes or developments within this network. Such a diachronic perspective on networks can be defined as a ResourceAssemblage that is both a state and a process (Bartelheim et al. 2021a, 15–16). Important actors in this perspective were the Phoenician populations, who may have brought about a change in the way the inhabitants of the southern Iberian Peninsula viewed crop cultivation and animal husbandry, acting as co-creators and drivers of this process of change and evolution of the ResourceComplex involving the resource cattle. These Phoenicians probably brought with them new ways of breeding cattle, or a new breed altogether, which would explain the differences in size and characteristics found in zoological studies carried out throughout the 20th century (Ulmansky 1918; Boessneck 1969; Bernáldez Sánchez et al. 2010, 372–373).

In fact, it was the Phoenicians who brought animals to the Iberian Peninsula that had not existed there before, such as the chicken, which is documented from the 8th century BC in sites in the south of the peninsula, such as Toscanos, Cabezo de San Pedro and Cerro de la Tortuga (Thesing 1977, 7–11). It seems that these Phoenician populations were also the bearers of fruits that were previously exotic and did not exist in the peninsula, such as the chickpea (Pérez Jordà/Peña Chocarro 2024, 10). Whether this eastern contribution was the driving force behind the exploitation of endemic resources such as the olive remains to be confirmed.

For the Greek population, cattle, especially bulls, were also animals with symbolic meaning and very present in their mythology. Therefore, the Greeks could have brought with them other ideas about cattle that were not present in the Iberian Peninsula, but which had an impact on the local population, as can be seen from the Iberian representations of bulls on ceramic vessels. This represents a further step in the development of a ResourceAssemblage centred on cattle.

The model that describes the resources themselves, in this case cattle, the networks around these resources (approached as a ResourceComplex) and the agents that diachronically influence the development and evolution of these networks (approached as a ResourceAssemblage) is what has come to be called ResourceCulture. This model aims to be able to compare complex contexts and relationships between resources and the different actors, objects, places and norms that interact with them (Bartelheim et al. 2021a, 16–18).

Power and knowledge are two closely related factors that directly influence ResourceCultures. Power determines who has access to knowledge, but those who have knowledge know how to gain power. Therefore, knowing what cattle need in order to be bred in the best productive conditions would also be related to the power that these animals might represent, giving a certain status to the people who breed them. Therefore, the ResourceCulture around cattle in the Early Iron Age in the south-west of the Iberian Peninsula would have been linked to who organised and distributed this knowledge, which in turn would have been linked to the arrival and knowledge of agriculture and animal husbandry that the Phoenicians brought with them.

The presence of *Bos taurus* bones as burial offerings in necropolises in the Lower Guadalquivir Valley shows that these bones were important to these people as a kind of bridge between the human world and the afterlife. Similarly, the sanctuaries and the cattle sacrificed there bear witness to the symbolic importance of these animals. These bones are something material, but in this case they represent something immaterial. The economic importance of these animals was great, as explained at the beginning of this case study. This importance is clearly linked to the ecological aspects and the symbolic value of these animals. These aspects point to cattle as a link in relationships and therefore with a social value. Thus, cattle seem to have played an important identity-constructive role in this context. The ResourceCulture built around cattle in the south of the Iberian Peninsula delimits all these cultural dimensions of practices around this resource, be they symbolic, economic or ecological, and therefore shaped the life of these populations in a distinctive way.

6.3 The Use and Non-use of Resources during the Bronze and Iron Ages in Menorca (Balearic Islands, Spain)

Marta Díaz-Zorita Bonilla, Edwin Mauricio Marciales Daza, Monice Timm

Introduction

The concept of ResourceCultures (Bartelheim et al. 2015; 2021a; 2021b) provides a framework for understanding tangible and intangible resources and their significance in shaping, sustaining or altering social groups and dynamics. By using resources as an analytical category, it becomes possible to explore how these social dynamics evolve according to spatial and temporal context (Danwerth et al. 2018). Resources cannot be studied in isolation, as their value lies in their role in the community. The use of a resource can trigger multidimensional processes that can change social groups, influence cultures in various ways and create a network of resources that can be approached as a ResourceComplex (Hardenberg et al. 2017). From this perspective, these resource networks can be analysed to observe how resources can shape social identities and differ from one culture to another (Ingold 1993).

Furthermore, these resources are never static; they embody agency that is influenced by the environment, space, technology and time. The diachronic transformation of these social networks extends to the development, movement and valuation of the social and cultural dynamics of communities, offering a new approach to understanding resources as assemblages of contexts across time, especially in the past. These network and dynamic models provide a means of understanding past cultures (Bartelheim et al. 2021a) and how cultures perceive and value these resources across time and space. This concept can be applied to both continental regions and islands, although the latter contexts are exceptional. In the case of islands, there are some physical limitations and the sharing of resources means that, to some extent, activities are shared. This can also lead to the development of unique cultural identities. Furthermore, it is important to recognise the specific focus of the project to explore past subsistence strategies, mobility, isolation, connectivity and colonisation in island contexts using the ResourceCultures framework. This approach allows for an in-depth examination of how resources, both raw materials and cultural assets, were available and used in the past, particularly in regions such as the western Mediterranean.

The Late Bronze Age and Iron Age burials from the Biniadrís cave in Menorca provide insights into these concepts. The exceptional state of preservation of the remains in the Biniadrís cave allows researchers to reconstruct palaeodemographic profiles, understand past identities and study the influence of social relations on the use of resources and funerary rituals. Specifically, in relation to the main research questions of this project, how diet and mobility patterns shape Late Bronze Age and Iron Age communities in the western Mediterranean. Through this examination at the Biniadrís cave, it becomes evident that resources were social constructions developed by Late Bronze and Iron Age communities in Menorca, shaping their cultural practices and reflecting their resilience.

Understanding the Use and Non-use of Resources

Resources are crucial for the optimal functioning and organisation of society (Bartelheim et al. 2015). Resources, whether tangible or intangible, are influenced by social relations and entities, creating, sustaining and altering identities (Hardenberg et al. 2017; Teuber/Schweizer 2020). They are more than mere physical objects; they are defined by cultural systems, including meanings, and vary from culture to culture (Burt 1992). The use of resources in each culture is unique and reflects the values, beliefs and traditions of its users. Socially connected people use this variety of resources and the specific interests of the group influence their use. It is essential to recognise that resources are social constructs with cultural connotations (Ash 2002, 32). These aspects, whether economic, political or social, influence the organisation and structure of social life. The way resources are used, controlled and distributed reflects the dynamics of a society and directly affects the lives of those who use them (Granovetter 1985). The meaning and value of resources vary considerably and depend on the individuals directly associated with one or more resources. Therefore, different approaches can be observed in coastal and mountainous regions, and may change over time as levels of industrialisation and technological development progress. Access to resources and their meanings are of great interest to the scientific debate, as they go beyond a basic materialistic understanding. It is therefore important to examine social networks in relation to cultural ideas and practices, and to consider entities and identities (Hardenberg et al. 2017, 14–15).

In the case of the Biniadrís cave, the study of funerary practices and resource use in Late Bronze Age and Iron Age communities on Menorca illustrates how resources, both tangible and cultural, were socially constructed and embedded in the group dynamics of ancient societies. This archaeological context, through the use of archaeological and bioarchaeological methods, highlights the use of resources and also the avoidance of some of them, particularly in Menorca, a particular aspect that can be replicated for other islands in the western Mediterranean.

The Application of the Resources Concept in the Western Mediterranean

The study of the prehistoric contexts of islands involves the exploration of the resources available to past communities, as well as how these resources were used and exchanged. Furthermore, they provide ideal case studies for examining social isolation and connectivity, colonisation and 'altering phases of cultural interaction' (Dawson 2014, 13). When it comes to acquiring and managing resources and navigating social relationships, islands offer a unique perspective that combines environmental and cultural approaches (Dawson 2019). In addition, transactional topics also provide valuable insights into understanding how resource structures have evolved through different stages of human use. The development of different levels of technology and knowledge in relation to specific resources, as well as the diachronic creation of power structures, can be observed through the historical material evidence provided by archaeological methods, which may vary depending on the context but collectively contribute to a broader understanding of island settings (Dawson et al. 2023).

In recent decades, several authors have redefined the use of resources and their definitions in various contexts (Baldacchino 2006; Stratford 2016; Connell 2011). While most definitions take an economic approach, there is an increasing emphasis on social engagement, highlighting the importance of resources as social constructs developed by communities to enhance their resilience. This concept of social constructs implies that resources are not simply physical entities with inherent value, but that their significance is shaped by cultural, social and historical contexts. Communities collectively assign value and meaning to resources and determine how they are used, managed and integrated into daily life. This social construction process reflects shared beliefs, norms and practices, which can vary widely between different societies and environments.

This study provides insights into the potential and limitations of different human habitats and offers a framework for understanding the complex relationships between people and their environment. From this perspective, the significance of resources in these environments is assessed not only in terms of their physical availability, but also in terms of how they are perceived and valued by the community. It can therefore be argued that the ways in which resources are developed and used represent shared characteristics that define the unique attributes of islands. These shared characteristics include the collective practices of resource management, cultural traditions and adaptive strategies that result from the interaction between communities and their environment. In this way, the social construction of resources

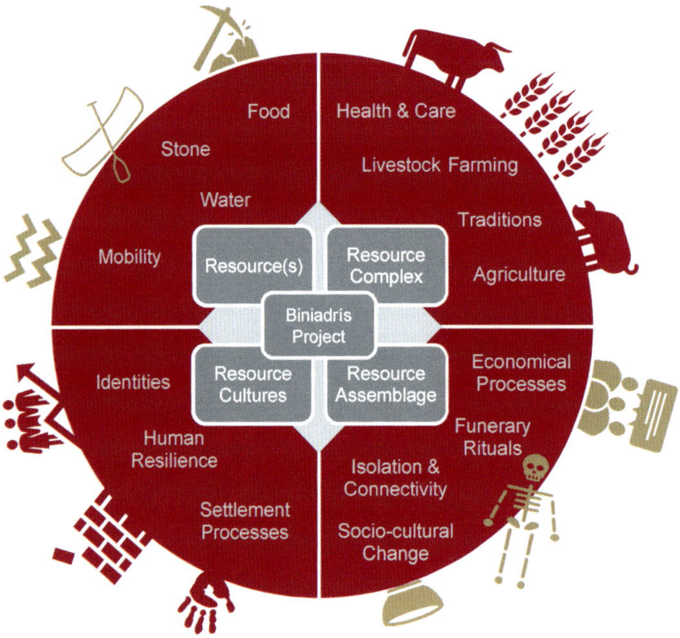

Fig. 28 Theoretical framework of the Biniadrís project according to the ResourceCultures concept (Graphic: Monice Timm/ Project Biniadrís).

becomes a central element in understanding the distinctive identity and resilience of island communities (fig. 28).

Differences in access to resources and their specific uses can find direct correspondences in physical appearance and thus provide valuable individualised but also general information about natural, economic and socio-cultural living conditions and shed light on interactions with cultural-historical dynamics and within the social structure to identify social (in) equalities. It is therefore crucial to combine materials in close relation to tradition and ritual within a community. This is important in order to gain insights into the daily lives of the deceased and to learn a little more about the 'hidden knowledge' of their community. On the other hand, different landscapes that are not distant from each other are suitable for clearly identifying the different ways in which resources are valued, accessed and used. Maritime and non-maritime societies can be studied because they offer a unique opportunity to reconstruct diet and health, as well as functional allocations and adaptations of use, especially during the Bronze and Iron Ages.

For a more detailed look, the burial cave of Biniadrís (Menorca) has been chosen because of its island location with a functional society, as well as its exceptionally well-preserved organic materials. One focus will be on food production based on regional dietary patterns, especially in relation to one of the main objectives of this project, which is investigating how resources related to food shaped Late Bronze and Iron Age communities. As seen from the ResourceComplex perspective, animal bones have been intensively studied, since the network of resources related to animal husbandry was one of the most important food sources on Menorca as a way of life during the Bronze Age. The factor of mobility, as another main objective of this project on how mobility also influences dietary habits and preferences, is used as an alternative hypothesis to explain the observed pattern in terms of certain forms of power structures expressed in culturally predetermined behaviour. The study of stable isotopes offers valuable insights into dietary patterns, allowing us to determine the likely diet

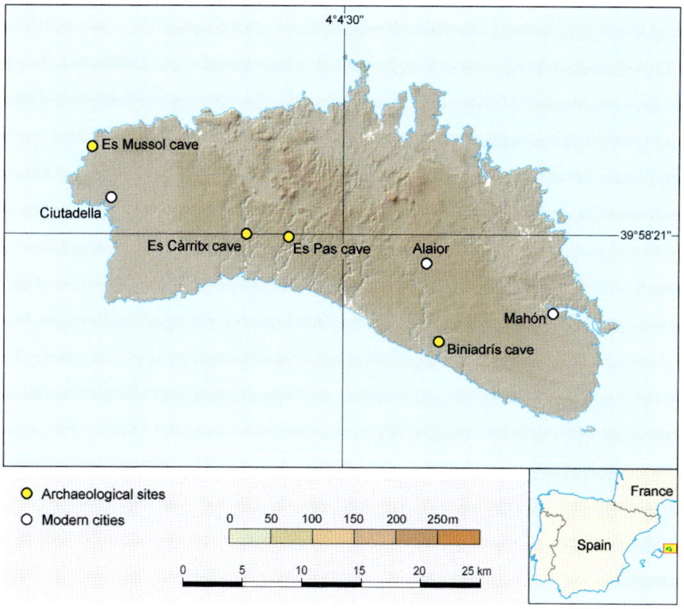

Fig. 29 Map of Menorca and the archaeological sites mentioned in the text (Data sources: CGIAR-CSI SRTM; Map: Monice Timm/Project Biniadrís; Graphic design: Richard Szydlak).

of the population and its relationship with the sea. In addition, this technique allows the detection of human and animal mobility in island contexts, providing a further understanding of residential and other forms of mobility. This analysis will contribute to a general understanding of the social and cultural structures of maritime and non-maritime populations. The observed dependencies in food consumption and migration events will be seen as processes of adaptation, integration and rejection.

The Western Mediterranean in the Late Bronze Age: The Biniadrís cave

The Balearic Islands represent some of the last Mediterranean islands to be inhabited (Dawson 2014; Ramis 2010). However, the prehistory of these islands is characterised by a unique and vast amount of monumental architecture and a large number of prehistoric sites. The island of Menorca, in particular, has been thrust into the limelight by its inclusion on the UNESCO World Heritage List for 2023 (UNESCO 2023).

The north-eastern part of Menorca is less populated due to several geological factors (salty marshes, strong northerly winds and barren soil formation). However, the southern coast (Migjorn) was more attractive to inhabit because it contains a greater number of archaeological sites in the massive cliffs, ravines or the softer limestone plateau, as well as in the flatter relief of the island with better soil quality (Pomar et al. 2002; Trias 2004; Van Strydonck 2014).

Menorca is a stony island, and the use of stone is not only visible in the archaeological context, but is still an important part of the cultural identity today. The social and economic structure of Menorcan prehistory fits well into the cultural context of the western Mediterranean, but there are still many missing links in the island's past in terms of social organisation and funerary behaviour. Thus, the Biniadrís cave is a good example of a Menorcan prehistoric burial cave (fig. 29).

Fig. 30 The reconstruction of daily life practices from the Biniadrís human community (Graphic: Lola Contreras/www.pastwomen.net).

This cave is the result of natural geological processes, particularly the dissolution of limestone over thousands of years. It is an excellent example of the cultural character of the inhabitants, who lived there from the late Bronze Age to the Iron Age for over 700 years (1200–500 BC). It brings together many material goods that reflect social identities that have also been found in similar contexts scattered around the island, such as other burial caves like Mussol, Es Càrritx and Es Pas (Lull et al. 1999a; 1999b; Fullola et al. 2008). The interior of this natural cave has a polylobular shape and encompasses approximately 18 m^2. The single chamber cave shows an organised space with a variety of funerary practices embodied by strong symbolism. This is represented by the traditional cyclopean masonry or dry-stone architecture ('*piedra seca*'), which formed a unique appearance at the entrance to the cave. The cave entrance was built with special care using limestone of different sizes. The upper rows of the wall did not contain any binding material, its presence was evident in the lower construction levels. A particular feature is the lintel, which consists of a large flat slab supported by some orthostatic jambs (Altamirano García/Alarcón García 2018; Alarcón García et al. 2020). Thus, the cave entrance is monumental and imposing. It is very impressive how this wall was built in the cave, 14 m above the ground, and only accessible by a ladder construction or from the platform above the cave. The purpose of this gate construction, whether as a 'house' for the deceased, analogous to earlier boat-shaped or '*Naveta*' structures, or to make the entrance to the cave less visible, remains unclear. However, it shows the strong relationship with the members of their community after their death and the preservation of the funerary memory as a social identity (fig. 30).

Fig. 31 Commingled human remains from the first stratigraphic units of the Biniadrís cave (Photo: Marta Díaz-Zorita Bonilla/Project Biniadrís).

Inside the cave, a high number of commingled human remains were visible, carefully organised and associated with several wooden, ceramic, metal and other objects (Moreno Onorato et al. 2019) (fig. 31). In particular, these were personal objects, such as the V-perforated bone buttons, designed for permanent use and deposited with the deceased. These objects combine the knowledge of their production, their careful manufacture and a symbolic character between the owner, their relatives and the object itself (Altamirano García/Alarcón García 2018). But they are also a recurrent object in the western Mediterranean, shaping identities in distant places (López Padilla et al. 2006). Within the cave, continuity within funerary rituals and the preservation of tradition are aspects that could be observed at Biniadrís, which are also representative of the period. It can therefore be assumed that knowledge of funerary procedures was important within this community, or even within these communities, and was passed on from one generation to the next. Since all age groups and both sexes were present in the cave, it can be assumed that they also had a similar status within the community and were included in the burial practices to keep their memory alive over a long period of time.

Understanding Resilience through Resources in Menorca

In the context of Menorca, and more specifically in the Biniadrís cave, an exploration may yield insights into the use and distribution of resources during the Bronze and Iron Ages in the western Mediterranean. In this endeavour, the theoretical framework of historical materialism is used to elucidate the dynamics of social relations within different periods in Menorca, as revealed through the study of archaeological artefacts found during excavations. Historical materialism posits that it is not the consciousness of individuals that primarily shapes their existence, but rather the prevailing social milieu that moulds their cognitive and perceptual capacities (Kohl 1981). By integrating the perspective of historical materialism, this research aims to gain deeper insights into resource use, allowing a more comprehensive understanding of human behaviour and social organisation in prehistoric western Mediterranean societies. This perspective makes it possible to discern the socio-cultural clues embedded in the material culture as a first-hand resource, discovered in both domestic and funerary sites in Menorca, as well as highlighting a continuity on the island. They serve as important channels for reconstructing everyday social interactions and relationships. In terms of subsistence strategies, the conspicuous avoidance or limited consumption of marine resources during the Bronze Age in the Mediterranean region can be attributed in large part to well-established traditions (Gilli et al. 2006; Rihuete Herrada 2003; Lull et al. 2013; Ramis 2010). The archaeological evidence of food consumption for earlier periods confirmed a significant decrease in marine food consumption during the Neolithic in Iberia, based on carbon and nitrogen isotope analyses (Fontanals-Coll et al. 2016; Salazar García et al. 2018; Cubas et al. 2019). Therefore, during the Bronze and Iron Ages, we can observe the inheritance of the non-use of marine sources and the exclusive strategy of relying on agriculture and pastoralism.

Archaeological research has revealed compelling correlations between religious and cultural practices and the discouragement of fish consumption, particularly in coastal and island areas of the Mediterranean. This phenomenon is clearly illustrated in places such as ancient Greece, where religious tenets initiated periods of abstinence and, in certain cases, outright bans on fish consumption (Dotsika et al. 2019; Hudson/Fernandez 2023). In this context, fish consumption was viewed with disdain by the elite, largely because of its association with lower social classes. This discernible trend underlines the prevailing preference for livestock as a more resource-efficient source of food within Bronze Age coastal and island communities.

Economic factors may also play a crucial role in explaining the avoidance of fish consumption during this period. Particularly in regions where fish was scarce, the cost of acquiring this marine resource was exorbitant, making it an unaffordable dietary option. For example, in places such as Crete and Cyprus, where food resources were insufficient, cereals and livestock took precedence as food products, relegating fish to the status of an occasional privilege or a culinary choice reserved for the upper echelons of society due to its preferred local consumption and possible high monetary value (Zohar/Artzy 2019).

Geographical determinants played an important role in shaping the dietary choices of Bronze Age communities, with proximity to the sea closely linked to the degree of dependence on fish and seafood. Coastal regions with direct access to the bounty of the sea would naturally and hypothetically show a strong dependence on fish. Conversely, in areas where fish resources were scarce or harvesting was impractical and dangerous, alternative sources of food were sought.

Rocky coastlines and restricted maritime access likely limited traditional fishing practices in the Balearic Islands, potentially leading to a greater reliance on livestock. The archaeological record of Menorca does not provide any discernible evidence of the exploitation of marine resources, with the scarce presence of molluscs representing less than 1 % in relation to the predominant livestock assemblages, as observed by both isotopic and zooarchaeological data (Ramis 2017; Van Strydonck et al. 2005). It is important to note that a high abundance of molluscs, coupled with meagre remains of fish bones, indicates their initial appearance during the late Iron Age, coinciding with an increase and more consistent interaction with Phoenician influences (Ramis 2017; Anglada et al. 2017). Approximately 78 % of Menorca's terrain is characterised by steep cliffs and elevated coastal features, exceeding three metres above sea level, which limits the accessibility of maritime contexts (Balaguer et al. 2017). Consequently, the diet of the island's prehistoric population was predominantly focused on terrestrial domesticated mammals, such as goats, cattle and pigs. Furthermore, this dietary pattern showed a neglect of non-domesticated species on the island, such as small game, including rabbits and bird species that were unintentionally introduced to the island (Ramis 2017; Gornés/Gual 2018).

Environmental variables also influenced fish and seafood consumption patterns, including factors such as water depth, coastal topography, seasonal variations and the availability of alternative food sources, such as crops and livestock (Evans 1977). On the other hand, the absence of arboreal vegetation typically observed along the island's coastline, as evidenced in the archaeological record, coupled with a deliberate preference for rainwater over river sources, reflects a discriminating and selective approach to the exploitation of local resources within these settlements (Picornell 2012; Pérez et al. 2018). This selection can be attributed to cultural taboos or anathemas that, together with the avoidance of marine resources, played an important role in the emergence and development of the Talayotic culture, implying a remarkable disregard for these resources.

Bronze Age Life in the Biniadrís cave (Menorca): Human Adaptation to the Environment

Access to raw materials in Menorca is limited by the geology of the island, as well as by navigation routes and skills. Specialisation is therefore necessary to adapt to the island's conditions and develop resilience. The individuals buried in the Biniadrís cave are only available as commingled remains, so that, with a few exceptions, no complete anatomical correlations have been found between the burials. This imposes certain constraints on the analysis, but also offers the opportunity to learn more about the lives of these people. One such possibility is given by the theoretical approach of the Bioarchaeology of Care (Tilley 2015), which focuses not on specific pathological cases of individuals, but on a population-wide change in the skeletal system, as well as on the practice of care and the structure of society. The use of some specific resources, such as stones and therefore mining, corresponds not only to the archaeological landscape of Menorca and the living conditions of these people, but also to the different activities in which the individuals were involved, which had a direct impact on their skeletons. Possibly due to intense physical activity, the bones of the individuals buried in Biniadrís showed a high degree of robustness, regardless of their sex and age. This is important as there does not appear to have been a sexual division of labour. The (over)loading over a long period of time was reflected in the bones through the reinforcement of muscle attachment zones (e.g. cortical defects/erosive lesions), the appearance of acquired traits as well as an increased general robust appearance and torsion of the bone morphology (fig. 32). This was observed in almost all bone elements and was independent of sex and age. The robustness and some cases of bone torsion were also

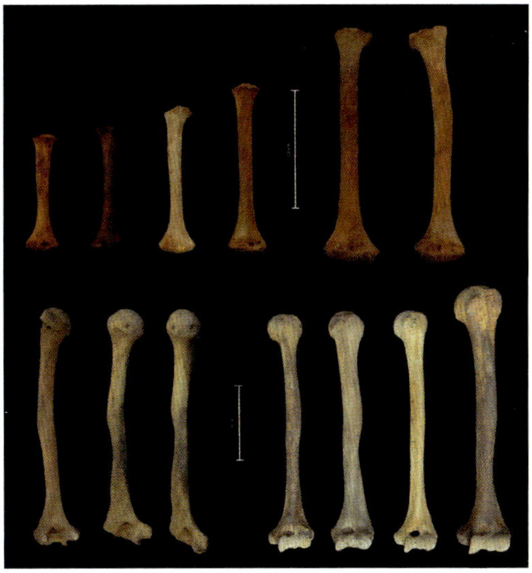

Fig. 32 Overview of Humeri from different sex and age categories from the Biniadrís cave (the scale reflects 10 cm) (Graphic: Monice Timm/Project Biniadrís).

seen in young individuals, which may indicate the involvement of the smallest members of the community in daily activities. The strong influence of certain activities also influenced the outcome of non-metric traits (e.g. squatting facets, Allen's fossa) and certain pathological conditions. In addition to healed fractures or traumatic injuries, the buried suffered from painful conditions of the joints and musculoskeletal system, which were seen in almost all age groups. The difficult living conditions caused by agriculture and animal husbandry, as well as the excessive use and processing of stone, left their mark on the bodies.

Recently published analyses of hair remains from the Es Càrritx burial cave (Guerra-Doce et al. 2023) show the use of various substances (alkaloid ephedrine, atropine, scopolamine) associated with shamanism, as they were used to alter states of consciousness and were found in specific containers with carvings of concentric circles. These carvings are found all over the world on objects containing hallucinogens (Guerra-Doce et al. 2023). Notably, the significance of these substances extends beyond their potential ritualistic use as deliriants, encompassing also their medicinal applications. They can also be used to increase physical activity and helps in cases of insomnia and respiratory diseases such as colds and asthma. It is therefore reasonable to assume that this society had solid medical knowledge and could afford to provide medical care for its people, even for those who needed full-time care. More 'general' care can also be seen in the daily life of these societies. An example of this is the relatively low rate of dental caries in both Biniadrís and Es Càrritx (Rihuete Herrada 2003). The development of caries is caused by variable and multifactorial processes, among which exogenous and endogenous factors, as well as environmental factors and pathogens, have a decisive influence on dental health (Roberts/Manchester 2010, 65–66; Powell 1985). The low caries rate in Biniadrís, as well as the increased robustness of the deceased, is very consistent with dietary factors in the cave. Marine resources could be important for a protein-rich diet, and their avoidance could therefore lead to a deficiency in the daily supply of nutrients. However, this deficiency was compensated for by animal husbandry, which provided the necessary protein. Proteins are important for bone and muscle (re)building and may have a positive effect on caries development (Hujoel/Lingström 2017).

Overview of the Application of the ResourceCultures Framework

The analyses of the individuals buried in the Biniadrís cave, as well as the results from similar burial caves, in particular Es Càrritx (Rihuete Herrada 2003; Lull et al. 2013), allows us to reconstruct the living conditions of the different communities and their subsistence strategies; in particular, how these populations were already organised and developed precise strategies to deal with resilience and adaptation to living conditions. The Biniadrís cave is an example of a stable and well-organised community with different occupational activities, such as stone mining, animal husbandry, agriculture and other subsistence activities (e.g. food preservation, ceramic production), which influenced their daily life and their general health. In summary, the occurrence of pathological processes of joint diseases, the high number of well-healed traumatic patterns, as well as the low occurrence of dental pathological patterns also indicate a community with specific care strategies, medical knowledge and probably specialisation, which positively influenced the health of this group.

The cultural and social significance of resources is crucial to understanding cultures. It highlights the diversity of resource use in different communities and the profound influence of values, beliefs and traditions on resource management. The role of resources is not static;

it evolves over time and affects the organisation and structure of social life. All these aspects can be seen as a dynamic within society, reflecting the power of both the individual and the community to influence the living conditions of their group members. It is also in constant interaction with tradition, knowledge and accessible materiality.

The use or non-use of resources can be seen as a controversial point, since (non-)use can occur for different reasons. It emphasises that resources are more than mere physical objects, they are deeply intertwined with cultural systems and have varying meanings in different cultures. The way resources are used and allocated is influenced by social dynamics and group interests, be they economic, political or social. This study emphasises the importance of recognising resources as social constructs with cultural significance. For example, previous research on dietary reconstructions has excluded a diet based on fish and seafood, which, as noted above, is common in the western Mediterranean from the Neolithic to the Bronze Age. Although the sea could be understood as a resource *per se*, it was not used for food consumption, but may have had other meanings than subsistence, such as connectivity or isolation. Alternative reasons could be economic and cultural. The study of resources in an island context offers a unique perspective that sheds light on how these resources have shaped the development of social identities and cultures. The use and avoidance of resources in Menorca and the Biniadrís cave during the Bronze and Iron Ages have revealed intriguing patterns related to the avoidance of fish consumption due to established traditions, economic factors and geographical determinants. It highlights the significance of environmental variables and cultural preferences in shaping dietary choices and, therefore, culture.

6.4 Islands and Water as ResourceComplexes – Interdisciplinary Reflections on Insularity

Laura Dierksmeier, Sophie Hüglin, Frerich Schön

Introduction

Islands frequently rank high on the freshwater scarcity index due to a combination of factors, including brackish groundwater, the absence of annually flowing rivers, low precipitation rates, aquifer pollution, deforestation, desertification and the exploitation of hydrological resources through water-intensive monocultures. In this case study, we adopt the ResourceComplex framework to examine how water on islands is influenced by their unique geography, exploitation and utilisation methods, governance structures, histories and religious practices. Conversely, we explore how the quantity and quality of accessible hydrological resources have shaped social relations and identities on islands.

Frerich Schön begins with an analysis of the Sicilian island of Linosa, focusing on Late Antiquity (5th–6th centuries) and the Modern Period (19th to mid-20th century). For the Middle Ages, Sophie Hüglin investigates the Balearic Islands, emphasising the role of monasteries before and after Muslim rule. Laura Dierksmeier examines the Early Modern Period (1500–1800) on the Canary Islands in the Atlantic, illustrating how water supplies both influenced and were influenced by island culture.

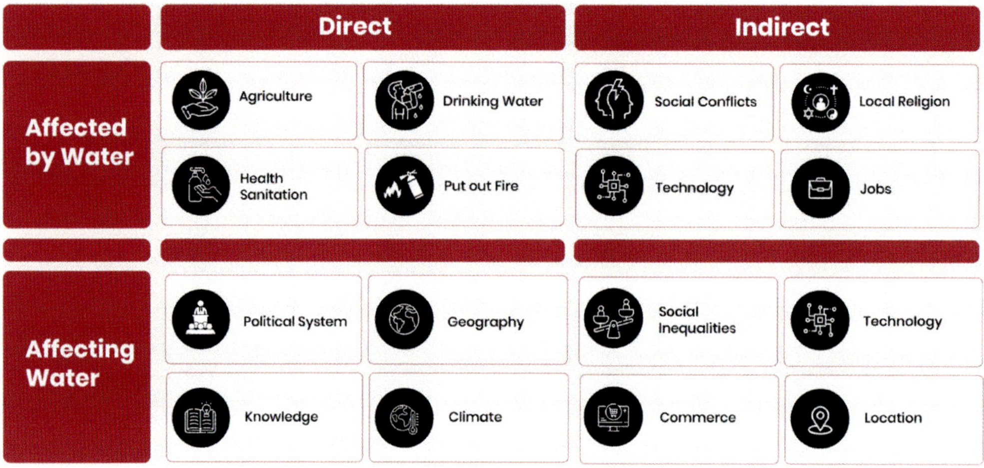

Fig. 33 The ResourceComplex Island Water (Data: Laura Dierksmeier; Graphic design: Fawad Saleem).

Water from the ResourceComplex Perspective

Water is embedded within an intricate resource network, comprising objects, people, knowledge and practices required for its use (Teuber/Schweizer 2020). The concept of the ResourceComplex for freshwater encompasses not only the source of water but also the infrastructure for its transport and methods of distribution. Examining freshwater through this lens highlights how its use and distribution shape local identities, influencing both self-perceptions and external perceptions of communities.

From the ResourceComplex perspective, water scarcity is not a fixed quantity but a culturally defined threshold that varies across regions (Schön/Dierksmeier 2021). Access to the quickest, cleanest and largest supply of freshwater is often tied to the power dynamics within a given territorial unit (see Förster/Bauch 2014, on water and power). Consequently, the total volume of water available globally is never equivalent to the amount accessible for practical use.

The study of water can be broken down into direct and indirect influences on culture and society, as well as the ways daily life in turn impacts water availability (fig. 33). Water, one of the few naturally occurring resources that falls from the sky, has shaped human civilisations over millennia, particularly in agriculture, which has been adapted to the availability of water (Bartelheim/Montero Ruíz 2009). Beyond its essential role for drinking, water scarcity significantly affects health and sanitation, limiting activities such as washing clothes and bedsheets – an issue of particular concern in hospitals with infectious patients. Furthermore, water shortages hinder efforts to extinguish fires.

Water also indirectly influences social cohesion, as conflicts over its unequal distribution have been a recurring issue throughout history. Religious practices involving water have been integral to daily life for centuries (Bradley 2017; Fagan 2011; Tvedt 2015; 2016). Technological advancements, such as water mills, rely on water availability, while occupations and economic activities are shaped by it. In premodern societies, predominantly agrarian economies were closely tied to the availability of water, which dictated the types of agriculture practiced.

The political system of a given region often determined how water was distributed and the extent to which water systems were maintained. Geography and climate influenced rainfall patterns, groundwater absorption and the potential infiltration of saltwater into freshwater sources. Power structures and social inequalities shaped access to water, directing distribution to certain segments of the population while neglecting others. Technological capabilities dictated the extent to which water could be purified, transported and stored, while local industries impacted the overall availability of water for residential use. The location of a household also played a critical role in determining water quality and whether it could be connected to a public supply system. In essence, water directly or indirectly affected many facets of daily life, while a multitude of factors simultaneously influenced its availability, usage and distribution.

Examples

Coordinated by Laura Dierksmeier and Frerich Schön, the DFG-funded Island Studies Network (2021–2023, project-no. 452312841) brought together an international and interdisciplinary group from the disciplines of history, archaeology, anthropology, digital humanities, geography, literary studies, sociology and philology to discuss, besides methodological issues (Dawson 2019; Schön/Dierksmeier 2021; Dierksmeier et al. 2021), three thematic clusters around 1) the impact of exchange on island communities and where social or cultural configurations and external relations have been enhanced or constrained (Christophilopoulou 2022); 2) environmental change (Depraetere/Dierksmeier 2023), island vulnerability and human responses to resource scarcity (Petzold et al. 2023); and 3) western and non-western, emic and etic perceptions of islands and islandness (Dautel 2021).

This case study presents water on islands through three examples from the perspective of ResourceComplexes. Starting with Late Antiquity and the Modern Period, Frerich Schön presents a study of Linosa, an island that, along with Lampedusa, is part of the Pelagic Islands in the Sicilian Channel of the Mediterranean Sea. For the Middle Ages, Sophie Hüglin analyses the role of monastic communities further west in the Mediterranean, on

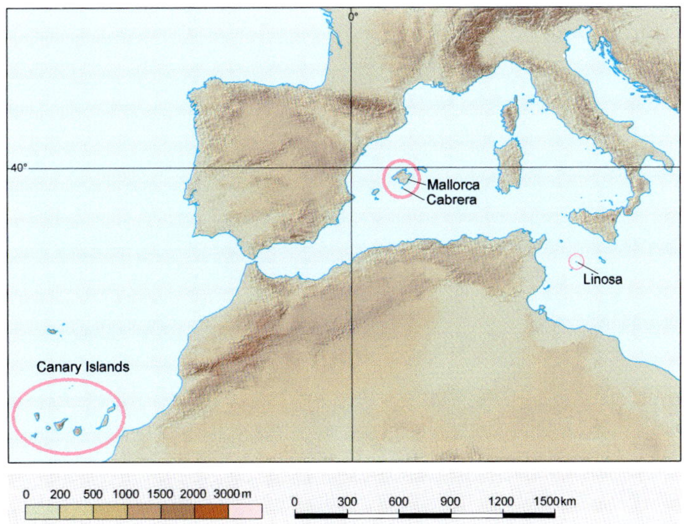

Fig. 34 Map with indication of the examples (Data sources: Open Topography; Map: Frerich Schön; Graphic design: Richard Szydlak).

the Balearic Islands – more specifically, on a site on Cabrera and another on Mallorca before and after the Muslim rule of Al-Andalus. For the early Modern Period (1500–1800), Laura Dierksmeier will show how water supply affected island culture in the Canary Islands in the Atlantic Ocean (fig. 34).

a) Bonded by Scarcity: Freshwater on Linosa Island in Late Antiquity and the Modern Period (Frerich Schön)

In September 1954, the Viennese anthropologist Fuchs visited the small Italian island of Linosa in the Strait of Sicily, 170 km south of Sicily, 165 km east of Tunisia, 140 km west of the Maltese Archipelago, in order to study the island society for 10 days (Schön 2014; 2020b; 2025). In his observations on the freshwater supply of the island, he stated: 'The water supply is exclusively provided by large cisterns in which the abundant winter precipitation is stored. The transport of water from the cisterns to the fields takes place in barrels carried by mules or donkeys and is the main work of the farmers' (Fuchs 1956, 25; translated from German by the author). This study examines 'Island Water' from the ResourceComplex perspective by comparing archaeological and historical sources on the island of Linosa (Pelagic Islands, Province of Agrigento, Sicily/Italy) from the 5th and 6th centuries and the 19th and first half of the 20th centuries on the extraction, distribution, use and perception of freshwater.

The settlement history of Linosa in relation to the island's water supply was the subject of an archaeological survey. According to archaeological sources, a permanent settlement was established on the island in Late Antiquity, but it was abandoned at the end of the 6th or beginning of the 7th century (Schön 2025, with references). It was not until the mid-19th century that Linosa was colonised by the Bourbon Kingdom of the Two Sicilies. Written sources from this period mention more than 150 cisterns of ancient origin that were repaired to supply the colony with freshwater (Sanvisente 1849; Calcara 1851, 10; Schön 2025, with further references). Since there were no freshwater springs on Linosa and the volcanic island had no usable groundwater table, the collection and storage of rainwater in cisterns was the only way of supplying the island with water for human consumption. The cisterns found by settlers in the 19th century are all of the same type: reservoirs dug into the bedrock with a bottle or cone shaped vertical section, 2.80–8.90 m deep, and a circular ground plan, 3.00–7.08 m wide (Schön 2014). The storage volume ranges from 12 to 90 m^3, with an average of 35 m^3. The majority of these cisterns are located on the mountain slopes and are fed by rainwater collected on rocky surfaces that have been levelled and cleared of vegetation and debris to act as an *impluvium*. Rock catchments to collect the rainwater, canals to convey it, sedimentation basins to purify the water and cisterns to store it form a rainwater harvesting system known in the local dialect of the island as the '*comprise* system' (from Latin *comprimere*; English compress or comprise) (fig. 35).

A high level of knowledge and planning is required to ensure that clean water can be collected and stored in these systems to provide a reliable supply: the cisterns must have sufficient storage capacity to cover regular consumption throughout the year, particularly during the dry summer period when there is no rainfall to replenish the reservoirs. The cisterns must also be lined with suitable hydraulic plaster to prevent leakage into the porous rock. The size of the catchments must be appropriate to the rainfall amount and the size of the cistern. The systems require regular cleaning and maintenance to prevent contamination of the stored water. Finally, the water has to be transported to the consumers – an activity that, as Fuchs noted, was still the main task of farmers in the mid-20th century (Fuchs 1956, 25).

The *comprise* systems of Linosa cover about 10% of the island's surface. These systems are found sporadically to supply scattered farmsteads or in clusters to supply the main settlement. On the outer western slope of Monte Bandiera, a 102 m high volcanic crater in the centre of the island, 30 cisterns with an average capacity of 32.1 m^3 have been documented, each connected to a rock catchment with an average area of 546 m^2. In one case, two cisterns are connected by an overflow channel to form a storage cascade. In contrast, 16 ancient cisterns with an average storage volume of 37.4 m^3 and one modern reservoir with a capacity of 1,200 m^3 have been documented in the Zona Gibbiuna in the north-western part of Linosa Island, connected to a single large rock catchment of up to 150,000 m^2 (fig. 36 a + b). Further ancient cisterns are suspected in both areas but could not be identified due to modern landscape transformation. The distribution of the cisterns and the relationship of the water supply systems to other remains of the Late Antique settlement, which have been documented as part of the Tübingen survey project, allow conclusions on the ResourceComplex around water.

The main ancient settlement of Linosa consisted of a series of at least 26 spacious caves built into the western slope of the foot of the Monte Bandiera volcano. The Late Antique settlement phase could be dated between the 5th and 6th centuries based on the pottery found in the survey material. It is possible that this was a monastic settlement, created as an island exile during the expulsion of monks and clerics from North Africa during Vandal rule (429–533) (Schön 2020b). The rocky slopes above the residential caves were used as catchment areas for rainwater harvesting. It is noticeable that the rainwater catchments and their respective cisterns were clearly parcelled out, roughly corresponding to the number of caves.

Fig. 35 Linosa, Monte Bandiera, comprise system: rock catchment and cistern (Photo: Frerich Schön; Design: Richard Szydlak).

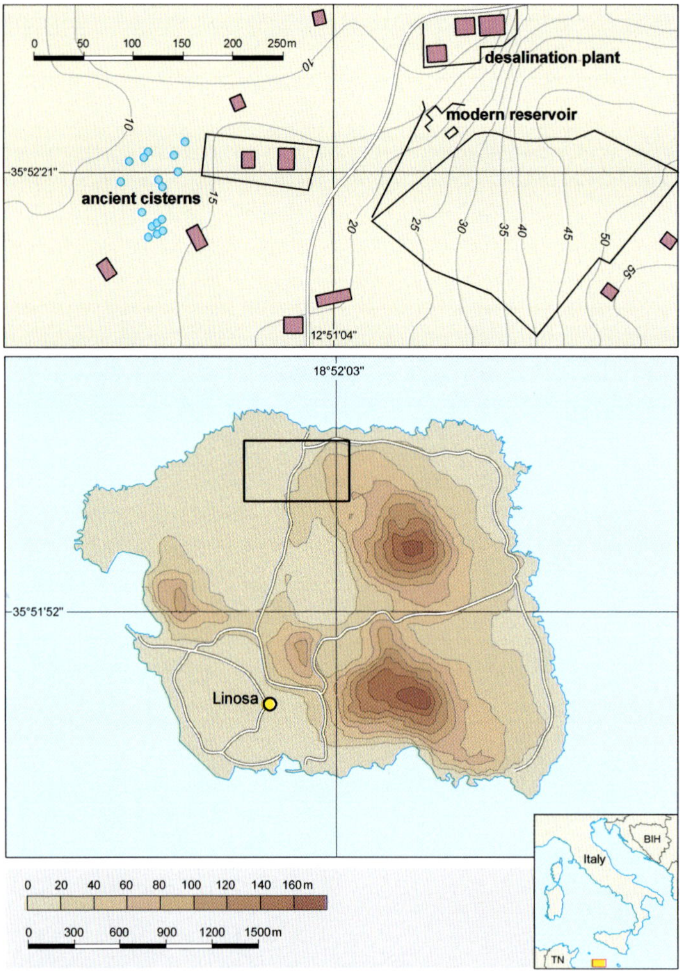

Fig. 36a Linosa, Zona Gibbiuna, catchment and cisterns (with ancient cisterns, modern reservoir and desalination plant) (Data sources: CGIAR-CSI SRTM; Data: Frerich Schön; Graphic design: Richard Szydlak).

Fig. 36b Linosa, Zona Gibbiuna, modern reservoir (Photo: Frerich Schön).

In the Zona Gibbiuna there seems to have been no parcelling out of the catchment area. All the cisterns documented here are located as a cluster in a small area at the foot of a large tongue of rock (see fig. 36a + b). No traces of ancient buildings have been documented in their vicinity, so the site may have served as an additional water supply for the settlement. The cisterns here have a higher average capacity (additional 5 m^3) than those on Monte Bandiera, and the common catchment area is many times larger than that of the Late Antique settlement, so that significantly more water could be collected and stored, and the cisterns here filled much faster during a precipitation event.

The design and distribution of the hydraulic systems allow further conclusions on their operation and the ResourceComplex water in the Late Antique settlement. The parcelling out of the catchment areas on the slopes of Monte Bandiera suggests that the *comprise* systems were run by smaller social units. So each cave dwelling would have had to look after a catchment area with one or two cisterns for the supply of freshwater. On the other hand, the use of a common catchment in the Zona Gibbiuna suggests that the maintenance of this facility was a communal task, while the existence of at least 16 individual cisterns seems to mean that the stored water was not a communal good, but rather was attributed to individual social units, such as individual households or other types of user communities.

Written sources from the 19th and 20th centuries show the problems and dependencies that water supply created for the population (Schön 2025, with references). The sources show that the *comprise* systems on Linosa Island were initially restored to their Late Antique use during the colonisation of the island. Some of the catchments were used as private property of individual settler families and others as communal property (*demanio*). The restoration of the installations was intended to make the colony, founded in 1845, independent of water supplies by boat from the neighbouring island of Lampedusa or from Sicily. According to the colony's governor: 'Of the considerable number of cisterns, eight of the most spacious have been cleared and five accommodated; these are now all filled and contain water for more than a year to serve the present colony of 116 souls' (Sanvisente 1849, 119; translated from Italian by the author). The colonial administration had already gained experience in reusing ancient cisterns on the neighbouring island of Lampedusa, which had been colonised since 1843. However, the failure of the hoped-for spring rains in 1844 had led to a serious supply crisis in the young colony, forcing the governor to ask for help: 'The colony, not excluding the military, has been rationing water from the cisterns for many, many days, since the hoped-for rains have not come. Each person is allotted two jugs of water a day; the same has been done with wine, giving a portion of half a quart per person per day, and for three days they have been without it' (Fragapane 1993, 423–424; translated from Italian by the author).

Just six years after the colony was founded, 136 ancient cisterns were reactivated (Calcara 1851, 9–10). However, water scarcity due to lack of rainfall remained a recurrent problem on Linosa Island. The settlement was regularly dependent on water supplies by ship from Lampedusa or Sicily. The American geologist Washington, who visited the island in September 1905, reported: 'Linosa is dry in the extreme, not a single spring being found on the island, and the water supply being dependent on the rains, which are collected from the flat roofs and preserved in cisterns. At the time of my visit, the island was favoured with the first rain in five months, and the inhabitants were reduced to a daily ration of two litres of water apiece, a supply having had to be brought by a government vessel from Sicily' (Washington 1908, 4).

The construction of a communal water supply by the Italian military at the end of the 1920s was an attempt to overcome this problem and finally made the island independent of external water supplies. To be able to store a much larger supply of water, the eastern part of the Gibbiuna catchment was separated from the ancient cisterns, enclosed separately and connected to a newly built reservoir with a storage capacity of 1,200 m^3. However, after the end of the Second World War, due to the stable connection with Sicily, there was little local need for these waterworks, especially as the island no longer had an institution to maintain them once the military had left. There may have been another reason for the abandonment of this large reservoir: its construction cut off the large Gibbiuna catchment from the numerous small cisterns to the west. This severely limited the functionality of the cisterns there, which certainly did not help the acceptance of the new facility. Perhaps even more serious was the fact that such a large reservoir required a different water distribution system. The small cisterns that previously served the Gibbiuna catchment were probably owned by individual families who could use the stored water individually. The new large cistern had a much greater storage capacity, but how was the water to be distributed after the military left? The islanders had no institution to take care of this. The sources cited unanimously show that the islanders systematically reinstated the ancient cisterns and their catchment areas and took great care to maintain them. However, independent supply was sought at family level, not at community level. Moreover, from the beginning of the colony there were close links with Lampedusa and Sicily, not least because of the lack of water. It was precisely this dependence on the elementary level of water, the incessant communication of water shortages, that was and is the guarantee that a continuous connection with Linosa will be maintained through regular water deliveries. The recurrent topos of the scarcity in the sources was the price the islanders were prepared to pay for their island life. This scarcity is constructed in very different ways from an emic and etic perspective and therefore has a socio-cultural dimension. This is illustrated in a report by the Swiss geographer Messerli from the 1950s, who notes with some surprise that, despite the narrow limits imposed on agriculture by the natural environment of Linosa, orange trees were cultivated, the growth of which was supplied by several cisterns (Messerli 1958, 238–239). Even the construction of a desalination plant in 1973, with an initial capacity of 50 m^3/day and, from 1986, 500 m^3/day of drinking water, did not make the island independent of external water supplies (Scifo 1989, 47–48, 65). Although water supplies by ship from Sicily are now less frequent, the drinking water supply is now almost exclusively provided by industrially bottled water delivered by daily ferry from Sicily. At the same time, the *comprise* on the slope of Monte Bandiera continues to be systematically exploited, no longer to supply people or livestock, but to provide water for forestry programmes.

b) Islands and Water, Monasteries and Muslim Communities – Changing Actors in Water Management on the Balearic Islands in the Middle Ages (Sophie Hüglin)

In the last decade, several archaeological investigations and historical studies of medieval monasteries have begun to be viewed from the perspective of ResourceComplexes. In parallel to dedicated papers in journals – such as the one by Vossler-Wolf on Bebenhausen (2016) – there has been an edited volume dedicated to 'Monasteries and their Resources' (Krätschmer et al. 2018). Within this collection, two contributions deal specifically with the resource water and its management in late medieval Christian – often Cistercian – monasteries such as Maulbronn (Gillich 2018), or highlight the water-related situation of many monasteries from a landscape perspective (Burkhardt 2018). Looking at the internal structure of a monastery, Burkhardt mentions 'the high symbolic value of the staging of a well-organised flowing water supply as an expression of a well-ordered community' and further notes that this 'can serve as an indication of how important it is also to consider the immaterial aspects of resources and the feedback effects of certain resource complexes and ResourceCultures' (translated by the author from Burkhardt 2018, 135).

More recently, in 'Waters' (Teuber et al. 2020), Vossler-Wolf's chapter 'Aqua Viva and the Monk in the Pond' demonstrated the multidimensionality of water use in medieval monasteries in south-west Germany. She uses 'written sources and the material remains of water use [...] to show how the connection of different meanings [...] leads to a Resource-Complex which is used by [...] the monastic community to establish and stabilise social connections.' Moreover, this ResourceComplex has an impact beyond the monastic precinct, because the 'remains of the ritual and economic use of water can be found both inside the monastic building and in the landscape' (Vossler-Wolf 2020, 73).

Islands as monastic sites have been explored by Nüllen in written sources from the British Isles and published in the 'RessourcenKulturen' book series in a volume of the Islands Studies Network (Dierksmeier et al. 2021). Nüllen speaks of the 'island/mainland dichotomy' with the island as the 'anti-field'; the aim being 'not the crossing of a boundary but the dissolution of this boundary' and describes 'the Christianisation of the islands [...] as a fulfilment of their paradisical potential' (Nüllen 2021, 237). Here we will try to approach all three aspects from the perspective of ResourceComplexes. This means looking at monastic institutions on islands from the perspective of water as a managed and spiritual element.[19]

The task ahead is to look specifically at island monasteries on a global and trans-religious scale. This means adopting an anthropologically inspired worldview of monastic life as a cultural invention that uses the natural context of an island landscape as a means and metaphor to enhance its social purpose (Bharati/Johnston 2023).[20] Here, the role of the resource of water – in terms of its physicality alone – becomes threefold: it surrounds the island as sea or freshwater, it is produced and available on the islands themselves in natural or artificial springs and streams and it is consumed and excreted by humans, animals and plants.

The Balearic Islands are the subject of a mid-term study, which looks at the different ways in which water was actively managed in three stages: the early and late medieval period by Christian monastic and in between by Andalusian tribal[21] communities. First, rainwater was collected in cisterns for drinking, then groundwater was tapped for irrigating agricultural land and finally technologies were used and further developed to collect and store all kinds of water dosing and distributing it over long distances to support urban development.

Between the 5th and 15th centuries on the Balearic Islands, water was managed in many different ways. In this example, the focus is mainly on monastic communities, before and

19 In the context of island studies, water on, in and around island monasteries is explored in a collaborative publication project led by Sophie Hüglin. This project brings together contributions from more than twenty scholars from across Europe and beyond.

20 Cf. <https://anthropology.iresearchnet.com/monasticism/> ; last accessed 15.03.2024.

21 The terms 'tribe' or 'clan' for the small Berber and Arab groups were introduced by Pierre Guichard (1977) in his analysis of the social structure of medieval Muslim Spain. They are still used today by leading archaeologists – without any derogatory undertone – to describe the small peasant groups responsible for most of the complex medieval irrigation systems in rural Al-Andalus (e.g. Kirchner 2024).

after the Muslim conquest: first, the early medieval period before the Arab and Berber conquest of Mallorca in 902; second, the subsequent period of Muslim rule in the Iberian Peninsula; and third, the period after the Christian conquest in the 13th century.

If Pope Gregory had not written to Defensor John in 603, urging him to control the behaviour of a male monastic community on Cabrera, no one might have searched for and discovered the remains of this monastery more than 1,400 years later. Before archaeologists searched for and uncovered the remains of stone buildings and graves of adult males in several campaigns, nothing was visible above ground (Riera Rullan 2014). The monastery on Cabrera belongs to a type of site that was typical of the Mediterranean in the early medieval period: communities tended to choose small islands; on these, they selected sites that were both close to a natural harbour and a reliable freshwater source (fig. 37). Also, these small islands should be reasonably close to the coast of the mainland or to a larger archipelago, as Cabrera is to the south of Mallorca, the largest of the Balearic Islands. The best-known example of a similar early medieval island monastery is Lérins off the coast of Cannes (cf. Riera Rullan 2017, fig. 2.1). A geographically much closer comparison is Ila del Rei in the natural harbour of Mahon on Menorca (cf. Riera Rullan 2017, fig. 2.4). There, the archaeological remains of a 'paleochristian' basilica with adjacent domestic buildings and related finds clearly indicate an ecclesiastical centre, but in the absence of written sources, the interpretation as a monastery is based on the living quarters uncovered next to the church.

The monastery on Cabrera was supplied with water from a perennial spring located a few hundred metres to the south-west of the remains of the possible church and graves. In the Muslim period that followed, the valley underwent a major transformation, and nothing remains of the early medieval installations. The monastery not only had to provide for its monks – and probably lay – communities as well as some agricultural – and possibly also eco-

Fig. 37 Modern cattle watering tank at the natural harbour of Cabrera. This small island – today a nature reserve – used to house a monastic community from the 5th to the 9th centuries. The monastery was ideally situated between the most reliable natural well and one of the best natural harbours of the Balearic Islands (Photo: Sophie Hüglin).

Fig. 38 This derelict *noria* or *sāqiyah*, from Arabic: ساقية, a scoop wheel that used to have buckets, belongs to the historic townhouse at Muro where the ethnographic section of the Museu de Mallorca is housed. This kind of animal-driven water-lifting device was invented in Antiquity and further developed by Muslim engineers like Ismail al Jazari who introduced the use of the carank in 1206. Until recently, this was the most commonly used water-raising device not only in Spain but also throughout the Middle East and India (Photo: Sophie Hüglin).

nomic – activities on the island, but in the summer, it also received guests, such as pilgrims, inspectors and perhaps also merchants. Therefore, a good supply of drinking water and other sources of freshwater of lesser quality was essential for the functioning of the network of resources made up of the island, the water and the monastery. In Cabrera itself, as well as in the smaller islands of the Cabrera archipelago, Riera Rullan has identified several sites that he interprets as hermitages from the same period. The most important element of these structures is always a watertight cistern built in masonry (Riera Rullan 2017, fig. 3.158). The cistern demonstrates the likely perennial nature of these dependencies, which could not rely on rain or freshwater supplies from other islands throughout the year. In winter, when there would be more rain, the sea would be too rough to sail or row to Mallorca or even the neighbouring islands. In the summer, when the weather could remain dry for weeks or even months, the demand for water would increase with the number of people and animals, as well as agricultural and economic activities. Cabrera, along with the south-west coast of Mallorca, Ibiza and Formentera, is one of the driest areas within the Balearic Islands (López Mayol et al. 2017). This made water the limiting factor for monastic and other settlements and may have been one of the reasons that led to the abandonment of the monastery even before the Muslim conquest at the beginning of the 10th century (Riera Rullan 2014, 83).[22]

With the Muslim invasion and the establishment of Al-Andalus – the incorporation of most of the Iberian Peninsula, including the Balearic Islands, into the Umayyad Caliphate – not only the faith but also the social structure changed profoundly. Family clans – mostly of Berber origin – took over valley by valley, transforming the landscape and agricultural practices through the creation of terraces (*marjadas* or *feixes*) and the installation of sophisticated irrigation systems (Foster 1952; Guichard 1977; Wozniak 2021). Coming from North Africa and Arabia, the new settlers were accustomed to arid climates and saline soils. To lift the groundwater, they used technologies – such as *qanât(s)*, a type of underground aqueduct, or animal-driven water wheels, *noria* or *saqiyah* (fig. 38) – that had been developed in desert regions. A whole range of new crops and vegetables were introduced, from durum wheat

22 From the beginning of the 8th century until 902, Cabrera may not even have been inhabited (Helena Kirchner, oral communication).

to watermelon and from sugarcane to artichokes (Decker 2009). Based on archaeological surveys and systematic descriptions of the Christian (re-)conquest of the mid-13th century, the so-called *Escrivania de Cartes Reials* (ECR), these transformations have been intensively studied by Kirchner and her team on the islands of Mallorca, Menorca and Ibiza (Kirchner 2009; Kirchner/Retamero 2016; Kirchner/Sabaté 2022). The ECRs can be compared to the Domesday Book, the 'Great Survey' that documented England and parts of Wales after the Norman Conquest; it was completed in 1089 after King William I ordered surveyors to list his possessions, and the dues owed to him. Similarly, the ECRs documented, estate by estate, the lands, canals and watermills that belonged to the Spanish crown.

The major changes in land and water management under Muslim rule in the Iberian Peninsula can be compared with the top-down reforms under Carolingian rule north of the Alps, where monastic communities played a pioneering role. In contrast to Charlemagne's reforms, Kirchner (2024) emphasises that in Al-Andalus, terracing and irrigation systems were not the result of central planning and implementation, but that of a bottom-up process, where Berber groups took the reorganisation and cultivation of each valley into their own hands, adapting technologies to local conditions. The network of resources would not be that of the island, the water and the Christian monastic community, but that of the island, the water and the Muslim farming group. This again raises the question of whether in this case it was the *zeitgeist* rather than the agency of the particular human community and its religious beliefs that led to a similar technological development.

The monasteries founded in the Balearic Islands after the Christian conquest, from the mid-13th century onwards, are the focus of the third medieval period chosen here. There are no more monasteries on the smaller islands. The monasteries of this period have many similarities with those of Central Europe in several respects: these are the motives and methods behind their foundation, their choice of location and their medium- and long-term survival factors. As they are located on the Balearic Islands, they are still in an island context, but other landscape or settlement features – such as hilltops, caves or towns – tend to be more important than the 'island factor'. This may have something to do with the lesser importance and even potential threat of water as a means of transport. Before a natural harbour had made long-distance travel possible, and boats had brought pilgrims as a main source of purpose and income; now such a location invited plundering and destruction, and had to be fortified and controlled by military forces and state actors. Freshwater and, above all, sources of drinking water remained a determining factor in the location of monasteries on the larger islands of the archipelago, as will be illustrated by the Cistercian monastery of La Real, north-west of Palma de Mallorca.

In the late medieval period, the Cistercian monastery of Santa Maria de la Real was famous for its library. This is why the famous Catalan philosopher Ramon Llull lived there at the turn of the 13th and 14th centuries. But the monastery is just as remarkable for its strategic location in relation to an important source of freshwater, which was developed under Muslim rule, and the urban layout of Palma de Mallorca, the new centre of power in the Balearic Islands. The Cistercian Order is known for its tendency to choose places of solitude, away from settlements or even urban centres. In fact, in 1236 the monks founded their monastery in Sa Granja, in the Sierra de Tramuntana mountains, but three years later they received a donation of land much closer to the town. There, at the Alpic farmhouse near Esporles, La Real was able to take advantage of an older water source and system that dated back to the time of Al-Andalus. In this way, they essentially controlled Palma's access

Fig. 39 The former Monastery of La Real is still being fed by an open water channel originating at Font de la Vila. The Cistercians chose this as the position for their convent in 1239 to control the water supply of Palma de Mallorca. The irrigation system was created in the previous Al-Andalus period, when the Iberian Peninsula was under Muslim rule (Photo: Albert Cassanyes Roig).

to freshwater. This system of canals, which supplies the monastery with drinking water, is still intact today (fig. 39) (Batet Company 2006; Soto Company/Mas Forners 2015). This example demonstrates the complex relationship between the control of water and power from the perspective of ResourceComplexes, by observing the interaction between water and monastic sites in a marine island context.

c) Freshwater on the Canary Islands from a ResourceComplex Perspective (Laura Dierksmeier)

The scarcity of freshwater on the Canary Islands during the early modern period led to direct or indirect legal, social, religious and cultural repercussions, all formed part of the network around water (see fig. 33).[23]

Following the Spanish conquest of the Canary Islands in the 15th century, water laws were initially modeled on those of Castilian Spain, specifically *el sistema ribereño*. Derived from the 13th-century Castilian legal code *Las Siete Partidas* (The Seven-Part Code), water was considered a public resource by default (Gómez Gómez 2000, 115–119; Guimerá Peraza 1960, 27; Fernández-Armesto 1997). Private ownership of water was granted only in specific cases, often as an incentive to encourage colonisers to emigrate from the Spanish mainland to the remote islands, located more than 1,000 km away (Macías Hernández 2005).

The governance and population densities of the islands influenced water rights and settlement patterns. The islands of Gran Canaria, La Palma and Tenerife, administered directly by the Spanish Crown (*islas de realengo*), were more densely populated. In contrast, La Gomera, Fuerteventura, Lanzarote and El Hierro – initially conquered by the Norman French and remaining feudal (*islas de señorío*) until the 19th century – retained communal water rights. Access to freshwater frequently shaped settlement patterns, as seen in La Laguna on Tenerife, where a lagoon was historically used by the indigenous Guanches for maintaining their goats and sheep (Pérez Herrero 1961; Gómez Gómez 2016).

23 This section is based on information published in Dierksmeier 2020.

Farm workers were granted private access to a specific amount of water per day, determined by the size of the land they cultivated and the type of crops they grew. A continuous and reliable water supply was crucial to incentivise farmers, as they would otherwise be unwilling to risk crop failure (Martín Martín 2003; Cruz García 2004). Additionally, exemptions were granted for the private ownership of substantial hydrological resources to attract new investors, particularly for the cultivation of sugarcane, known as 'white gold'. These exemptions were deemed essential due to the high water demands of sugarcane, which requires significantly more litres of water per kilogramme to produce compared to other crops. Consequently, sugarcane became the dominant monoculture of the Canary Islands (Pérez Herrero 2006).

Primary sources indicate that local islanders, given the choice, preferred a diversified agricultural landscape, enabling them to grow their own staple foods and reduce reliance on imports. Louis Sicking cites a similar example in Corsica, where locals petitioned the Genoese authorities for a closed economy to avoid monoculture and exploitation (Sicking 2014). Foreign investors with prior experience in sugar production – including Italians, Flemings, Portuguese, British and Germans (notably the German banking family of the Welsers) – played a significant role in the Canary Islands' sugar industry. These investors, who controlled the water supply, held special status and were referred to as waterlords (*señores de agua*). On Tenerife in the 16th century, water supply control was concentrated in the hands of four individuals, who collectively owned approximately 80% of the island's total water resources (Macías Hernández 2005). A similar pattern of concentrated water control was observed on Gran Canaria (Hernández Ramos 2004).

Each time new water sources were discovered, sugar production increased, but this growth was ultimately unsustainable as these sources were rapidly depleted (Hausen 1954). The resulting water scarcity forced many inhabitants without access to water to emigrate in large numbers to the Spanish Caribbean. There, they applied their expertise in managing sugarcane plantations in Jamaica, the Dominican Republic and Cuba (Hernández González 1995; Rodriguez Morales 2010; de Luxán Meléndez/Viña Brito 2006). The productivity and agricultural knowledge of Canarian emigrants in the Caribbean ultimately contributed to the decline of the sugarcane industry in the Canary Islands.

By the second half of the 17th century, the dominant crop in the Canary Islands shifted from sugarcane to wine, which required significantly less water. This trend continued until the late 18th century, when improved water supplies and the liberalisation of trade under Charles III facilitated the introduction of new crops such as cotton, tobacco and silk. Among wines, Malvasia – a sweet dessert wine – was particularly prominent. However, it faced strong competition from Portuguese and Madeiran wines throughout the 17th and 18th centuries (Macías Hernández 2005, 190).

Entire communities were able to obtain licences to convert public water into private shares, which could then be sold to sugarcane or vine growers for profit (Macías Hernández 2005, 183; Hernández González 1986). This practice occurred, for example, in 1520 in the Orotava Valley, home to Tenerife's primary water source (Macías Hernández 2009, 736). Contracts governing these transactions not only outlined the transfer of ownership but also designated a person responsible for maintaining the pipes, as timely repairs were essential to prevent leakage and waste (Gómez Gómez 2016). The privatisation of public water, initially intended to generate revenue for public use through sugarcane sales, ultimately initiated an irreversible process (Macías Hernández 2005, 185). Water, once considered a public good, was transformed into a commodity for sale.

One of the most significant changes in water ownership occurred when Canarian law mandated that land and water rights be sold separately. As a result, individuals could own land without having the right to access the water beneath its surface. This distinction underscores the value of water as a privilege that needed to be purchased separately, rather than as a resource inherently tied to land ownership. Water reforms were initiated after lawsuits exposed landowners for exceeding the authorised amounts of water from their holdings (for examples, see lawsuits in the Archivo Histórico Provincial de Santa Cruz, 1518: Her 1–23; 1558: Her 1–27). However, municipal reform efforts failed due to the opposition of politically influential elites (Gómez Gómez 2000, 342–349; Suárez Grimón 1987). In place of municipal reforms, water commissions (*heredamientos de aguas*) were established in the 16th century, accompanied by water regulations (*ordenanzas de las aguas*). Two new roles were created to oversee water management: the water mayor (*alcalde de aguas*) and the water allocator (*adulado*).

The water mayor was responsible for monitoring the total availability of freshwater, ensuring its quality and regulating waste discharge to prevent contamination (Archivo Municipal de La Laguna, cartas diversas, C-V, 1785. Alcalde de aguas; Archivo Municipal de La Laguna, C-VIII, 1780. Abastecimiento de aguas; Archivo Municipal de La Laguna, C-VIII, 1783. Desagüe a las calles de aguas sucias). Elected by municipal authorities, the water mayor had police powers to enforce the law. In contrast, the water allocator oversaw the distribution of water. Allocators were nominated by the island councils and confirmed by the king, reflecting the monarchy's interest in maintaining direct control over individuals managing such a vital resource. This distinction in the appointment of water officials highlights the king's recognition of the importance of water control and the need to personally approve those in such influential roles. Individuals with authority over water had lasting impacts on the island's finances and politics, effectively placing the fate of the broader population in the hands of a few.

To safeguard their precious and limited water resources, communities deliberated on the best methods to conserve and revitalise mountain areas and gorges abundant in water. Pine forests in the mountains were particularly valued for their role in fog harvesting, as water was collected not only through the trees' roots but also via their pine needles (Marzol Jaén 2005; Martín Medina/Gouzévitch 2008). Pine trees were so esteemed that they inspired the creation of a religious cult in their honour. The Virgin of the Pine Tree (*la Virgen del Pino*) became the patron saint of Gran Canaria.

On the island of El Hierro, the inhabitants venerated a specific tree known as the *Garoé*. This tree's remarkable ability to collect enough water on its leaves to supply the island's entire population earned it sacred status (Sánchez García 2007). The *Garoé* had served as a vital water source since indigenous times but was destroyed by a hurricane in 1697. Today, the tree's legacy as a natural water collector of horizontal rain is honoured through a 16-km trail featuring information about the water history of El Hierro.

Water was also collected from snow on the peak of Mount Teide in Tenerife, the highest mountain in Spain at 3,700 m (the pits for snow collection are described in the *datas por el adelantado y el cabildo* in the Archivo Municipal de La Laguna). When measured from the ocean floor, the Teide volcano rises 7,500 m, making it one of the highest volcanoes in the world from seabed to summit. In addition to snow collection, mountain gullies directed melted snow and rainwater into cisterns. These cisterns utilised sand for decontamination and to reduce evaporation. Underground cisterns were particularly valuable, as they provided a cool storage environment that inhibited the growth of bacteria – an issue more common in water stored aboveground in warm climates.

The significance of cisterns is exemplified by the town hall of San Miguel de Abona, Tenerife, which was constructed over a large 17th-century cistern. This architectural integration of a government building with a water collection facility underscores the critical importance of ensuring local communities had access to and control over reliable water supplies.

When technology failed to alleviate water stress, communities turned to *rogativas* – public prayer processions – to petition God for water (Archivo Municipal de La Laguna, Asuntos eclesiásticos. A-IX, 1761. Rogativa por la sequía; Archivo Municipal de La Laguna, Asuntos eclesiásticos. A-IX, 1779. Acción de gracias por haber llovido). As Mariano Barriendos has demonstrated, *rogativas* were performed throughout early modern Spain to address a variety of societal and environmental challenges, including epidemics, torrential rain, drought, storms and extreme cold (Barriendos 2005).

These *rogativas* were notable as one of the few religious rituals documented by secular authorities and formally presented to the Spanish monarch with an official seal. At a time when commoners lacked direct channels of communication with the central government, such processions may have served as a means for residents to convey their water shortages to the *cabildo insular* (island government) and, ultimately, to the King of Spain.

Extensive sources suggest that most rogation ceremonies were performed to address droughts and floods. In the context of frequent conflicts over water, *rogativas* served an important social and cultural role, uniting townspeople in their shared hardships. During these ceremonies, large statues of saints and Marian sculptures, such as the *Virgen de Remedios* (Virgin of Remedies), were carried on the shoulders of participants (Archivo Municipal de La Laguna, 1761, A-IX, 7).

Rogativas for water scarcity were widely regarded as effective remedies for communal needs and were often followed by ceremonies of gratitude after rain had fallen (Archivo Municipal de La Laguna, A-IX, 13). These rain processions occurred not only during the dry season, from April to September, but also during months typically associated with reliable rainfall, such as February. This irregular timing suggests that these church records may provide evidence of unexpected climatic and environmental variations in certain years.

The case of the Canary Islands illustrates how political dynamics further restricted already scarce water resources, as water shifted from public to private control. A small number of foreign investors, driven by profit through monocultures, leveraged colonial structures to exploit local resources for financial gain. This colonial exploitation had both direct consequences, such as reduced availability of water for local use, and indirect consequences, including social conflict and the adaptation of cultural practices.

Communities employed innovative strategies to cope with diminished water resources, such as establishing a dedicated water police force, implementing extensive water cistern systems and utilising fog and snow harvesting techniques. Religious and cultural traditions also played a role, serving both as an outlet for communal suffering and as a means to communicate local needs to secular authorities. The frequency of public rogation ceremonies for rain was meticulously recorded by these authorities, further highlighting the intersection of cultural and resource management practices.

In summary, access to freshwater on the Canary Islands was determined by a complex interplay of factors beyond mere rainfall. The distribution of water often reflected underlying power structures, with certain islanders enjoying greater access based on their social or political influence. Those who lacked sufficient freshwater relied on diverse gathering techniques, supplemented by water-specific cultural and religious traditions to navigate their challenging environment.

Conclusion

This case study has presented the ResourceComplex perspective on water-related resource networks. The ResourceComplex serves as a valuable analytical framework for examining how a resource such as freshwater influences society both directly and indirectly. The technologies available to a population, such as cisterns on Linosa or *norias* on Mallorca, depend on two critical factors: the availability of water and the knowledge required to store and recover it effectively. Consequently, the amount of accessible water is not merely a result of natural abundance but also of the systems in place for collecting, transporting and storing freshwater.

Agricultural practices, such as sugarcane cultivation on the Canary Islands, significantly impact the availability of water for non-agricultural uses. Furthermore, political systems – whether centralised governments or localised clan structures – shape water distribution and the maintenance of supply facilities. Geology, geography and climate also play a critical role; for example, small volcanic islands like Linosa and the Canary Islands face distinct challenges compared to limestone archipelagos with seasonal rainfall and mountain ranges, such as the Balearic Islands. These factors influence rainfall patterns, groundwater absorption and the risk of saltwater intrusion into aquifers.

Power structures and social inequalities further dictate access to water. For instance, monastic communities, such as the Cistercian monastery of La Real in Palma de Mallorca, were often granted strategic control over water supplies, while marginalised groups, such as Muslim or Jewish communities, were dispossessed following the Christian conquest of Spain in the 13th century. Similarly, the availability of specific technologies, such as rainwater harvesting or groundwater extraction systems, determined the techniques that could be employed to purify and transport water. Local industries, like sugarcane production on the Canary Islands, also influenced the overall availability of water for the general population.

In summary, many facets of daily life on islands are directly or indirectly shaped by freshwater, and the ResourceComplex framework offers a comprehensive tool for exploring the interconnected factors that influence or are influenced by freshwater.

6.5 Landscape and Settlement Patterns and the Concept of Resources and ResourceCultures. Two Examples from Iran and Northern Iraq

Mohammad Karami and Benjamin Glissmann

Introduction

Landscape archaeology is a distinctive discipline that explores the complex interactions between human societies and their surrounding environments. The field encompasses the reciprocal influence of human agency on the environment and, conversely, the profound impact of the environment on its inhabitants. Beyond physical objects or changes in the landscape that can be observed and measured (tangible elements), landscape archaeology recognises the role of non-physical elements that influence cultural and social dynamics (intangible aspects), collectively referred to as resources, a focus that numerous scholars have emphasised in shaping human behaviour and cultural expression (e.g. Tilley 1994; Bender 1993; 1998).

Landscapes have played a central role in shaping human societies and cultures, serving as reservoirs of essential resources such as water, timber, minerals and wildlife. They have provided the basis for human settlement and economic activity. At the same time, human societies have developed diverse strategies and technologies to extract and manage these resources in a sustainable manner, adapting to the specific characteristics of their local landscapes. In addition, the landscape has had a profound influence on cultural traditions, with indigenous cultures exemplifying a deep understanding of local ecosystems woven into their cultural practices (Turner 2005).

In response to the environmental challenges and opportunities presented by landscapes, human societies have ingeniously developed technologies and systems for sustainable resource management. These solutions, diverse and tailored to specific environments, range from intricate irrigation systems that prevent soil erosion in agricultural communities to sustainable timber harvesting techniques that prevent deforestation. Such practices are imperative for conserving resources for future generations while minimising environmental damage (Ellis et al. 2010).

The conceptualisation of landscape and settlement as resources has attracted attention across a range of academic fields, including anthropology, archaeology, geography and environmental studies. Scholars have explored the many ways in which societies have adapted to and interacted with their local environments, and how this has influenced cultural practices and traditions.

This perspective emphasises that landscapes are not passive settings for human activities, but dynamic entities that are actively shaped and transformed by human societies' practices of resource use and management. It emphasises the dynamic and reciprocal relationship between humans and their environment, and highlights the social and cultural dimensions of resource use. Key concepts within this framework include 'resources', ResourceComplexes, ResourceAssemblages and ResourceCultures (see Chapter 2).

The concept of ResourceCultures provides a robust framework for comprehending the significance of landscape archaeology as a method for understanding the dynamics of resource use and management over time. This approach involves the systematic study of the material remains of past resource use practices, as well as the landscape features that reflect these practices, such as agricultural terraces, irrigation systems and settlement patterns. It allows for a more comprehensive and nuanced understanding of the relationships between people and their environment, within the historical and cultural contexts that shape these interactions (see Bartelheim et al. 2021b).

Landscape and Settlement as Dynamic ResourceComplexes

In anthropology and archaeology, the conventional distinction between natural and cultural resources has long been the cornerstone of understanding landscapes. The ResourceCultures perspective, however, challenges this dichotomy and proposes a broader and more interconnected comprehension of how resources and landscapes intricately shape human societies and their environments. This paradigm postulates that all resources are inherently culturally embedded, deriving value from the social practices and symbolic meanings attributed to them. It emphasises that resources are not static entities, but dynamic processes that evolve through integration into cultural practices and social relations (Hardenberg et al. 2017).

A central facet of this paradigm shift is the concept of 'key' resources, which have the transformative power to reshape social structures. These resources, which vary across time and space, acquire cultural connotations that determine their strategic importance. The significance of a resource in one society may not have the same resonance in another, underscoring the influence of cultural context on resource perception.

Bartelheim, García Sanjuán and Hardenberg, for instance, examine landscapes as complex resource networks in southern Spain. Through a careful analysis of Bronze Age resource use in the south of the Iberian Peninsula, the authors discuss the fundamental role of both material and immaterial resources in shaping societies. They introduce the concept of ResourceAssemblages, combinations of resources that evolve through complex relationships over time, and how these assemblages shape cultural landscapes and influence the organisation and establishment of social groups during the Bronze Age (Bartelheim et al. 2021b).

However, landscapes transcend mere physical entities and serve as expressions of collective work, repositories of memory and tools for establishing social relationships. These landscapes possess agency, influencing identity formation and becoming sites of political contestation (Filippucci 2023). This concept also accentuates the role of the environment in shaping the co-occurrence of resources within specific regions, with geological, climatic and ecological factors influencing the distribution and availability of resources, and thereby influencing the strategies and behaviours of communities (Teuber/Schweizer 2020).

Methodology

To investigate the relationship between environment and settlement in the SOJAS (South of Jiroft Archaeological Surveys) and EHAS (Eastern Habur Archaeological Survey) sub-projects, a multidisciplinary approach was adopted, drawing on a range of methodologies and theoretical frameworks from archaeology and related fields: extensive fieldwork, including archaeological surveys, excavations and geophysical surveys, documented and analysed physical features, settlements, buildings, hollow ways and material culture, including pottery, stone and soils.

Environment and Settlement

The symbiotic relationship between environment and settlement is profound, with natural landscape features influencing the location and layout of settlements. At the same time, settlements often leave a significant imprint on their surroundings. Understanding this relationship provides insights into how human societies have interacted with and shaped their natural environment in the past. This approach is especially useful for deciphering archaeological landscapes, as it allows researchers to reconstruct historical settlement patterns, resource management strategies and the socio-cultural dynamics of ancient communities. By examining both the tangible and intangible traces left behind, archaeologists can better understand the complex interplay between human activity and environmental change over time.

Adaptation

The adaptation of settlements to their specific environmental contexts is a nuanced process that reflects the ingenious responses of human communities to different geographical challenges. In mountainous regions, settlements often sit on terraces or cling to steep slopes, demonstrating an adaptive strategy that maximises habitable space while minimising the impact of elevation. Conversely, in low-lying areas, settlements are strategically located on elevated ground to escape the threat of flooding. In desert landscapes, settlements tend to gravitate towards oases or other water sources, demonstrating a strategic reliance on available resources. These adaptive practices not only ensure the survival of communities, but also shape the cultural traditions of societies by harmonising with and exploiting the unique features of their environment.

Technology and Infrastructure

The evolution of human settlements is closely linked to the development of sophisticated technologies and infrastructures tailored to the demands of their respective environments. Ancient civilisations, such as those of southern Mesopotamia and the ancient Maya civilisation of Central America, exemplify this relationship. In these regions, intricate networks of canals and reservoirs were meticulously designed to manage water resources for agricultural use. Similarly, the Romans in Europe built an extensive network of roads and aqueducts, not only as conduits for transport but also as critical components of a broader infrastructure that supported urban centres and facilitated trade. These technological marvels underscore the dynamic synergy between human ingenuity and the environmental constraints that shaped settlement practices.

Social Organisation

The topography and ecological characteristics of a landscape play a pivotal role in shaping the social organisation of settlements. On the one hand, isolated or harsh environments, such as mountainous regions or deserts, often give rise to close-knit communities characterised by strong social ties. In these environments, survival requires collective effort, which fosters a sense of communal interdependence. On the other hand, settlements in more fertile or accessible areas, such as river valleys, often develop complex social hierarchies and political systems. The abundance of resources in such areas facilitates specialisation and division of labour, contributing to the emergence of intricate social structures. The geographical setting thus becomes a silent architect, shaping the social dynamics within settlements.

Environmental Impacts

While settlements adapt to and shape their environment, it is imperative to recognise the reciprocal impact of human activities on landscapes and ecosystems. Deforestation, mining and intensive agriculture, among other extractive practices, can cause significant environmental change. These activities can lead to changes in soil composition, erosion, loss of biodiversity and other ecological challenges. Recognising the potential environmental impacts of human endeavours is crucial in the pursuit of sustainable coexistence. Finding a delicate balance between the utilisation of resources for human needs and the preservation of the environment is essential for the longevity of settlements and the well-being of future generations.

Examples

In this case study we present two examples that are geographically and chronologically differentiated. These examples allow us to explore and compare the intricate relationship between people and their environment in different contexts.

a) The South of Jiroft Archaeological Surveys (Mohammad Karami)

Geography and Geology

The research area is the region in the south of Kerman Province, south-east Iran. The archaeological part of an interdisciplinary project focusing on this region is the so-called SOJAS (South of Jiroft Archaeological Surveys). The aim of the SOJAS was to understand the dynamics of social development in the region over a long time span, from the earliest occupations to the late Islamic periods (Pfälzner/Alidadi Soleimani 2017; Pfälzner et al. 2019). The research area was divided into different zones, including the southern part of the Jiroft Plain (Zone G), the Boluk Plain (Zone A), the Faryab Plain (Zone B), the western edge of the Rudbar Plain (Zone C), the Kahnuj Plain (Zone D), the Goolashgerd Plain (Zone H), the corridor between the Boluk Plain and the Esfandaqeh Plain in the west (Zone I) and the intermountain areas of Bagh-e Borj (Zone X) (fig. 40).

The Jiroft basin is oriented in a north-south direction and contains the longest river in south-eastern Iran, the Halilrud. The Jiroft basin of the Halilrud valley is surrounded by the Hezar and Dalfard mountains in the north, Jebal-e Barez in the east, Esfandaqeh and Bagh-e Borj mountains in the west and the Jaz Murian depression in the south-east.

Geologically, the Jiroft Plain was formed as a result of the tectonic collision of the Zagros Mountains to the north and west and the subduction of the Makran Plate to the south and south-east. The Makran Plate corresponds to the accretionary prism formed by the subduction of the Arabian Plate beneath the Iranian Plate behind the deformation point, creating a back-arc basin, the so-called Jaz Murian depression (Byrne et al. 1992). Also in response to the tectonic collision of the Zagros and subduction beneath the Makran, the Jiroft basin was formed by the combination of a system of two important right-lateral strike-slip faults; the Sabzewaran fault in the west of the basin and the Jiroft fault in the east, both of which strike from north to south (Jolivet 1995; Fouache/Garçon 2007, 136) (fig. 41).

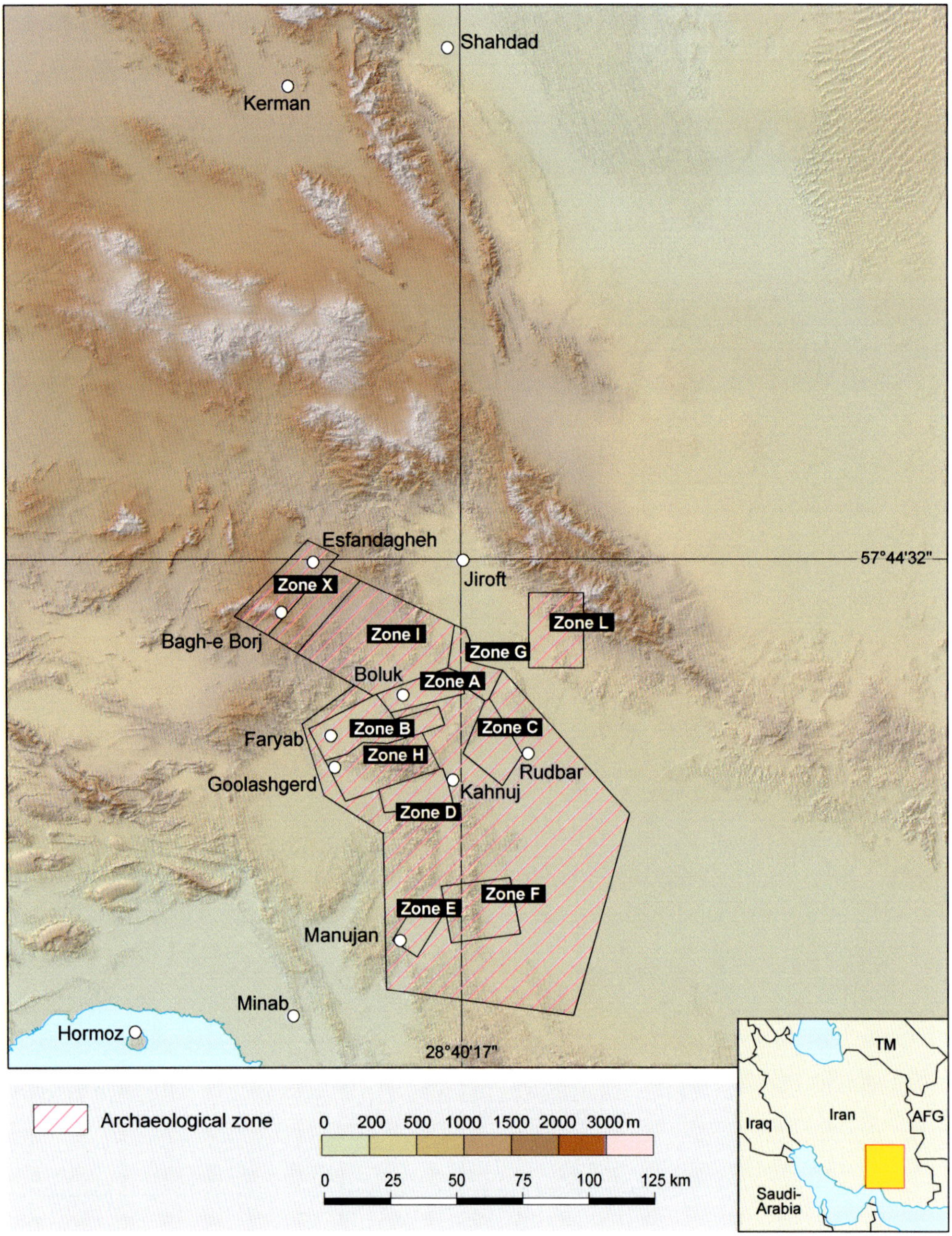

Fig. 40 Digital Elevation Model map of SOJAS and the defined archaeological zones in the south of Jiroft (Data sources: ESA COP30; Map: Mohammad Karami; Graphic design: Richard Szydlak).

The closest available climatic data refer to the city of Kerman, located at an altitude of about 1,650 m, 150 km further north of Jiroft. At this station, annual precipitation ranges from 168 to 203 mm. In the Halilrud valley, the Water Service of Jiroft estimates that the average annual rainfall is 375 mm in the mountains north of Jiroft and 125 mm in the Jaz Mu-

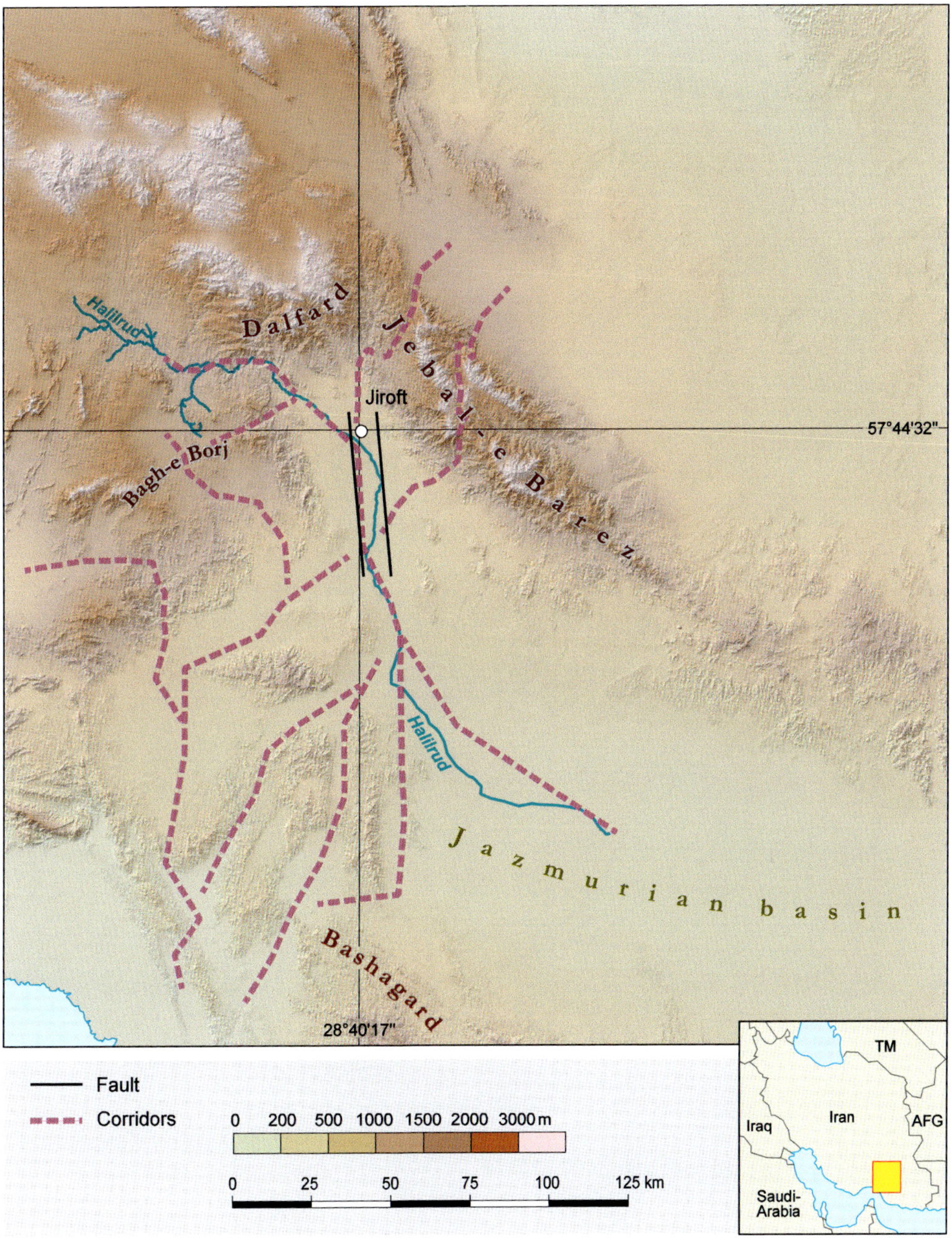

Fig. 41 Location of the city of Jiroft and the surrounding mountains, two faults of Sabzewaran and Jiroft, the long river of Halilrud as well as potential corridors that connect the Halilrud basin to the southern shores of the Persian Gulf and to the south-east (Data sources: ESA COP30; Map: Mohammad Karami; Graphic design: Richard Szydlak).

rian depression to the south. In fact, the Jiroft basin is located in a transitional zone between the semi-arid area in the north and the arid area in the south. This transitional area is very sensitive to the consequences of climatic fluctuations (Fouache/Garçon 2007, 136–137).

Fig. 42 One of the seasonal flash floods inside of the city of Jiroft on April 14, 2019 (Photo: Mohammad Karami).

Fig. 43 One of the rarely active artesian wells near the site of G67, i.e. Cheshmeh Shah, in the south of the city of Jiroft (Photo: Mohammad Karami).

Geomorphology and Potential Resources

The Halilrud basin is characterised by a unique geological setting that has given rise to significant geomorphological features. These include distinctive lithological patterns, seismic activity and a dynamic hydrological regime. Understanding the geomorphology of this region is pivotal to comprehending its historical development and the factors that have influenced settlement patterns and resource distribution over time.

Lithology

The intermediate position between the Zagros and Makran systems explains the lithological distribution observed in the Halilrud basin. In the west of the basin there is a mass composed mainly of metamorphic rocks, and in the east crystalline and volcanic rocks dominate. This geological context is crucial to understanding the distribution of mineral resources around the Jiroft basin. To the east, i.e. Jebal-e Barez, there is copper; to the west, zinc; to the southwest, chlorite; and locally, gold-bearing veins. The pebbles found in the Halilrud basin and surrounding areas provide a rich sample of the rock types present in the river basin (Fouache et al. 2005, 111; Fouache/Garçon 2007, 136–137; Fouache 2008).

Seismic Activity

Seismic activity plays a crucial role in the selection of settlements and the provision of water resources in arid zones. The Zagros region is characterised by a high level of seismicity, which affects the Jiroft basin. In recent decades, six earthquakes have been recorded in the Jiroft basin, five of which were epicentred along the Sabzewaran fault. The lateral displacements along the faults have an impact on the aquifer resources, which could indirectly influence the choice of settlement locations (Fouache et al. 2005).

Hydrology

The Halilrud, the longest river in south-eastern Iran with a length of more than 400 km, rises in the mountains to the north and north-west of Jiroft and flows in a south-east direction, crossing the bed of the Jiroft basin before emptying into the Jaz Murian depression. The Halilrud River has a strong dynamic in terms of lateral movement and alluvial deposits. This river has the hydrological regime of a wadi with intermittent flow during the winter months and very high interannual variability (Fouache 2010, 10). The size of the river watershed, the precipitation in the upper reaches and the occurrence of flash floods triggered by intense, short-duration rainfall, which can be potentially catastrophic, explain why the riverbeds are pronounced and water sources are abundant in the region (Fouache et al. 2005, 115–116) (fig. 42). In addition, the region benefits from an abundance of aquifer resources in the alluvial fans. The water level is high and flows close to the surface. The natural pressure of up to eight bar in the aquifer forces the water to rise to the surface, resulting in artesian wells (Fouache/Garçon 2007, 140–141) (fig. 43).

Corridors and Pathways

There are several natural pathways and corridors that connect the Jiroft basin to the south, leading to the coast of the Persian Gulf (see fig. 41). These routes, along with other resources, have played a significant role in the development of societies in this region throughout history. For example, these connecting routes enabled the Jiroft basin to become an impor-

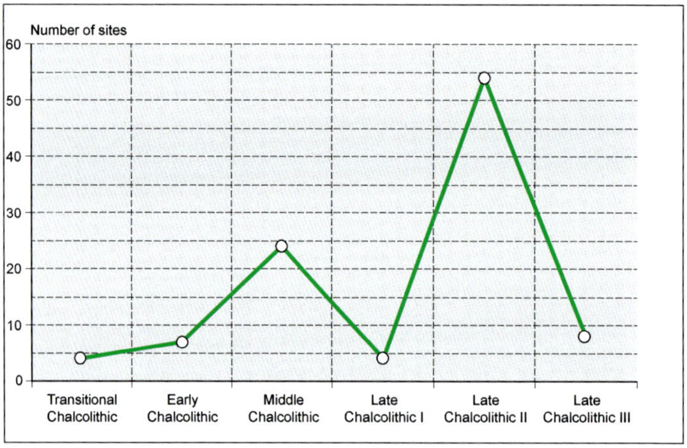

Fig. 44 The settlement pattern and the number of archaeological sites in the south of Jiroft during the Chalcolithic period (Data: Mohammad Karami; Graphic design: Richard Szydlak).

tant supply region for chlorite and diorite/gabbro stones, which were in high demand by the city-states of southern Mesopotamia during the Early Bronze Age (approximately 3000–2000 BC) (see Karami et al. 2021). Notably, these same routes retained their importance thousands of years later when Marco Polo, the Italian traveller, used one of these corridors from Jiroft to Hormoz and Minab along the shores of the Persian Gulf during his journey through south-eastern Iran (see Mir Ahmadi 2008, 10–11).

Example: Chalcolithic Period in the Halilrud Basin

The Chalcolithic period in southern Kerman covers a long time span and is divided into the Transitional Chalcolithic (ca. 5200–5000 BC), Early Chalcolithic (ca. 5000–4500 BC), Middle Chalcolithic (ca. 4500–4000 BC) and Late Chalcolithic (ca. 4000–3200 BC). The Late Chalcolithic is further subdivided into Late Chalcolithic I (ca. 4000–3700 BC), Late Chalcolithic II (ca. 3700–3400 BC) and Late Chalcolithic III (ca. 3400–3200 BC).

As a result of the SOJAS investigations, four archaeological sites provided pottery evidence for the Transitional Chalcolithic, seven for the Early Chalcolithic, twenty-four for the Middle Chalcolithic, four for the Late Chalcolithic I, fifty-four for the Late Chalcolithic II and eight for the Late Chalcolithic III. The above-mentioned numbers of settlements show that the survey zones were gradually settled from the late Neolithic/Transitional Chalcolithic into the Middle Chalcolithic. During the Middle Chalcolithic there was a rapid increase in the number of settlements. From the Middle Chalcolithic to Late Chalcolithic I a strong decrease in the density of settlements could be observed. The second rapid increase in the number of sites was determined from Late Chalcolithic I to II, where the number of settlements from the Transitional to the Late Chalcolithic periods reached the highest number in the whole study area. Again, after the Late Chalcolithic II, we observed a significant decrease in the number of archaeological sites in the survey zones (fig. 44).

The Chalcolithic Settlements and the Resource Concept

The heterogeneity of the settlement pattern during the long Chalcolithic period was also observed in subsequent periods, including the Bronze Age, the Iron Age, Historical Periods and the Islamic period in the survey zones (see Pfälzner et al. 2019). In order to understand the reasons for this heterogeneity, we can examine the development and expansion of settlements in different periods within the geographical and geomorphological conditions of the research area.

As mentioned above, the Jiroft basin is located in an almost arid region. Its geomorphological position testifies to continuous surface changes as a result of earthquakes, flash floods and changes in the aquifer. The presence of two major faults in the Halilrud basin could both provide underground water sources and pose a significant seismic risk to surrounding settlements.

In addition, surface flooding has played a significant role in the location of settlements as well as in agriculture and animal husbandry. This is especially evident in the western half of the plain and to the west of Halilrud, where surface flooding has led to the formation of another river, the Chil, which runs parallel to the Halilrud.

The settlement patterns of the different phases of the Chalcolithic in the south of Jiroft demonstrate the resilience of the occupation within the geomorphological characteristics of the region. According to the survey results, the main concentration of Early and Middle Chalcolithic settlements was not far from the Halilrud in the Jiroft basin and the Chil in Boluk (fig. 45). During the Late Chalcolithic, especially Late Chalcolithic II, a change in the choice of settlement areas could be observed. In this phase, in addition to the areas near the Halilrud, a remarkable number of settlements were established along the southern parts of the Sabzewaran and Jiroft faults. Moreover, the higher areas, such as the alluvial fan to the north and west of Boluk, which had not been settled before the Late Chalcolithic, were densely populated by the Late Chalcolithic II (fig. 46). This shows that during the Late Chalcolithic II artesian wells and aquifer resources, in addition to surface water, had a decisive influence on the development of societies. In fact, natural disasters such as seismic activity and flash floods did not act as a deterrent to settlement. The SOJAS studies clearly show that the choice of higher terraces along the faults and higher points on the alluvial fans during the Late Chalcolithic II was not accidental. It seems that not only were people able to reduce the risk of hazards by choosing higher points, but they were also able to introduce new water resources into the system, i.e. aquifer resources.

A geomorphological study in the southern region of Jiroft focuses on the beneficial role of flash floods in the area. In fact, flash floods deposit a significant amount of silty clay downstream, replenishing the soil and leaching the salt-saturated soil, making it more fertile. The study shows that the palm tree, which remains the main product of the orchards, is well adapted to these hydrological and edaphic conditions (Fouache/Garçon 2007, 143). In addition, the existence of natural artesian wells in certain regions, together with water flowing on the ground through artificial waterways, can be used as irrigation channels (Gentelle 1989). However, the adaptation of human societies to the dynamic geomorphological conditions of the semi-arid to arid region was not limited to subsistence economic systems such as agriculture and animal herding. Almost half a millennium after the Late Chalcolithic, the inhabitants of the Jiroft basin began to exploit specific resources, including chlorite and diorite/gabbro stones. They also took advantage of natural corridors to the south and south-east, integrating themselves into the complex socio-political system of South-west Asia. This integration had a significant impact on the settlement patterns and socio-political situation of the Jiroft basin during the 3rd millennium BC.

In short, the interdisciplinary research of the project, which focused on the archaeology of southern Kerman Province in south-eastern Iran, unravels the intricate relationship be-

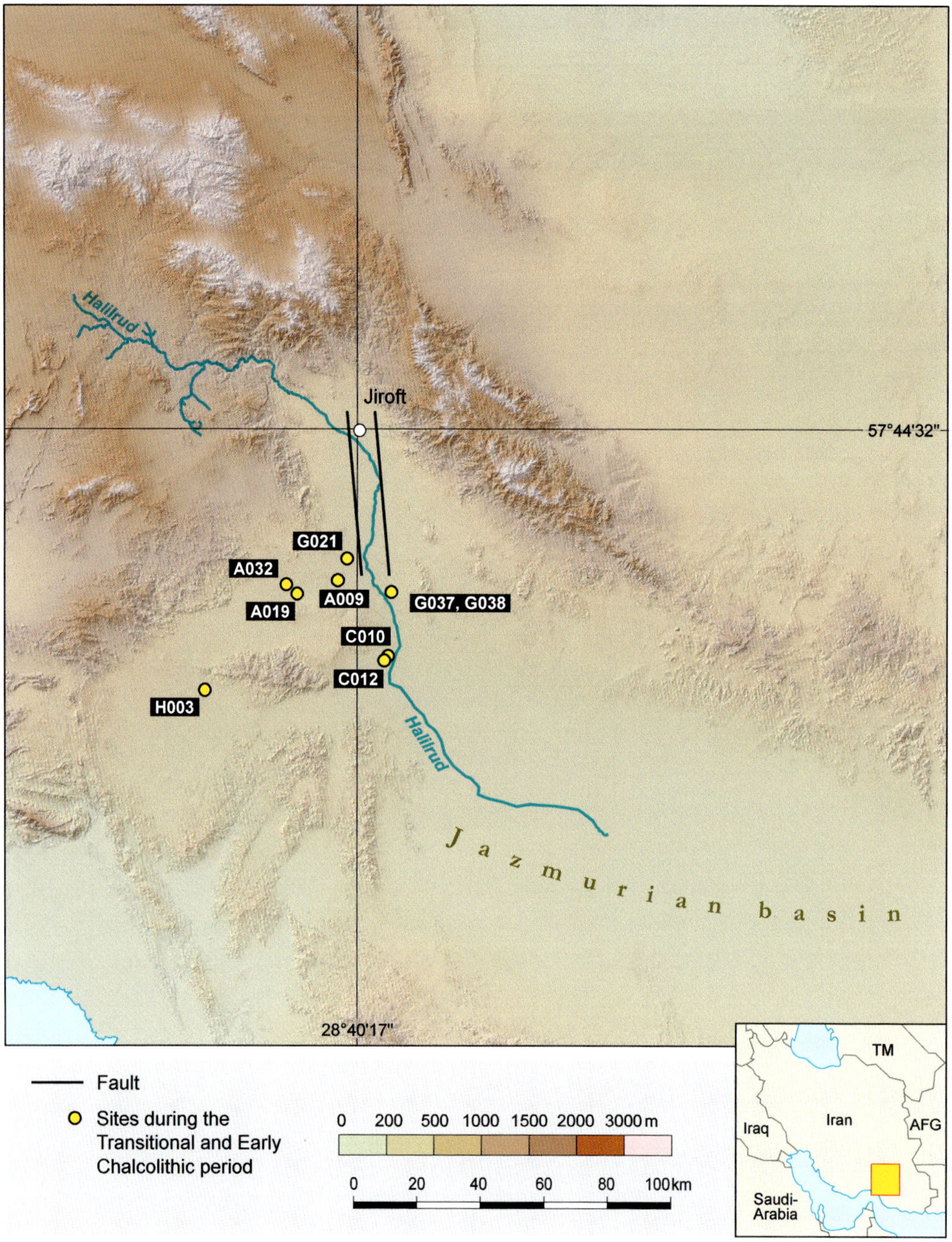

Fig. 45 Settlement pattern map of the study area showing a small number of the sites in the plain of Boluk and near the river of Halilrud during the Transitional and Early Chalcolithic periods (Data sources: ESA COP30; Map: Mohammad Karami; Graphic design: Richard Szydlak).

tween the region's unique geological and environmental features and the development of human societies over millennia. By considering the impact of geological context, seismic activity, hydrology and natural corridors, the study offers profound insights into the resilience,

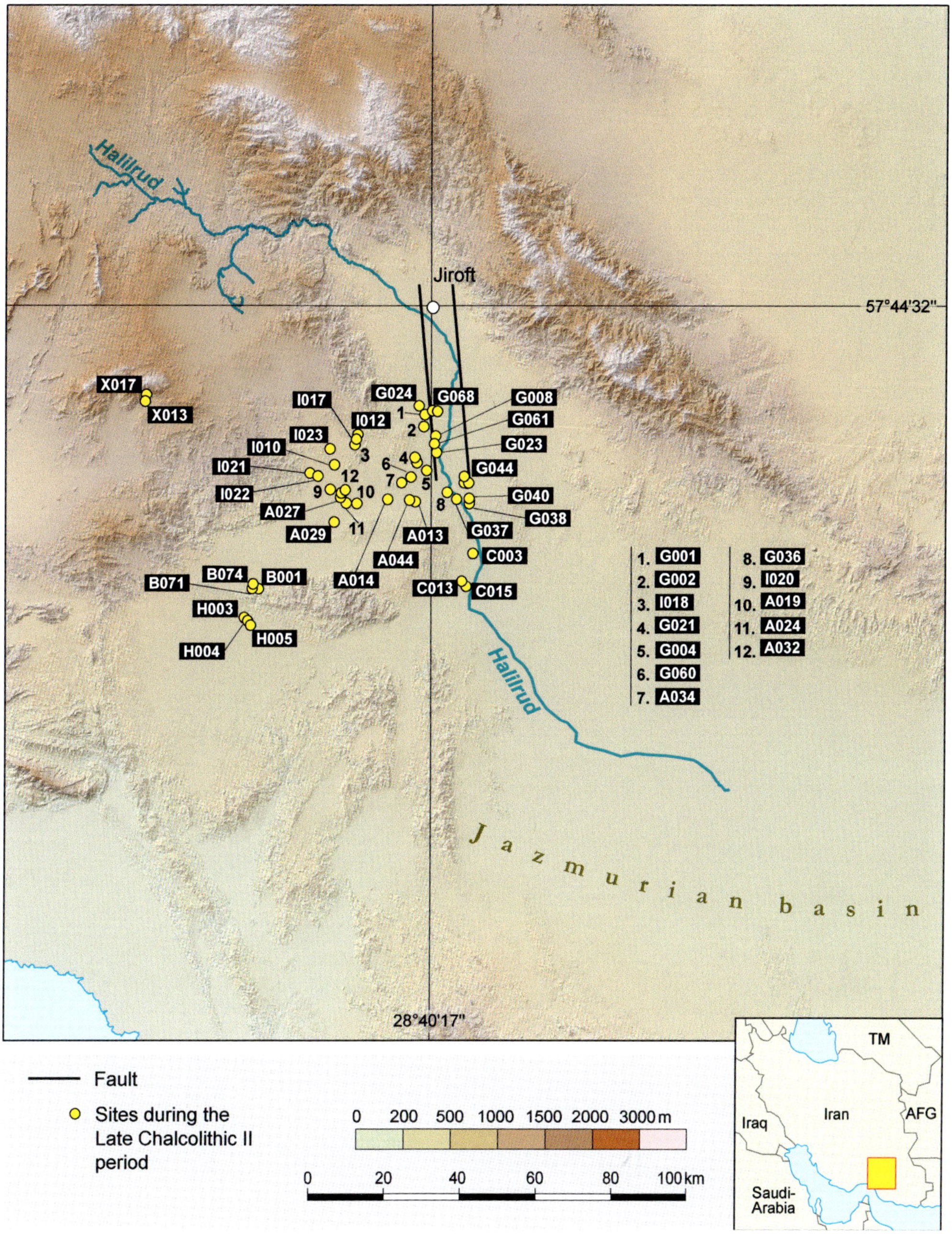

Fig. 46 Settlement pattern map of the study area during the Late Chalcolithic II showing a dense settlement not only in the Halilrud basin, but also along the faults and on the higher areas of the alluvial fans in the west of Halilrud (Data sources: ESA COP30; Map: Mohammad Karami; Graphic design: Richard Szydlak).

adaptability and resource use of past communities. It shows how these factors influenced settlement patterns, resource distribution and socio-political dynamics in this historically rich and ecologically diverse area of south-eastern Iran.

Dynamic Interaction:
Unravelling ResourceComplex and ResourceAssemblage in the Halilrud Basin

In exploring the example located in the south of Kerman Province, the intricate dynamics of resource use emerge, harmoniously integrating the perspectives of ResourceComplex and ResourceAssemblage with three aspects – materiality, knowledge and potentially power.

From the perspective of ResourceComplexes, the geological and environmental features of the Halilrud Basin contribute to a spatially diverse network in which distinctive mineral resources are distributed across different zones. This intricate spatial network includes not only physical elements such as minerals, mountains, rivers and faults, but also human settlements.

Temporal shifts in settlement patterns during the Chalcolithic represent a set of dynamics that can be approached as a ResourceAssemblage. Adaptation to changing climatic conditions, seismic activity and technological advances reveals a constant negotiation within the assemblage. This adaptability is particularly evident during the Late Chalcolithic II, where deliberate choices of settlement locations along faults and the use of higher zones demonstrate a continuous response to the evolving environment.

Materiality plays a pivotal role in shaping this assemblage, as evidenced by the diverse distribution of mineral resources and the deliberate use of natural features. The lithological composition of the landscape, seismic activity and hydrological dynamics directly influence material conditions, affecting settlement locations and resource availability.

Knowledge as a central element is manifested in the adaptive responses observed throughout the Chalcolithic. The shared understanding of the environment, evident in responses to flash floods and the use of natural artesian wells, reflects a collective knowledge base. The integration of natural corridors into socio-political systems during the 3rd millennium BC represents a strategic application of knowledge, influencing trade routes and connections within the wider region.

Although power dynamics are not explicitly inferred, the strategic choices of settlement locations, especially during the Late Chalcolithic II, imply a potential influence on resource use. The use of natural corridors and the historical significance of resource supply suggest a strategic power play, emphasising the interconnectedness of power within the ResourceAssemblage.

In conclusion, the integrated analysis of the example demonstrates a harmonious connection between the ResourceComplex and ResourceAssemblage perspectives, encapsulating materiality, knowledge and potentially power. The landscape of the Halilrud Basin unfolds as a dynamic canvas where human societies negotiate and adapt to the ever-changing interplay of resources, knowledge and strategic decision-making over time.

b) The Eastern Habur Archaeological Survey (Benjamin Glissmann)

The Project and the Area of Investigation

The second survey project presented here investigates an area of approximately 4,400 km^2 in the north-western part of the Kurdistan region of northern Iraq, bordering Syria and Turkey. Due to the historical development of the field of Ancient Near Eastern Archaeology, with early excavations in and a strong focus on southern Mesopotamia as the cradle of early states and advanced civilisations, the regions in the northern part of modern Iraq have long been regarded as peripheral and little explored until recently. Moreover, the military campaigns and expansion efforts of the southern Mesopotamian empires into these regions, as documented for the study area, for example, by the Akkadian statuette base dedicated to Naram Sin from Bassetki or the Neo-Assyrian rock relief of Tiglath-pileser III at Mila Mergi, have so far been interpreted as a hunt for natural raw materials by the materially poor south. The Eastern Habur Archaeological Survey (EHAS) project aims to test this theory through systematic investigations and the study of key resources from the perspective of ResourceComplexes, used to identify ResourceCultures in the northern fringes of Mesopotamia.

The core element of the project is the EHAS-Survey conducted between 2013 and 2019 and led by Paola Sconzo and Peter Pfälzner (Pfälzner/Sconzo 2015; 2016a; 2016b).

The EHAS survey area is bounded by the Tigris River to the south-west and the Zagros Mountains to the north, and extends from the modern state border in the west to the area of Amedy. The survey area was divided into five zones based on topographical features: the low and meandering Tigris Valley (Zone A), the adjacent Selevani Plain (Zone B), the lower Zagros foothills with the eastern Habur River (a tributary of the Tigris) north of the Selevani Plain (Zone C), the Zagros chains in the far north (Zone D) and the intra-mountainous corridor of Amedy (Zone E) (fig. 47).

A total of 529 sites and archaeological features were recorded, spanning several millennia – from the Palaeolithic to the Islamic periods. As part of the documentation process, each site was photographed, mapped (in most cases using aerial drone imagery to produce topographic maps) and its former size and extent estimated from the distribution and scatter of surface pottery. In addition, nearby resources and landscape features were documented.

Furthermore, several spin-off projects and excavations carried out so far at the archaeological sites of Bassetki (e.g. Pfälzner/Qasim 2020; Pfälzner/Faist 2020), Muqable (e.g. Puljiz/Qasim 2020), Kemune (Puljiz/Qasim 2019) and Jubaniyeh (Sconzo/Qasim 2021), as well as geophysical surveys in the lower town of Bassetki, provide detailed insights into the layout and function of individual settlements in the area.

General Settlement Dynamics in the Survey Area

Based on the location and periodisation of the surveyed settlements, broad trends and dynamics of the settlement systems can be determined in a diachronic manner. It is striking that the Tigris Valley and the Selevani Plain (Zones A and B) are densely populated throughout all periods and that many of their sites show a certain continuity based on the surface pottery finds. In contrast, the hilly Zagros foreland (Zone C) shows only sparse and isolated settlements, mainly in the vicinity of the eastern Habur River. A major expansion of the settlement system into this zone is only observed in the Middle Bronze Age, the Neo-Assyrian period and finally the Islamic period. This suggests that the natural environment and existing resource networks in zones A and B were more attractive for settlement than those in zones C, E and especially D (see fig. 47).

While simple plotting of sites according to superordinate epochs provides only a rough picture of settlement dynamics, more detailed developments in settlement patterns can be determined at the meso and micro levels. As various R-based modelling analyses for zones

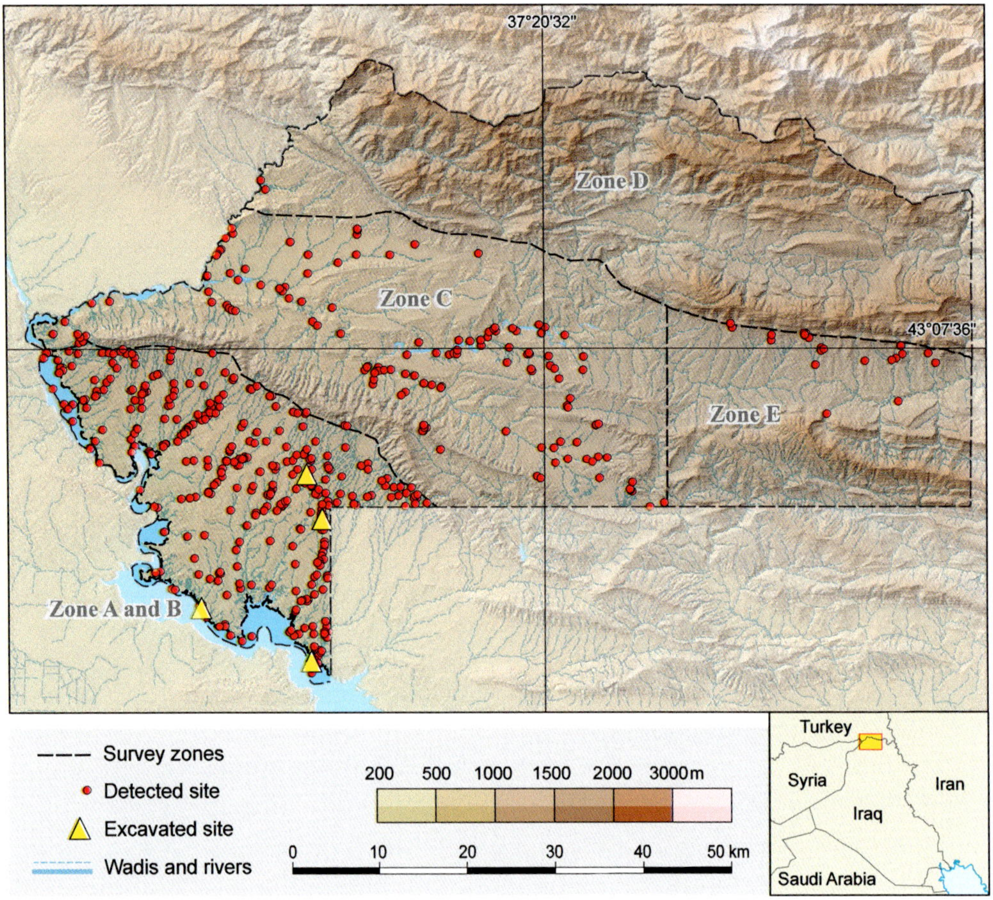

Fig. 47 Map of the survey area with the survey zones and detected sites of all periods (Data sources: CGIAR-CSI SRTM; Map: Benjamin Glissmann; Graphic design: Richard Szydlak).

A and B in the 2nd millennium BC show,[24] hot spot analyses and core density estimates can be used to identify finer settlement trends: while the site of Bassetki, located in the middle of the Selevani Plain, played a prominent role throughout the entire 2nd millennium BC, core density estimation reveals a striking concentration of settlements along an east-west axis running south of the Djebel Be'khair for the Middle Bronze Age. In contrast, the subsequent period between 1600 and 1350 BC shows a shifting settlement focus along a north-south axis, with a cluster near the present-day Eski-Mosul reservoir and the site of Kemune (fig. 48). While this axis continues to exist between 1350–1100 BC, the earlier east-west connection regains importance.

The Resources

The specific concentration of settlements in the Selevani Plain as well as the field mapping of natural features and resources in the immediate vicinity of the identified sites showed early on that, unlike in the SOJAS project, natural raw materials such as stone or possible metal

24 These analyses are carried out by Mona Walter.

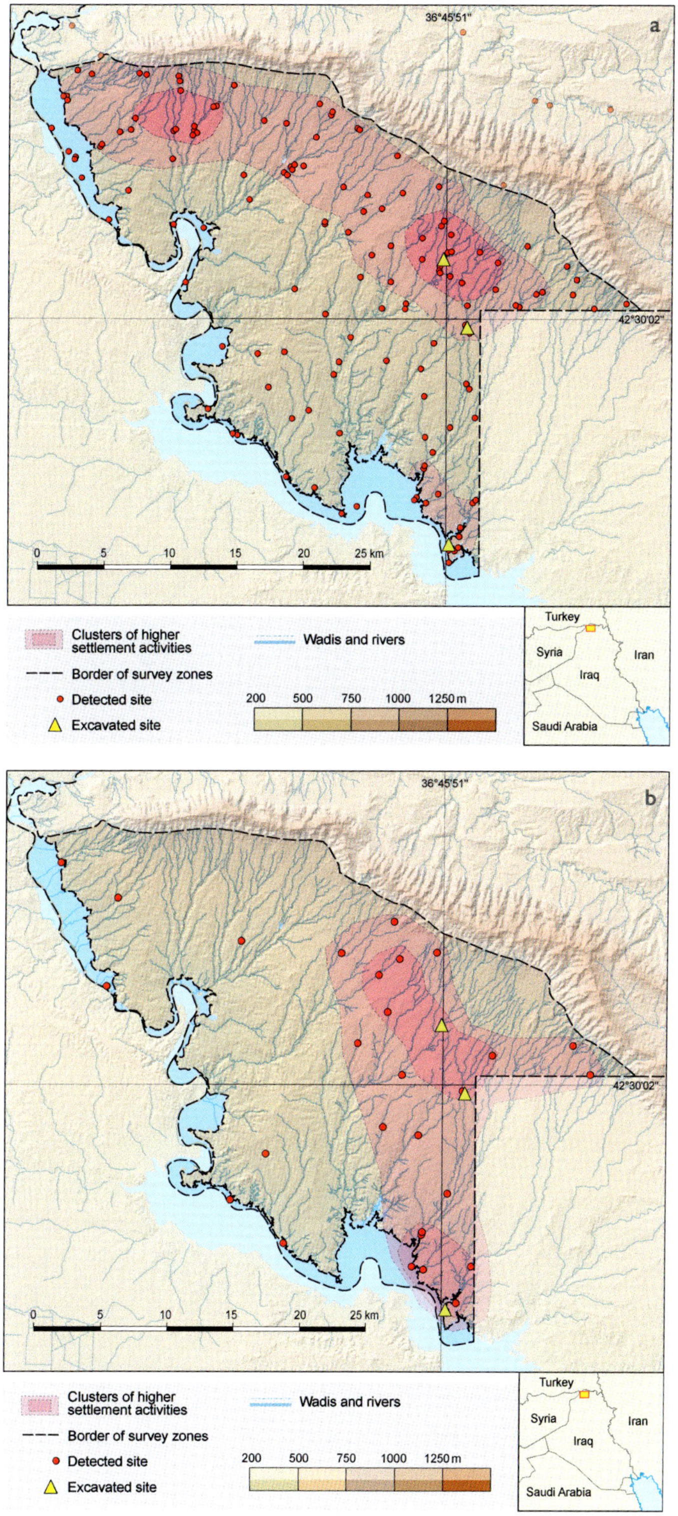

Fig. 48a + b Rough clusters of higher settlement activities in periods 8 (a) and 9 (b) detected by unweighted hot spot analyses and core density estimations for the zones A and B in the 2nd millennium BC (Data sources: CGIAR-CSI SRTM; Analyses by Mona Walter; Map: Benjamin Glissmann; Graphic design: Richard Szydlak).

deposits did not have a noticeable influence on the observed indigenous settlement dynamics, nor could they be regarded as a thriving force stimulating expansion from the south. Rather, soils, agricultural potential, water, roads, paths or trackways and settlements were identified as key resources, interwoven in different endogenous networks, and are studied in a diachronic approach.

Water and Soils

Water is one of the most important resources for this semi-arid region, especially during the hot summers, and is the most basic prerequisite for survival and livelihoods, including agriculture and livestock. Given this vital importance, it is not surprising that almost all of the sites identified are located in close proximity to rivers or smaller or larger wadis. In some cases, the settlements are also fed by local springs, as in the case of Bassetki, where this feature was certainly an important founding factor and contributed to the long and continuous settlement history from the Ninivite 5 period (mid-3rd millennium BC) to the Hellenistic period (330–150 BC) (Pfälzner/Faist 2020, 349–350).

Another important site-related factor is the accessibility of land for agricultural use. This resource is especially abundant in the vast and largely flat areas of the Selevani Plain, which are still characterised by agricultural activities, including livestock farming. Preliminary results of soil sampling – carried out by the Department of Geosciences of Tübingen University as partner in the project – show that the calcitic, slightly alkaline soils of zones A and B contain low levels of organic matter, indicating that the soils have been in continuous use over a long period of time. The less densely populated hilly foothills of the Zagros Mountains (Zone C) still provide good conditions for the cultivation of fields and orchards, but also include some areas of poor soil and a more pronounced relief, leading to higher rates of erosion. The question of whether soil composition, more limited access to land, slightly different environmental conditions or other (geographical or cultural) factors have influenced the different settlement patterns between zones A and B compared to zones C and E can be answered with the help of detailed soil quality maps currently being produced by soil scientists.

While access to water is crucial for any settlement, the resources of soil quality and access to agricultural land may be particularly important for smaller, rural and self-sustaining settlements and may influence their spatial location. While these resources support basic livelihoods, other, more specialised settlement types also show a strong connection to more complex and socio-culturally related resources, such as route systems.

Routes and Pathways under the ResourceComplexes Perspective

Routes, streets, pathways and tracks represent features of movement and communication and constitute essential components of settlements and settlement systems. In general, these features can be studied in the EHAS project at different scales:

1) at the interregional scale, with overland routes connecting different regions in a broader (commercial) network;
2) at the regional scale, with routes and paths connecting contemporaneous settlements of different sizes and hierarchical order or with off-site features; and
3) at the scale of individual settlements, where streets and paths organise communication between different areas and connect the settlement with its immediate surroundings, such as fields or pastures.

In addition to being simple architectural elements that structure and facilitate movement, routes and pathways can be understood as resources that are integrated with other material and immaterial components (such as building materials, the surrounding landscape, knowledge or power) into a broader resource network. From the ResourceComplexes perspective, the links and components of the network are not fixed or static and may vary slightly depending on the specific types and scales of the route systems. This will be illustrated briefly by a few examples from the Bassetki site.

Most excavations focus primarily on the study of public buildings or residential structures, while streets or pathways are usually of less interest and are often treated as 'by-products'. In the case of Bassetki, large-scale magnetic prospections carried out between 2015–2017 and 2021–2023 detected several linear anomalies representing possible routes in and around the lower town. To verify these features as streets and routes, and to investigate their structure and date, some of them were excavated.

The test trenches revealed various types of streets and routes. A short strip of a single-phase street dated to the Ninivite 5 period (first half of the 3rd millennium BC) was uncovered about 120 m north-east of the high mound. The narrow street, up to 1 m wide, consisted of small stones, pottery fragments of various sizes, some animal bones, pottery slag and fragments of burnt mudbricks mixed with ashy loam material. The heterogeneous nature of the pavement, as well as the discovery of two clay sealings, indicates that the street was an integral part of a settlement area and accumulated through settlement deposits. Thus, the resource 'street' is linked to a broader ResourceComplex that includes different aspects of knowledge and daily life practices such as house building, dwelling, pottery production and use, storage, waste management, nutrition habits and animal husbandry. Furthermore, the cumulative construction of the narrow street shows that it developed in concordance with the surrounding settlement, rather than being planned and executed by a higher power.

On a regional scale, route systems linking individual settlements are more difficult to identify, as they are largely located outside settlements and therefore mostly beyond the scope of excavations. Such regional road systems can be modelled using a least cost path analysis. Based on a digital elevation model, the shortest path between two points or (main) settlements can be calculated using the natural slope of the terrain and then compared with the distribution of smaller settlements not included in the model to refine the course of the pathways. Further indications of such systems are the so-called hollow ways. These are shallow, linear depressions, sometimes visible as colour differences or crop marks in aerial photographs or older satellite images, such as the declassified Corona satellite images, and often correspond to ancient trackways connecting settlements with the surrounding areas (Ur 2003). Apart from this function, hollow ways may also be ancient roads linking different settlements.

Nevertheless, two partially excavated streets in the north-eastern part of the lower town run in an east-west direction and can be dated to the Middle Bronze Age. Like the street from Ninivite 5, they are paved mainly with small stones and pottery fragments, but contain less settlement debris and are generally wider (> 3 m). According to the geophysical prospections, these streets appear to be extra-urban streets connecting the city of Bassetki with the surrounding settlements. This is most obvious in the case of the northern most of the two streets, which first runs parallel to the eastern boundary of the city, then turns at right angles to the south-west and crosses the wadi to the east of the tell via an artificially constructed ford. Especially the connection to the ford, as well as the more elaborate construction of the streets, suggests that they were not built by chance, but were planned and executed by some higher, institutional power. Nevertheless, from the perspective of ResourceComplexes, such types of roads are not only associated with building materials and techniques, but also integrate further aspects of knowledge about the landscape, the environment, waterways, other settlements, agriculture, trade, communication and exchange, as well as the component of centralised power and its representation.

The distinction between regional and interregional routes is in most cases limited by the fact that the latter often followed or used pre-existing regional route systems. A major road passing by the site of Bassetki, 40 m south of the lower city boundary, was detected by magnetic prospections in 2015. Excavations in the following years revealed that this street was 5.5 m wide and carefully paved with flat-lying, large-sized Middle Bronze Age pottery fragments resting on a layer of stones (Pfälzner/Qasim 2017, 33–36, fig. 23). Several cremation urns containing burnt human bones and some artefacts such as beads or toggle pins were dug into the pavement and adjacent to the street, while an extra-urban cemetery with stone-built cist graves was discovered immediately south of the road (Pfälzner/Qasim 2017, 36–37, figs. 25–26). Although only partially excavated, the course of the street can be traced on magnetic imagery for a distance of 180 m in an east-west direction before it disappears.

The identification of Bassetki as the ancient city of Mardama(n), mentioned as one of the stops on one of the trade routes linking the Assyrian capital of Assur with the trading colonies in Anatolia in Old Assyrian times, possibly identifies this road as an interregional trade route (Pfälzner/Faist 2020, 362–363, 377–378). Its course near Bassetki, instead of following the deep and meandering Tigris Valley, can be explained by the flat and easy-to-travel natural conditions of the Selevani Plain. Even today, the modern motorway leading to Turkey via the ford at Feshkhabur follows exactly this course, covering the ancient interregional route south of Bassetki (Pfälzner/Faist 2020, 348; Pfälzner/Qasim 2017, 38).

The carefully constructed and maintained Middle Bronze Age trade route testifies to the existence of a centralised institution participating in the interregional trade system. While the street can be understood as an expression and demonstration of (local) power and belonging to a broader system of trade and exchange of goods and ideas, from the perspective of ResourceComplexes, it also integrates further components into the network of resources, such as funerary traditions and belief systems or knowledge of the wider region with possibilities for easy travel and suitable places for crossing rivers as well as natural boundaries.

The Link between Settlements, Landscape and ResourceComplexes

The spatio-temporal distribution of the settlement sites detected in the EHAS, as outlined above, highlights important trends and dynamics of settlement strategies in relation to the surrounding landscape and the given resource networks observed under the ResourceComplexes perspective. Resources in the sense of natural raw materials such as stone or metal do not seem to have played a decisive role in the expansion of the Mesopotamian great powers, and their intrusion does not significantly change the general settlement patterns. Rather, it becomes evident that the endogenous key resources of water, soil and route systems played a major role in the formation and dynamics of settlement systems, and that resource networks were assigned different meanings and values in different periods.

While the overall higher density of settlements in the Tigris Valley and the Selevani Plain (Zones A and B) may be related to more favourable environmental conditions such as easy access to water and wide, flat plains for agricultural activities and easy travel, the general settlement systems appear to have been stable and not radically affected by the expanding powers from southern Mesopotamia. Nevertheless, the preliminary modelling of zones A and B for the 2nd millennium BC has revealed minor changes that can be interpreted as differences in the importance and valuation of resources and ResourceComplexes in the sense of ResourceAssemblages.

From the perspective of ResourceComplexes, the extra-urban street south of Bassetki, an intermediate station of the interregional trade route system, as well as the contemporaneous dense settlement concentration along an east-west axis south of the Djebel Be'khair mountain range (see fig. 48), corresponding to the presumed course of the interregional trade route to Anatolia, highlight the route system resource as an important key factor of the Middle Bronze Age. Since interregional trade routes already crossed this region in the 3rd millennium BC, this idealistic attribution of value to route systems as a means of establishing and maintaining (commercial) power may also have stimulated the earlier expansion of the Akkadian Empire into this region and is indicated by a similar settlement pattern as in the first half of the 2nd millennium BC.

In the Late Bronze Age, at least for some sites, the route system resource seems to lose some of its importance and other landscape factors play a flourishing role. While the previous distribution of settlements along the east-west axis is significantly less pronounced, a divergent distribution of sites along a north-south axis now emerges, with a cluster of sites close to the Tigris (see fig. 48). This can be linked to the expanding Mitannian and Middle Assyrian realms. The location of the site of Kemune, with its Mitannian palace and pronounced fortification walls on the edge of the Selevani Plain overlooking the Tigris Valley, seems to have been determined by geopolitical factors such as control of the river valley and the natural protection of the settlement by the surrounding slopes. While such factors were particularly valued for sites associated with territorial expansion and political power, other key resources – such as water, soil and route systems – continued to play an important role in the distribution of settlements. This is evidenced by the resurgence of settlement density along the east-west axis in the Middle Assyrian period, indicating an increasing importance of the (interregional) route system, with Bassetki and the site of Derabun, close to the modern borders of Syria and Turkey, as possible gateway communities.

The site of Muqable, a few kilometres south of Bassetki, has been identified as a possible Middle Assyrian *dunnu* and highlights different aspects of landscape valuation and use. These types of settlements were fortified farmsteads specialising in the production and processing of agricultural goods and were granted by the Assyrian king to loyal elite families or individuals (Puljiz/Qasim 2020). In this way, the landscape and various resources, such as soil, were used to establish and consolidate social hierarchies and loyalties, and to extend the king's power within and beyond the core empire.

The results of the EHAS-Survey, refined by the excavations at the sites mentioned above, together with the application of the ResourceComplex perspective, show how the landscape and the various resource networks were endowed with different values and idealistic meanings, and served – apart from mere subsistence – the establishment and maintenance of political and commercial power. The identified resource networks, like the route systems, can be studied diachronically as ResourceAssemblages and help to better understand specific settlement patterns and the distribution of sites. Although this region has been little explored, ongoing investigations involving the production of detailed soil quality maps, the modelling of route systems and more detailed investigations of specific sites have the potential to improve our understanding of the role and use of resources on the northern edge of Mesopotamia.

Conclusion

The study of landscape archaeology, situated within the paradigm of understanding environments and settlements as resources, reveals a profound understanding of human-environment interactions. This emphasises the need for holistic approaches that recognise the dynamic nature of resources and settlements, the interplay between cultural and environmental factors and the significance of balancing resource use with environmental conservation. The central theme of landscapes and settlements as resources resonates as a core concept, emphasising the significant impact of landscapes on the shaping of societies and, conversely, the profound impact of societies on their landscapes. This highlights the depth and complexity of human-environment relationships across time and space.

The case study presented, with its two examples, serves as a complex vignette contributing to this understanding, focusing on different societies, natural environments, climatic regions and epochs. The exploration of south-eastern Iran vividly illustrates the intricate interconnections between geological features, environmental conditions and the development of human societies over time. It highlights the resilience and adaptability of communities in response to the unique conditions of their environment, and demonstrates the interplay between landscape and societal development.

Research in northern Iraq shows how different settlement patterns and locations are linked to natural landscape conditions such as access to water and agricultural land. On the other hand, shifts in settlement distribution are also linked to and shaped by geopolitical and economic factors, as well as different valuations and uses of resources. For example, many Middle Bronze Age settlement sites are associated with interregional trade routes and the exchange of goods and ideas, whereas some specific sites in later periods are more associated with aspects of territorial control and natural fortification. While environmental conditions, for example, influence the establishment and trajectory of interregional trade routes, such features, as well as other resource networks, influence the inhabitants and authorities of these regions and were used to gain and maintain political and commercial power.

Such insights further validate the concept of landscape and settlements as material and immaterial resources.

Landscape archaeology stands as a testament to the interconnected relationship between human societies and their surrounding environments, emphasising in particular the unique perspective of landscape and settlements as resources. This framework presents an intricate dualism of influence between human agency and the environment, highlighting the interactions and complex dynamics that shape cultural expressions and societal development. The central theme of landscape and settlements as resources is echoed throughout the subsequent examination of various cases, highlighting the multifaceted nature of human-environment interaction in different historical and geographical contexts.

The underlying premise of landscape and settlements as resources highlights the multiple ways in which landscapes are not passive backdrops but active participants in the shaping of human societies. This is consistent with the framework of dynamic ResourceComplexes and ResourceCultures, which elucidate the intertwined and evolving relationship between societies and the resources they use. These concepts emphasise the fluidity and adaptability of resources, depicting their evolution as integrated elements within social practices and cultural contexts.

Our research across different geographical areas and time periods has identified similar factors that have played an essential role in the development of societies:

- Water resources not only played a crucial role in the choice of settlement locations, but also contributed to the knowledge and techniques of the inhabitants, thus ensuring the sustainability of societies over time.
- The importance of pathways, whether natural corridors in Iran or communication networks between settlements in Iraq, is evident in shaping settlement dynamics and resource exchange.
- Geological and climatic features are significant in both study areas, from the unique geological and climatic features in Iran to the divergent distribution of settlements in different topographic zones in northern Iraq.
- Geomorphological factors such as the influence of lithology, seismic activity and geomorphological conditions are recurring themes that influence settlement locations and historical development.
- The adaptability and resilience of societies to their environment is evident in all areas of research. They adjusted their agricultural practices, settled in response to natural phenomena such as earthquakes, floods or climatic changes, as well as to geopolitical or commercial needs and adapted to available resources.
- Taken together, these projects highlight the multidimensionality of human-environment dynamics and reaffirm the pivotal role of landscapes and settlements as resources. The adaptability of ancient settlements to diverse landscapes, the sophisticated technologies developed to support these societies and the societal impacts on the environment all underscore the intricate web of human-environment interactions throughout history in the context of resource use.

6.6 Pathways as ResourceCultures. A Perspective from a Late 19th Century Colony in Paraguay

Attila Dézsi

'... I was awakened by Manuel around ten o'clock with the news that the moon had risen and the horses were saddled. [...] We continued in constant alternation through fields, forests, and lagoons throughout the long, long night; the moon had bathed everything in a silver glow. [...] About a legua before San Pedro, we turned back onto the path we had taken during our journey there; I noticed it immediately by the farms along the way, which the moon illuminated distinctly.'[25]

Klingbeil 1889, 134 (translated by the author)

Introduction

Landscape archaeology has always been a significant approach in the archaeological sciences to understanding the interrelationships between human communities and natural or man-made landscapes. Landscape archaeology is a specialised term that encompasses different approaches, scales and perspectives (see the history of the discipline in Historical Archaeology: Branton 2009).

'Landscape is the spatial manifestation of the dynamic relations between humans and their environment' (Crumley/Marquardt 1990), and landscapes are not simply passive backdrops to human activities. Places and spaces as parts of landscapes, together with their characteristics, play an active role, both enabling and constraining human behaviour. As dynamic participants in a relationship, landscapes shape communities just as they are shaped and reshaped by humans. They necessarily contribute significantly to the economic, political and social expressions of societies and to identity formation. The importance of landscapes for identity and well-being becomes particularly apparent when they are suddenly lost due to war or natural disasters (van Duijvenvoorde et al. 2023; Bagwell 2009, 284; Schlanger et al. 2016).

The study of landscapes allows for the analysis of the material components of place, identity and power (Branton 2009, 51). Historical archaeology has in many places explored the symbolic and political dimensions of landscapes that have left their mark and remnants in a particular place and continue to influence it to this day. This includes, for example, a long tradition of work on former plantations in North America (Delle 1998; 1999) or studies of the impact of contemporary protests on the perception of landscapes (Schofield 2005; Badcock/Johnston 2009).

The ambivalence of the term 'landscape archaeology' brings diversity to research questions and allows critical enquiry to bring underrepresented perspectives to the foreground (Hicks/McAtackney 2007, 15, 23). It is to be highlighted that a landscape is not only used, perceived and valued by one group, but that different groups have different relationships

25 In the original: '[... als] ich von Manuel gegen zehn Uhr mit der Nachricht geweckt wurde, daß der Mond aufgegangen und daß die Pferde gesattelt seien. [...] Fort ging es in steter Abwechslung durch Cämpe, Wälder und Lagunen die lange, lange Nacht; der Mond hatte Alles mit silbernen Glanze übergossen. [...] Ungefähr eine Legua vor San Pedro lenkten wir wieder in den von uns bei der Herreise eingeschlagenen Weg ein, ich bemerkte es sofort an den am Wege liegenden Gehöften, der Mond ließ diese deutlich erkennen.' (Klingbeil 1889, 134).

within the same landscape. Landscapes are thus multivocal: places can be given different meanings by social groups (even by individuals within a group) that co-occupy a landscape (creating 'nested' landscapes) or share it collectively (Branton 2009, 52). Additionally, landscapes do not exist in isolation and can overlap and/or be interconnected. Over time, social relations to landscapes can change, leaving behind remnants of previous uses and being constantly overwritten by new relations, creating a palimpsest of past and present features existing in one space (Lucas 2012, 115–118).

To understand such a complex relational system, in which features of a landscape are valued by different scales and people through social practices over a period of time that are themselves intertwined and changing, the concept of ResourceCulture can be an interesting tool (see Chapter 2; Bartelheim et al. 2021b, 13–14). In this case study, the resources pathways and the plant yerba mate are discussed as potential ResourceCultures within a regional subtropical landscape in northern Paraguay: how they have been used, valued and linked to social practices by different communities in a very specific local historical context of a colonial settlement.

Nueva Germania – An Intentional Dis-/Utopian Settlement in Paraguay

In order to examine the relationship of both resources to the cultural landscape, we first need to understand the historical and social context of the emigrant colony Nueva Germania, founded and collapsed in the late 19th century.

Migration has been a pivotal aspect of human history, driven by a variety of factors. The second half of the 19th century saw a significant European exodus for a number of reasons. Crop failures associated with the Tambora volcanic eruption of 1816 (Behringer 2018, 179–185) and the socio-economic turmoil of rapid industrialisation, accompanied by the global financial crises of 1857 and 1873, with far-reaching economic repercussions on grain prices, forced many Europeans to seek a new life abroad. Economic opportunities, but also political dissatisfaction, were the main reasons for the increased emigration of German-speaking people since the 19th century. The majority of German emigrants chose North America as their destination, while a minority set their sights on Eastern Europe, Africa and Latin America (Bergmann 1994, 5).

This migration away from industrialisation, urbanisation and social upheavals ushered in a phenomenon that has recently attracted the attention of historical archaeology: the establishment of 'utopian' communities in response to perceived social injustices (Van Bueren/Tarlow 2006; Hewitt 2007; Tarlow 2002; Kozakavich 2017). These 'utopian' visionaries consciously adopted alternative societal structures and sought to manifest ideals such as religious freedom, socialism, gender equality and spirituality in order to persuade others to adopt their political principles (Van Bueren/Tarlow 2006, 1–2). In particular, archaeological research has been conducted on 'utopian' settlements, including cases such as New Harmony (Strezewski 2015a; 2015b), Feltville (Tomaso et al. 2006), the Koreshan Unity Settlement (Tarlow 2006), the Llano del Rio Cooperative (Van Bueren 2006) and Shaker communities (Starbuck 2004) in North America.

In the genesis and organisation of these 'utopian' settlement initiatives, the practical experimentation with visionary and alternative political or religious ideologies played a central role, suggesting that their beliefs were likely to have left tangible imprints on the physical aspects of the settlement landscape. As a result, their ideologies may be discernible through the layout and pattern of the settlement, its relationship to the surrounding landscape, architectural features, household material culture and the physical remains of the utopians, making them subjects of archaeological investigation (Van Bueren/Tarlow 2006, 4).

This case study presents the 'utopian' settlement of Nueva Germania, a short-lived and ultimately unsuccessful private colonial venture that developed in Paraguay between 1887 and 1889. This colony has been a focus of historical research (Kraus 1999; 2008; Fischer 2015; Schneppen 2001; also Williams et al. 2001) and journalistic interest (Macintyre 1992). It is also the subject of an ethnographic case study examining the social identity of German, Paraguayan and Germanino residents in the vicinity of the former utopian settlement (Kurzwelly 2019; 2024).

Important historical sources include contemporary pamphlets and books that both promoted and criticised the project (Förster 1883; 1886; Förster-Nietzsche 1891; von deutschen Freunden 1889; Klingbeil 1889), as well as written materials from the planning phase authored by the founders and archived at the Goethe-Schiller Archive in Weimar (GSA).

Founded by the anti-Semitic activist Bernhard Förster (Kraus 2009) and his wife Elisabeth Nietzsche (Sieg 2019), Nueva Germania departed from the norm of contemporary socialist, progressive and religious 'utopian' settlement projects. Instead, it adhered to an anti-modern and Germanic ideology rooted in Social Darwinism and anti-Semitic beliefs. Paraguay was considered an ideal location for this experiment for several reasons. First, the country was in dire economic straits following the Paraguayan War (1864–1870) (Reber 1988), and to address labour shortages and the search for economic investment, the government actively encouraged European immigration (Sieg 2019, 101–102). Second, Förster believed that only in such a sparsely populated country could German culture be 'preserved' (Kraus 1999, 140; Förster 1883, 54–55). This vision of a better life with a new social order in a 'utopian' colony was compelling enough to convince 40 families, totalling at least 160 people, to emigrate from Germany to the colony by July 1888 (Förster-Nietzsche 1891, 3–5).

However, the 'utopian' project soon came to an end. The newly established colony could not sustain itself economically, and Förster passed away in June 1889 in seclusion under unexplained circumstances (Kraus 2009, 237). Elisabeth Nietzsche returned to Germany in 1893 and attempted to raise funds for the colony project (Förster-Nietzsche 1891), but did not return to continue her life in Paraguay. The town of Nueva Germania still exists today. For its current inhabitants, the challenging legacy of the early settlement history is of little relevance, although echoes of the German heritage are visible in the city, such as murals, streets and garbage bins painted in German national colours, an annual Oktoberfest (beer festival 'El Choppfest') and city celebrations with parades in Paraguayan and German national colours. Since the founding of the settlement, the town has grown considerably, and some material remnants of the early years of the town's discordant history have survived – the pattern of streets and pathways in the wider landscape and some of the settlers' first buildings still remain and are the homes of families.

A Planned and Experienced Dis-/Utopian Landscape

To gain a full understanding of the diachronic development of the colony – before and during its deliberate settlement, with a particular focus on its aftermath – it is imperative to

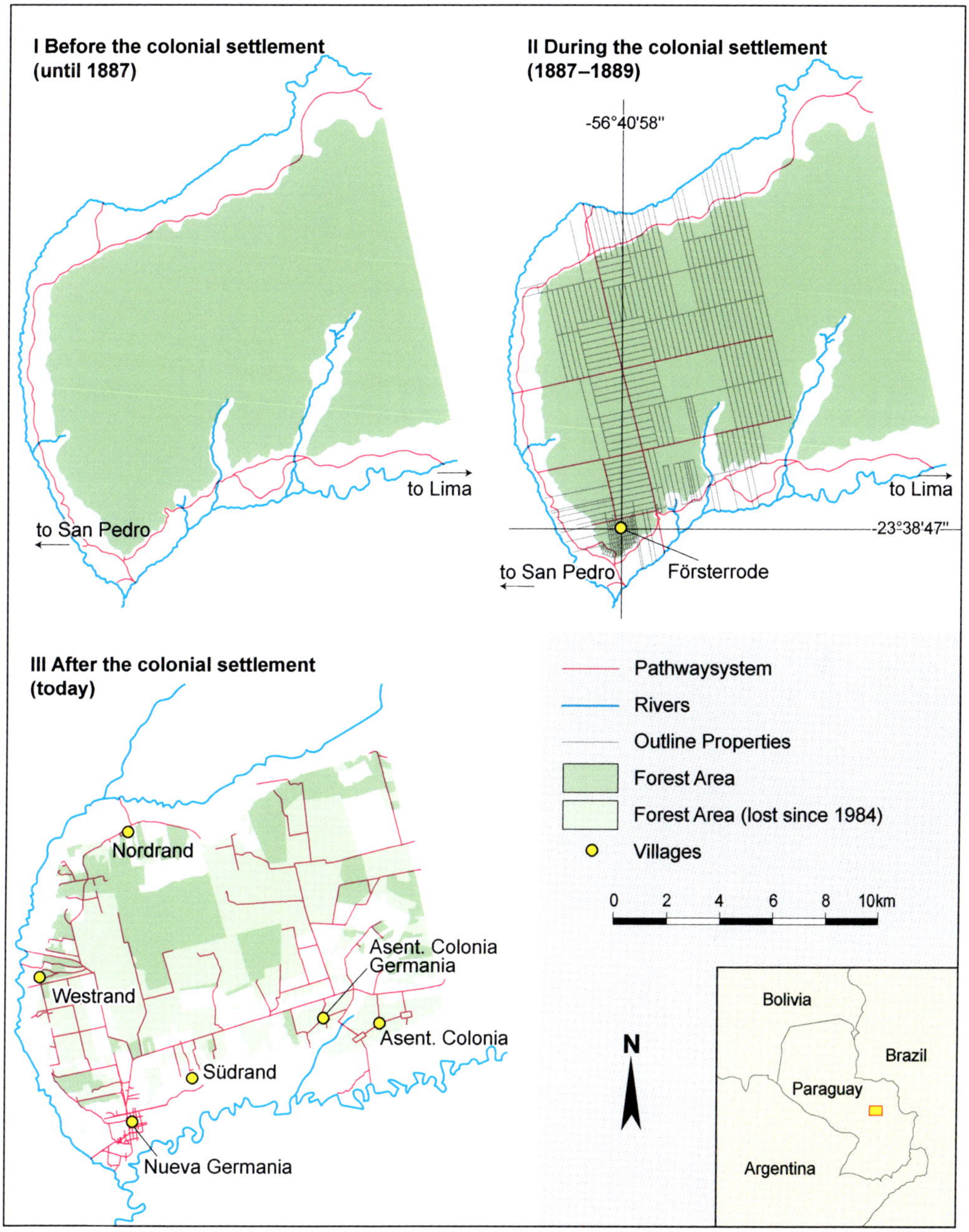

Fig. 49 Diachronic comparison of the study area of Nueva Germania in regard of pathway systems. Upper left: indigenous pathways and settlements, river and forested area before the start of the colony, depicted in GSA 72 947-27. Upper right: intended pathways and property outlines for the colony and main settlement Försterrode, depicted in GSA 72 947-27. Bottom: current pathways, settlements and forest outlines from 1984 and 2023 (Data sources: GSA; INE Paraguay; Map: Attila Dézsi; Graphic design: Richard Szydlak).

consider the dynamic interplay of different resources that have shaped the experience and livelihood of people in the area (fig. 49). The following discussion focuses on several key resources of interest – primary pathways and secondary yerba mate – and highlights their transformations and perceptions over time by indigenous people, colonial planners and migrating families, bringing us closer to understanding this interplay as a ResourceCulture.

Phase I – Pathways before the Colonial Settlement

Between 1883 and 1885, Förster travelled through Paraguay on horseback and recorded his experiences while scouting for a suitable site for his colony (Förster 1886). Although Förster was largely alone on this journey, he occasionally encountered Paraguayan locals who provided him with food and shelter. He was particularly interested in a particular area in the present-day municipality of San Pedro, in the north-central region of Paraguay. The colony's territory covers an area of roughly 200 km², framed by the meandering Aguaray-Mí, Aguaray-Guazú and Arroyo Zanjita rivers on its southern, western and northern flanks. The associated floodplains are extensive and subject to periodic flooding. The settlement area was located on a plateau approximately 132 m above sea level and was therefore protected from flooding. The Goethe-Schiller Archive (GSA 72/1508 sheet 19) has preserved a fascinating hand-drawn map entitled 'Neu-Germanien', apparently by Bernhard Förster. This map shows intricate details of the area. In particular, the terrain between the Aguaray-Guazú and -Mí rivers is covered in forest. A commentary on the forested region describes it as '600 km² zu ¾ Urwald, zu ¼ schöner Weideland [600 km², ¾ jungle, ¼ beautiful pastureland; translated by the author]'. A note in the lower right corner of the map acknowledges that the forest has been slightly understated in the representation. Importantly, to the east of the 'Neu-Germanien' area, Förster identifies another distinct region using different symbols, which he refers to as '*Yerbal*' (yerba mate plantation, see below).

South of the *yerbal* and Aguaray-Guazú river, the map shows the only existing settlement in the vicinity: Lima. Curiously, Förster's documentation makes no mention of any other settlements or pathways in the area – giving the impression of an almost empty, uninhabited area.

A second printed map entitled 'Karte des Herrn Dr. Bernh. Förster in Paraguay erworbenen Landes' (GSA 72/1508 sheet 5), also sheds light on the planners' perception of the landscape prior to the establishment of the colony. Once again, the main focus is on the forested area, referred to as 'Monte del Aguaray Guazu', meaning that the area between the rivers was elevated and thus protected from flooding. Once again, Lima appears on the map as the only Paraguayan settlement in the vicinity during this period.

A system of pathways stands out in comparison to the first map. These winding paths appear to follow the rivers and the edge of the forest, with a few instances of river crossings. Identified as the 'Camino a San Pedro' and connected to Lima, this system of paths suggests the existence of a significant resource for the indigenous population, with whom Förster had only occasional contact – a well-maintained network of pathways between Paraguayan villages. Some of the paths end at the edge of the map, suggesting that they were part of a wider, long-distance network of pathways connecting the countryside and Lima to the wider Paraguayan landscape and other settlements.

This system was intricately linked to topographical features and the natural landscape, avoiding the crossing of forested areas. As the opening quotation shows, the routes were often traversed at night to avoid the subtropical heat. But navigating in the dark required local

Fig. 50 Old pathway in the western part of Nueva Germania, following the river Aguaray-Guazú (Photo: Timon Zirlewagen).

and spatial knowledge of the pathways. Klingbeil, one of the colonists of Nueva Germania (who quickly abandoned his plans to join the colonial enterprise after a few weeks) was navigating the pathways with the help of local guides (see also Klingbeil 1889, 95–97) (fig. 50).

The pathways were connected to the aforementioned *yerbal*. Yerba mate is an evergreen tree native to the subtropical rainforests of South America. Revered for its invigorating properties and its integral role in fostering social bonds, the cultural history of yerba mate transcends geographical boundaries. Its origins can be traced back to the Guarani people, who recognised the magical properties of the leaves of the *Ilex paraguariensis* and collected them from wild, mature trees.

The way in which yerba mate is prepared and shared, often referred to as the 'mate ceremony', remains a pivotal cultural expression to this day. It involves the communal sharing of the infused beverage using a hollowed-out gourd (*mate*) and a metal or wooden straw (*bombilla*). Beyond the act of sipping, the ceremony fosters a collective experience and serves as a conduit for unity and intimacy among participants. In Paraguay, it is usually served ice-cold as '*Tereré*', a practice that has been recognized as an official UNESCO Intangible Cultural Heritage (UNESCO 2020).

Yerba was traditionally harvested from older wild trees. Its agricultural domestication was discovered twice – in the 16th century, when Jesuit monks settled in South America and created *reducciones* (McNaspy 1987), missionary settlements established for the evangelisation and (forced) assimilation of indigenous populations. Here they developed a technique for germinating seeds and domesticating the plant for orchards (Folch 2010, 13–14). With the expulsion of the Jesuit order in 1767, the yerba plantation system continued to force the indigenous population to work on the so-called *yerbales*.

The area marked on the historic map of Förster's first visit shows that he was aware of the yerba plantation in the vicinity of his proposed settlement. At present, there are no other sources that tell us more about the daily life of the people working in the *yerbal* near (and during) the colony, but since the area is connected to the pathways, it is conceivable that workers and crops were transported along these pathways. The traditional means of transporting goods in Paraguay is the use of a *carreta*, a large ox-drawn cart with high wooden wheels that can cross swampy areas (Klingbeil 1889, 26–27).

Phase II – Pathways during the Colonial Settlement

In the subsequent phase of Förster's deliberate settlement project, after negotiations with the Paraguayan government, he secured the colonial territory in return for an initial payment. However, this agreement came with a condition: Förster had to attract at least 140 families to relocate within two years. However, he was unable to fulfil this condition throughout the colony's existence. Instead, he opted to sell plots to interested investors and prospective settler families without actually owning the land.

Seven printed and hand-coloured maps have survived from this phase, providing an insight into the intended layout of the colony (GSA 72/947-27 and -47; 72/1482-3; 72/1508-7, -9, -15 and -17) and thus into the founder's ideas of the colony's spatial and social organisation. In addition, four smaller-scale maps were produced exclusively for the planned village of 'Försterrode'. All of these maps show the intended transformation of the landscape and shed light on how the landscape and its features were assessed and valued by the colonists.

Of particular note is the newly (to be) established network of pathways, which followed strict north-south and east-west orientations. The planned pathway system traversed the wooded area and imposed a grid of rectangular sections that would later be divided into parcels for sale to settler families. Each parcel was given a number, sometimes accompanied by a family name, probably indicating reserved or sold plots. The maps show variations in the parcelling system, with family names both crossed out and replaced by others, reflecting the evolving planning and reorganisation of colonists' spaces.

These maps reflected an approach that valued land primarily for its spatial and mathematical properties, carving up the landscape without regard to natural topography or inherent features for profit. However, other elements played a significant role. Most of the plots had a relatively uniform size of 1.3 ha, but some were considerably larger than the standard parcels. In particular, the founders and their close associates were allocated plots 12.5 times the size of the standard plots, while few others received multiple plots, probably in proportion to the capital they invested in their future homes. This suggests that arriving colonists were not treated equally, suggesting the deliberate establishment of a social hierarchy from the outset.

In addition, the newly designed pathway system was treated differently from the pre-existing pathways, not only reflecting the introduction of a new spatial order, but also illustrating the multiple uses of the same resource, the pathways, within the same landscape. This planned space is in direct contrast to the spaces experienced by the first settlers.

The colonial territory was first surveyed between June and September 1887 by a surveyor called 'H. Ibbecken' from Asunción (GSA 72/949). Later, the land was surveyed by Maximilian Podleck, who was also a settler in Nueva Germania (GSA 72/7962; GSA 100/533; Kraus 1999, 169). However, the challenges of translating the planned space into reality became apparent early on, as noted in one document:

'Moreover, the entire surveying appeared illusory... all the small markings made solely for surveying purposes were quickly overgrown by the lush vegetation, rendering boundaries and markers invisible. What was once divided could only be discerned by the forest itself.'[26] (GSA 72/962; translated by the author).

The rapid growth of the tropical forest was not the only obstacle overlooked by the planners. The colony had built a port at the confluence of two rivers south of the village, the only infrastructure linking the colony to larger towns like San Pedro for the transport of essential materials and food. During the scorching summers, the water would evaporate, leaving land routes as the only option to access the local market. Daniela Kraus argues that this lack of connectivity for buying and selling goods was the main reason for the colony's economic decline (Kraus 1999, 174–179).

The new pathway structure, however, brought with it another set of problems. The distribution of goods was centralised in the village, a local store controlled by Förster and colonist Georg Freitag sold goods to the colonists, forcing families to travel long hours from their plots to the town or even to visit their neighbours. The Neumann family's property, for instance, was more than 8 km from the centre of Försterrode within the colony. A settler wrote home in 1889 describing the situation: 'My plot is a good two hours' walk from the so-called town, and in doing so, you have to wade through three lagoons up to your hips in water'[27] (cited in Kraus 1999, 209; translated by the author). These considerable distances meant that the first settler families lived in relative isolation from each other.

Bernhard Förster was well aware that most of the land for sale was deep in the jungle and that the colonists would have to undertake the arduous task of clearing and cultivating the land, as well as building their own houses. In several writings, Förster and Nietzsche acknowledged the necessary efforts and challenges that settler families would face (Förster-Nietzsche 1891, 111–122). The daily challenges of dealing with wildlife, such as big cats hunting cattle and parrots devouring crops, and unknown diseases caused by parasites, provide a glimpse of the hardships these families endured. One farmer interviewed, who has lived on a plot of land 7 km from the town for three generations, recalled how his grandfather had spoken of this relationship with the founders – the evocative phrase 'they brought us here to die! [sie haben uns hier zum Sterben hergebracht!]' was still fresh in his memory and underlines the hardship experienced by the families.

This experience, more dystopian than utopian, reflects the settler families' perception of the landscape. It was the result of poor planning and a disregard for the inherent conditions of the landscape. Another interpretation of these living conditions can be seen by taking the

26 In the original: 'Noch dazu ließ das ganze Messen als illusorisch [...], nach alle die kleinen Grenzungen, welche nur der Vermessung wegen gemacht wurden, durch die üppige Vegetation, in kürzester Zeit überwuchert wurden, so dass von Grenzen und Grenzsteinen nichts mehr zu sehen war. Was einmal eingeteilt gewesen ist, könne dem nach der Forst sehen.'

27 In the original: 'Mein Grundstück ist von der sogenannten Stadt [...] zwei gute Stunden Weges entfernt und dabei muß man durch 3 Lagunen bis an die Hüfte durchs Wasser gehen.'

planners' ideology seriously. In Förster's worldview, only strong Germans who could withstand the harshest conditions should form the basis of his new Germanic utopia, rooted in his belief in social Darwinism (Mehler/Dézsi 2024, 224).

At the time of his first arrival, Förster noted the consumption of yerba mate, and even in the map mentioned above he documented the existence of a *yerbal* to the east of the area of Nueva Germania. However, he did not write any further about the economic potential of yerba in his prospects, and the new maps no longer mention the *yerbal*. The settler families themselves tried to cultivate fruits brought from Europe, such as apples, potatoes and wheat, which could not grow in the tropical climate. Traditional crops of the region, such as Maniok, must have been unknown to the arriving families from Saxony. Leaving behind the habits of European diets and adapting to the local subsistence economy was an important change that made it possible to live independently of the local colony store.

Phase III – Pathways after the Initial Colonial Settlement

Following the departure of the founders, the remaining families attempted to reorganise themselves as a society. Unfortunately, there are few sources that can tell us about this phase of the settlement. However, an examination of the current layout of the area and its pathways reveals a new spatial organisation, partly based on the colonial layout. A motorway built at the end of the 20th century connects the eastern town of Santa Rosa with San Pedro, passing through Nueva Germania and partially dissecting the old historic centre. The high volume of lorries and agricultural machinery on the road underlines its importance in the current regional economic infrastructure.

Fig. 51 Processed *Ilex paraguariensis* leaves and stems from the area of Nueva Germania, at Yerba Fortin in San Pedro (Photo: Attila Dézsi).

Interestingly, a local story about a yerba mate processing and distribution company in San Pedro claims that the founder of the company rediscovered the germination process needed to cultivate the plant in Nueva Germania, which is supported by a report of Nueva Germania's increasing yerba mate exports in 1901–1902 (Fischer-Treuenfeld 1903, 48–49) (fig. 51). Indeed, one of Förster's closest associates when he took over the administration of the colony, Oscar Erck, became one of the leading industrialists of yerba mate production in Paraguay at the end of the 19th century and was also operating yerba mate plantations in Nueva Germania in 1900 (Kraus 1999, 192–193).

Several companies in both Argentina and Paraguay claim to have been the first to rediscover the technique, and by the early 20th century in Argentina yerba was produced in an industrial scale. Despite this economic success in South America, yerba mate did not penetrate the European market (Folch

2010, 16–17). Farmers interviewed during fieldwork in Paraguay in 2022 stated that cultivated mate was a major commodity and that tending and harvesting yerba was an important experience on their family farm. Public murals and art in the town of Nueva Germania serve as reminders of the historical importance and heritage of yerba mate production in the region.

After Phase II, families began to replant yerba mate, and for some this venture proved economically successful. Today, however, soybeans have become more economically viable. The new and improved motorway connects the region to the wider world market.

Smaller roads connect the town with satellite villages called Nordrand, Westrand and Südrand. These villages appear to be remnants of a decentralisation of social organisation following the abandonment of the planned centralised layout of the former colony – these newly formed satellite villages were not included in the founders' plans and had once established their own schools. Aerial photographs also show the continued existence of long rectangular plots and some of the original pathway systems.

Some of the old pathways have been abandoned in favour of modern tarmac roads and the motorway, but the grid of the colonial layout still connects households and families, linked more closely by cars and motorbikes than by horses and *carretas*. But all three pathway systems still exist in the landscape, creating a palimpsest of the remains of different connectivity needs and social practices.

Conclusion: ResourceCultures Pathways and Yerba Mate

In a short period of time, different pathway systems were established in the landscape of the area to meet the needs of the local population, each of which was valued differently. Prior to the arrival of the colonists to establish their intentional settlement, the pathway system appears to have been used and maintained by the indigenous population to link settlements with each other, while maintaining a deliberate distance from forested areas and their inherent dangers. The design of the resource pathways was adapted to the natural and topographical characteristics of the landscape. With the arrival of the colonists, a rigid grid system was introduced and cut into the forest. This new concept of pathways was not only introduced into the landscape to calculate and unequally distribute the value of the land with plots, but the pathways system also became a resource to project power, hierarchy and perhaps even a deliberate isolation of families. After the end of the political project, new pathway systems developed on top of the old ones, adapting to the need to decentralise social life and later to open up economic opportunities to the capitalist market with the long-distance motorway system. Like a palimpsest, the old systems remained in the landscape and were visible, but newer layers were created and valued differently.

'Landscape increasingly began to be seen as engaged socially and culturally as much as it is engaged environmentally, and it is this engagement that defines the lie of the land, what a landscape looks like. Landscapes are topographies of the social and the cultural as much as they are physical contours. To understand a landscape one has to outline its means of engagement, the way it is understood, codified, and lived in social practice; and each of these, along with the landscape itself, have history. Engagement gives and is defined by the way we give cultural meaning to the location of our existence – so that even the trees and the rocks mean different things to different people.' (David/Thomas 2008, 35–36).

The pathways, indigenous, colonial and modern, have been and are important resources for each period of engagement with the landscape, but are intertwined with different ways of perceiving and using the landscape for their needs and in relation to the characteristics of the area. As these ResourceCultures do not necessarily replace each other and continue to coexist, although of course with a different meaning than in the intended concept, it might be possible to understand this temporal sequence of ResourceCultures itself from the perspective of ResourceAssemblages. As the use and valorisation of the resource pathways and associated social practices change in the different phases from enabling connectivity with local knowledge of travelling on horseback, with guides and beyond daylight (I), to creating control, power and isolation with a colonial grid system being pushed back by the landscape itself (II), to a decentralised use of land connected to the world market (III), we see a rapid change within a short period of time.

From the perspective of ResourceAssemblages, different dynamics of development can be drawn in relation to yerba mate within the study area. *Yerbales* show that the extraction of yerba leaves was already established in the region before the colony was initiated (I), the colonists and families tried to start agriculture, but did not implement yerba mate as a crop from the beginning, as it is not mentioned as a possible prospect in Förster's text and probably did not fit in with his ideology (II). But after the colony, some entrepreneurs started or invented the cultivation of yerba and became successful, for other families it became an important crop and important heritage of the region (III). The opening up and adaptation to yerba mate can be seen as an important shift from a colony based on an exclusionary ideology to one that is at home in the Paraguayan landscape and society. However, it should be noted that yerba has been harvested in the area from the beginning, which is linked to a longer tradition of yerba mate consumption.

In this case study, historical sources are available and can show different synchronous perspectives on resources. For the phase during the colonial period, they showed how differently people related to the existing and newly built pathways – for some of the arriving colonists they became an opportunity to materialise ideology and create an intentional social order, for others they caused damage and isolation. All these different relationships to the pathways as a whole can be interpreted as ResourceCultures, changing over time within a short period of just 150 years.

The experience of the people living in the area was inextricably linked to the dynamic interplay of these resources. The relationship with these resources changed significantly over time and their value was perceived differently by different groups within the historical context. In analysing this historical site, it is essential to consider the perspectives of different stakeholders, including the indigenous population, the colony's leaders and planners and the German families who migrated. By examining these perspectives dialectically, we can gain a more nuanced understanding of how different communities validated and used these resources over time.

In summary, the use and perception of resources in the area underwent profound changes during the three different periods of the 19th and 20th centuries. This multi-dimensional approach is essential for a comprehensive analysis and facilitates discussion with other similar studies. It highlights the transformative nature of landscapes, resources and the adaptive capacity of communities in response to changing circumstances.

6.7 Gone with the Wind? Why the Palatines of Tübingen Have Not Entered Cultural Memory

Christian Kübler

Introduction

At least since Aleida and Jan Assmann's research on collective and cultural memory in the 1980s (Assmann 1988; 1992), this topic has aroused increasing interest in historical scholarship. This has been particularly true of those scholars who have focused on the nobility of the pre-modern period in all its various facets. This seemed only logical, since Maurice Halbwachs himself had already cited the nobility as a prime example of the existence of a specific group memory:

'Two nobles who meet without ever having seen each other must be able, after exchanging a few words, to recognise each other as two members of the same extended family who have rediscovered their relationship of kinship or alliance. This presupposes the perpetuation within the noble class, down through the generations, of a whole set of well-linked traditions and memories. As nothing similar was found in other groups, it has to be said that the noble class was for a long time the medium of collective memory.' (Halbwachs 1925, 167; translated from French by the author).

In the years that followed, the consensus among historians was that the aristocracy's self-image was largely based on an awareness of its own (family) past, primarily through the cultivation of its *memoria*.[28] In 1986, the historian Otto Gerhard Oexle proposed a concise definition of the noble families of the High Middle Ages that is still widely recognised:

'"Lineage", in the language of the sources: genus, stirps, origio, genealogia or the like, is the community of descent of the family belonging to the noble house (domus).' (Oexle 1986, 47; translated from German by the author).

He distinguishes between the *domus*, i.e. the house of the noble family, as the historically present manifestation of the noble family, and the noble lineage, i.e. the memory of one's own ancestors. The noble lineage is thus a mental construct that is decisively shaped by the noble families themselves (Hechberger 2007, 47).[29] In 1995, Oexle concretised his thesis:

'Memoria, memory is the decisive moment that constitutes "nobility". [...] Without memoria there is no "nobility" and therefore no legitimisation for aristocratic rule. This is why the cultural production of commemorative rituals, texts, images and monuments, which constitute and represent the "culture" of the group, is particularly diverse in the noble "houses" and "lineages".' (Oexle 1995, 37; translated from German by the author).

But to what extent is this paradigm of a close connection between the culture of remembrance and the noble self-image to be followed (Krieb 2001, 58)? The *opinio communis* of the historians' guild developed primarily around a few important noble families of the High

28 In medieval studies, the term *memoria* refers to the ritual commemoration of the dead by (aristocratic) families in the Middle Ages. This was mainly done in the form of funeral masses and religious foundations.

29 Hechberger rightly points out that such a distinction cannot be based directly on the source terminology, but that it is certainly justified in the theoretical context, which must always be taken into account, due to the now long-standing research tradition on this topic.

Middle Ages – the Welfs and the Staufers. But are they representative of the entire medieval noble world? Some scholars, notably Spieß, have rightly expressed doubts. In his pioneering research on the high noble families of the late Middle Ages, Spieß was unable to establish a distinct culture of *memoria*; this was rather a product of humanistic scholarship (Spieß 1993, 490–492). In the following, I will ask what sources support the assumption that a collective memory and a distinct culture of remembrance existed among the medieval nobility, and to what extent such sources are known for the Palatines of Tübingen. Finally, I will discuss the extent to which this noble family has – or has not – found its way into the cultural memory of posterity.

First of all, however, it must be pointed out that the concept of collective memory as developed by Jan Assmann does not only refer to written sources, but includes a much wider range of different source genres. Rituals, portraits and monuments should be mentioned first and foremost.[30] Of course, it is impossible to deal with this vast field of the most diverse testimonies to the culture of remembrance of the high medieval nobility here. We will therefore concentrate on those sources in which genealogical information is most likely to be found. These offer the best clues to the self-image of the nobility and their ideas about their own family lineage, as well as a possible collective memory.

Sources on *Memoria* of the Nobility in the Middle Ages

What are the sources of a distinct culture of remembrance of the nobility that have come down to us? First and foremost, there is the noble family tradition. The famous 'Welfenchronik' ('Guelph Chronicle'; König 1938, 2–74), written around 1170, is almost always cited as a key witness, as it is the oldest known noble family chronicle in the Holy Roman Empire. However, it is not possible here to go into the complicated history of its creation, its authenticity and the background to its writing. Only one point, which is relevant to the question posed here, should be briefly mentioned. If one agrees with Oexle's considerations on noble consciousness and *memoria*, this necessarily presupposes for noble house traditions that the respective noble family at the time of the writing of the house tradition actually understood itself as a noble lineage, i.e. as a common descent community (Graf 1993, 128). That this should not be the case may seem surprising at first, but is indeed so regarding the Welfs. In the 12th century there were several 'Guelph houses' and they certainly did not have an identical self-image. Henry the Lion and 'his *domus*' referred to different ancestors than his uncle Welf VI. Henry's son Otto IV, who was the only member of the 'Guelph family' to become emperor, was not referred to as a 'Guelph' in contemporary sources (Hechberger 2007, 55). Thus, if one asks about the noble self-image and the noble lineage associated with it, one must also ask about the corresponding 'knowledge of memory'. What could the respective *domus* remember at all? What did they want to remember and what did they deliberately sweep under the carpet (Schneidmüller 2000, 17–40, 288–300)? This shows that even with supposedly certain, objectively recorded data, the author's intention must always be questioned. The high medieval 'House of Guelph', as it is generally known today, is a product of the 19th and 20th centuries (Hechberger 2007, 48). A further complicating factor

30 '[…] the stock of recurring texts, images and rites peculiar to every society and every epoch, in the cultivation of which it stabilises and conveys its self-image, a collectively shared knowledge, preferably (but not exclusively) about the past, on which a group bases its awareness of unity and uniqueness' (Assmann 1988, 15; translated from German by the author).

is that the Guelph Chronicle, as a noble family tradition for this period, is an almost singular phenomenon and cannot therefore be taken as representative of the entire nobility of the High Middle Ages.[31] Noble family tradition for the High Middle Ages is therefore a difficult matter. But even for the late Middle Ages, the written record is not much better. Although genealogical testimonies, such as the family book of the Lords of Eptingen (Andrina 1992) or the biography of the knight Georg von Ehingen[32] (Ehrmann 1979), have come down to us more frequently, such records are still rather rare until well into the 16th century and can hardly be taken as evidence of a 'total social phenomenon' (Oexle 1995, 39; translated from German by the author). Interestingly, in these late medieval records one repeatedly finds that a reasonably secure and detailed knowledge of one's own family's ancestors is limited to three, at most four generations, i.e. exactly the 80 to 100 years that Jan Assmann has also worked out for oral societies (Assmann 1992, 48–50).

Closely related to these late medieval family books are the tournament books, which have become more common since the 15th century (Krieb 2001, 69–74). These were often used to provide the often required proof of nobility for those wishing to take part in a tournament (Vok 1997, 2035). The coats of arms used and recorded there draw attention to one of the most important contemporary source genres, which like no other represents the collective memory of individual noble families, but also of the entire nobility as a specific group. From the late 11th century onwards, coats of arms and seals as their predecessors served individual families and groups (e.g. knightly confederations) as individual symbols of recognition (Scheibelreiter 2006a, 12–15).[33] These were displayed on a shield according to certain principles and consisted of personally compiled characters and images. From the 13th century onwards, the shield as the original carrier of the coat of arms was extended by the helmet and its ornamentation to a full coat of arms (Scheibelreiter 2006b, 9). A coat of arms can be compared to a modern identity card. Wherever the coat of arms of a nobleman or his family was affixed, it served to identify persons or to indicate (property) legal claims (Eckart 2012, 354).[34]

Finally, liturgical *memoria* should be mentioned. Many noble families tried to establish a common burial ground for their families. Those who could afford it even endowed their own monastery. Family tombs and monasteries were places where especially holy people (monks, nuns, canons) prayed for the salvation of the family and their relatives. Medievalists have assumed that the high medieval nobility in particular had a special urge to found monasteries for themselves and their families, and have called these monasteries 'house monasteries'. These are defined by 1) the founding of a particular noble family or at least a family that had a large share in the founding process, 2) the existence of a founder's grave in the monastery church or on the monastery grounds, which could be extended to a family burial ground, 3) that the monastery communities prayed for the founder's family and celebrated masses and 4) that the family of the monastery founders secured the bailiwick over the institutions and kept it in the family (Störmer 1980, 148). In addition, when monasteries were

31 Two further source texts of that period, which can also be attributed to this source genre, come from Gislebert of Mons: Monumenta 1868, 481–601 and Lambert von Ardres, see Monumenta 1872, 550–642.

32 Ehingen is now a district of Rottenburg am Neckar.

33 Even in 1958, Ahasver von Brandt described coats of arms, or heraldry, as '[...] Study of a characteristic medieval symbolic form of social and genealogical groups [...]' (von Brandt 1958, 23; translated from German by the author).

34 This is why coats of arms can be found not only in historical written sources, but also on all sorts of historical remains such as churches, epitaphs, coins, city gates, stove tiles, carpets and all sorts of portraits.

founded it was not uncommon for a separate monastery chronicle to be written, in which the founder and his family were also mentioned (Patze 2002, 109–249). However, research into the 'house monasteries' of the high medieval nobility has become more cautious. Rarely do all four of the above criteria apply to the founding of a monastery, and the development of individual institutions was often marked by disruptions (Dendorfer 2014, 17–38).

Sources on *Memoria*, Focusing on the Palatines of Tübingen

To what extent can these different genres of historical sources (house lore, coats of arms, house monastery) provide information about the collective memory of the Palatines of Tübingen? The ancestors of the Palatines of Tübingen probably came from the northern Black Forest. Towards the end of the 11th century, they achieved an enormous social rise, which brought them great (territorial) influence between the Black Forest in the west, Stuttgart in the north, Ulm in the east and the Swabian Alb in the south. They also founded numerous monasteries. However, the family's rise was matched by a rapid decline towards the end of the 13th century. Due to the relatively short period of about 120 years, a period for which there are still very few sources, only a few testimonies of the family have survived.[35]

As already mentioned, family chronicles of high medieval noble families are the exception rather than the rule. It is therefore not surprising that such a chronicle for the Palatines of Tübingen has not survived and probably never existed. However, the coat of arms of the Palatines of Tübingen is known and is one of the oldest known German coats of arms (Liesching 1984, 1–5). It shows a vertical three-pointed flag, a so-called *gonfanon*, with three rings. Nothing is known about the original tincture, as the oldest images were colourless. It is generally believed that the original image consisted of a red flag on a golden background (Liesching 1989, 69). The first known illustration dates from 1181. It is a so-called equestrian seal. It belonged to the Palatine Hugo II of Tübingen, who had it affixed to a document to prove a legal transaction (WUB 2, no. 423). Since a coat of arms served as a sign of recognition, the Tübingen *gonfanon* was not only attached to deeds as a seal, but was certainly widely used in the sphere of influence of the Palatines of Tübingen. This coat of arms was painted or affixed to many places and objects in order to demonstrate their dominion and splendour. However, it is also true that when the influence of this once powerful family collapsed and their lordship and possessions were taken over by other noble families, the coat of arms was also replaced by that of the new lordships. As a result, it has almost disappeared from public view, and with it the memory of the Palatines of Tübingen. A rare example still adorns a tomb plate in the monastery of Bebenhausen. Unfortunately, it does not bear an inscription and there are no other sources to indicate who was buried there (Schiek et al. 1989, 28–30).

Bebenhausen brings us to the question of a possible 'house monastery' of the Palatine counts of Tübingen. The monastery was founded around 1185 by Count Palatine Rudolf I in the Schönbuch forest a few kilometres north of Tübingen and was initially assigned to the Premonstratensian Order, but for unknown reasons was entrusted to the Cistercians only a few years later (Setzler 2003, 184). If one now tries to apply the above criteria for the definition of a 'house monastery' to Bebenhausen, disillusionment quickly sets in. The monastery

35 The history of the Palatines of Tübingen has not yet been adequately researched. Sönke Lorenz has presented several detailed studies, but has not been able to produce an overall assessment. It is therefore necessary to refer to the work of Ludwig Schmid (1853), which also compiled the most important sources. A concise overview is provided by Lorenz (1999, 11–32).

can be regarded as a foundation of the Palatines. This is certain. But even its function as a family burial ground calls for caution. As far as we know from historical records, only the founder of the monastery, Count Palatine Rudolf I, with his wife Mechthild and two children and one of his grandsons, Count Rudolf III, with his wife Adelheid and their daughter of the same name, were buried there (Sydow 1984, 35). However, as the Cistercian Order's rules did not allow founders or other secular persons to be buried in the monastery grounds, the Bebenhausen abbot received a severe rebuke from the General Chapter (Schiek et al. 1989, 19–23). Rudolf I deliberately renounced the bailiwick of the monastery (WUB 2, no. 466) – as was customary among the Cistercians – and instead the protective function was assumed by the empire (WUB 2, no. 482). Even the most important point for this study, the question of the monastery's prayers for the founder's family, can only be answered in the negative. This does not necessarily mean that there was no *memoria* for the Palatines of Tübingen in the monastery, on the contrary.[36] However, this no longer exists today and apparently was not important enough for the family to have it recorded in writing.

Outsourcing of *Memoria*

This conclusion touches on a very important point that is now increasingly being discussed in medieval studies. Presumably, for many noble families in the High Middle Ages, knowledge of their ancestors and thus of their own noble family was not as important as Otto Gerhard Oexle postulated and presented at the beginning. Bernhard Jussen asked what role kinship played in the organisation of the care of the dead in the Middle Ages, especially with regard to spiritual *memoria*. Whereas in ancient Rome this task had been the sole responsibility of the *pater familias*, who was responsible for the entire *memoria* of the ancestors in the Roman cult of the ancestors, Christianity de-familialised the care of the dead and thus separated it from kinship. Instead, specialised institutions (monasteries, convents, parishes) took over this task (Jussen 2009, 314–316). Jussen concludes 'that Christianity is a religion of memory, but that it does not reward lineage' (Jussen 2009, 315; translated from German by the author). It is no longer surprising, therefore, that knowledge of kinship and genealogy among the medieval nobility was for the most part less pronounced than had long been assumed.

Scattered Evidence in the Sources for a Communicative Memory among the Palatines of Tübingen

This brief overview illustrates how difficult it is to interpret the scant source tradition on the Palatine counts of Tübingen. There is no known family chronicle, nor is there a founder's chronicle of the supposed 'house monastery' of Bebenhausen, which, strictly speaking, did not serve as such. However, it is also worth asking to what extent a house chronicle and a founder's chronicle[37] were regarded as an indispensable prerequisite for proving a family

36 Martin Crusius, professor of Greek and Latin at the University of Tübingen, reported in the third volume of his historical work 'Annales Suevici' in 1596 that some 50 years after the presumed last burial of a member of the Tübingen family, prayers were still being said for the family members in the monastery, but it was no longer known exactly who was buried in which grave and where exactly they were located in the monastery (Crusius 1596, part 3 chapter 8).

37 At least one of the oldest sources of the monastery, the so-called 'Bebenhäuser Annalen', does not contain a history of the founder's family. The 'Bebenhausen Annalen' were written at the beginning of the 16th century by the Bebenhausen monk Ulrich von Baldeck (+ 1518), but he seems to have used original records that can no longer be found today (Zagolla 2002, 1).

culture of remembrance in a society that was still largely oral. What remains are the seals and coats of arms, and it is here that the communicative memory of the Palatines of Tübingen is on display in the truest sense of the word. By means of this self-chosen symbol, the Palatines of Tübingen communicated both internally and externally their very own conception of their own family. Even at a time when the Palatines had lost their power and had sunk into insignificance, their coat of arms was still listed in the armorial and tournament books of the late Middle Ages. At this point I only need to mention the famous Zurich coat of arms roll of around 1335/1345 (Zürcher Wappenrolle, 2760)[38], the 15th century coat of arms book of the Abbot of St. Gallen Ulrich Rösch (Codex Sangallensis, 1084, 209)[39] or Siebenmacher's (Apphun 1994, 34) famous coat of arms book of 1605. And of course they were not alone. Almost all the other major noble families of the High Middle Ages used this particular medium, so it is reasonable to assume a collective memory of the nobility as a specific group, as well as of their individual families as 'noble houses'.

Even if it is difficult to prove that the Palatines of Tübingen had a generally widespread collective or even historical memory and identity, there is at least one unique, extraordinary piece of evidence, and it goes back once again to the Palatine Rudolf I, the founder of Bebenhausen Monastery. It is a fragment of an inscription stone from the ruins of Königswart (castle) in the Black Forest near Baiersbronn. The castle was built in 1209 by Palatine Rudolf I of Tübingen. During the excavation of the ruins in 1974, this part of the inscription stone was uncovered, which reads as follows:

'Rudolf, Palatine of Tübingen, had this house built in 1209, the year of Christ's incarnation, so that all who hunt here may remember him and pray for the salvation of his soul.' (Wein 1979, 90; translated from German by the author).[40]

This inscription, which has only been partially preserved, shows that at least the Palatine Rudolf I wanted to commemorate his *memoria* in a very individual way (hunting and remembrance). Unfortunately, almost nothing is known about the inscription and the ruins of the Königswart. Why Rudolf had it built and how it was used remains a mystery.

Reasons Why the Palatines of Tübingen Have Not Entered Cultural Memory

In the end, one can speak of a collective memory of the Palatines of Tübingen. Why, then, did they not succeed in entering the cultural memory of future generations, which I would like to call, in a somewhat simplified way, the public or general historical memory of the respective contemporary environment? The answer has probably already been given. It must be assumed that the desire to preserve the family in the memory of future generations was simply not considered particularly necessary at the time. The liturgical *memoria* in the various ecclesiastical institutions were first and foremost for the salvation of one's own soul, usually that of the closest family. Nothing was more important to a medieval man than the fear that

38 The coats of arms of the Tübingeners are on page f. 2r. Five of the six coats of arms depicted show the characteristic *gonfanon*, each in its own individual form. These are the Tübingen lines of Werdenberg, Feldkirch, Tettnang, Chur and Asperg. Only the Bregenz line had adopted a different coat of arms motif.

39 The Tübingen coat of arms is on page 209.

40 The original inscription reads: [+ RUDOLPH]US . P(ALATINUS) . C(OMES) . DE TUWING[EN .] D(OMUM) . I(STAM) . P(ROCURAVIT) [.] F(IERI) . ANNO [INCARNATIONIS (CHRISTI) M] CC . IX . VT . OMNES . HIC . VENATURI SU[I . S]INT ME[MORES ET SALU]T(EM) . ANIME EIVS . INPRE-CENTVR +, quoted after Lorenz 1999.

he would not enter heaven after his death. So it may seem plausible that worrying about whether one's family would be preserved in the minds of (secular) future generations would not have been of much interest. But if it seems that the family itself had no interest in preserving its own history for posterity, it is still possible to wonder why the rest of society did. It is not the case that the Palatines of Tübingen were completely forgotten. Otherwise, as shown above, Siebenmacher would not have included the Tübingen coat of arms in his coat of arms book in 1605. Early modern historians also wrote about the history of the Palatines. Only the chronicle of Sebastian Küng from 1554 (Sommer 1971) should be mentioned here. However, heraldry and historiography were, and still are, subjects for specialists and not for the general public. The Palatines did not take root in the public sphere and thus in the public memory.

A major factor in this was undoubtedly the dramatic decline of the Palatine dynasty in Tübingen from the end of the 13th century. The power vacuum they left behind was quickly filled by other noble families. The counts and later dukes or kings of Württemberg, who dominated the German south-west from the 14th century until the early 20th century, are particularly worthy of mention, as are the Margraves of Baden and the Hohenzollerns. These three families not only managed to establish themselves as major players in the German south-west (and to some extent beyond) during the Middle Ages, but also to continue their dynasties into the present day.[41] A glance at the names of the individual states in South-west Germany after the defeat of National Socialist Germany and the founding of the Federal Republic of Germany makes this clear. In 1945/46 the military governments created the states of Württemberg-Baden in the American zone and Württemberg-Hohenzollern and Baden in the French zone. Württemberg, Baden, Hohenzollern, not Tübingen! In July 1948, the three states were ordered by the occupying powers to reorganise themselves into a new south-western state. At the end of this difficult and controversial process, the new state of Baden-Württemberg was founded on 25 April 1952. Again, there is no reference to the once powerful noble family of Tübingen (Wehling 2006, 34).

There is another noble family which, although it also died out very early, was of such importance for regional, national and ultimately European history that it is firmly anchored in the cultural memory of our time. This is the dynasty of the Staufers. In the 19th century, the idea of the nation increasingly entered the consciousness of the state, of the public and thus also of its own historiography. The period of Staufer rule could be understood as an example of a united German empire, even a universal empire under German rule (Luhmann 2021, 62). Under the Staufer dynasty, court culture, *minnesong*, chivalry and castle building flourished – in other words, everything that was understood, or wanted to be understood, as 'typically German'. Even after the defeat of National Socialist Germany, the Staufers were able to serve as a welcome projection surface. But since it was no longer possible or desirable to look at the Staufer past from a national perspective, the focus shifted to a local historical level. Since the Hohenstaufen were a 'Swabian noble family', this reflex was very strong in the German south-west. However, this new local cultivation of the Hohenstaufen tradition served primarily to (re)establish an identity as well as a certain escapism from the present, which at the time was often perceived as a crisis, by preserving those aspects of the Staufer rule that were considered non-political and therefore not suspect, such as castle building, chivalry, etc. (Luhmann 2021, 74). Because of its cultural and historical significance, the

41 For the individual noble families, see the respective contributions at Schwarmaier/Schaab 1995, 1–163 (Württemberg), 164–246 (Baden), 360–378 (Hohenzollern).

Hohenstaufen period continues to serve as a reservoir for a stereotypical image of the Middle Ages, from which modern popular culture is all too happy to draw, be it in the form of historical novels, music, films, etc. (Graf 2010, 305–306). As a result, the Staufers were in direct and, at the same time, outsized competition with the Palatines of Tübingen on a regional level in terms of local historical tradition. Under these circumstances, it is quite understandable that the counts of Tübingen have disappeared from public memory.

Conclusion – Gaining Knowledge with the Help of the Resource Concept

The Palatines of Tübingen can certainly not be regarded as 'losers of history', but the step from their own collective memory to the superior, diachronic cultural memory has nevertheless been denied them.

This case study has clearly shown how difficult it is for historians to localise the family of the Palatines of Tübingen in terms of their *memoria* as an individual identity. This problem can easily be extended to other aspects of this noble family. Their significance, rank and influence stand in stark contrast to the poor tradition of historical sources. For this reason, the resource concept offers a welcome opportunity to significantly expand the scope for interpretation, and thus the knowledge gained from this small number of sources, through consistent application. In this sense, ResourceComplexes describe specific contexts of things, environments, actors, monuments and landscapes, as well as knowledge, techniques, infrastructures, practices and orders (see Chapter 2).

Using the ResourceComplexes approach, it is possible to observe how the Palatines of Tübingen managed to create a resource network through a skilful combination of monasteries and castles (which could not be included in this case study) in order to use it to significantly expand their rule. The material and immaterial components observed under the ResourceComplex perspective, and their interrelationships and dynamics, can be understood in a larger context as ResourceCultures, in the example discussed here as a specific AristocraticCulture. This has manifested itself in various ways, such as the tournament books discussed above, coats of arms, seals, burial vaults, or even buildings such as the Königswart. These tangible and intangible resources are clear indications of an aristocratic cultural memory among the Palatines of Tübingen.

6.8 Karbalāʾ Bārānī – An Ancestor as (Re-)Source of a Collectively Shared Identity in the Mountains of Kermān, Iran

Wulf Frauen

Introduction

Pierre Nora once described the *lieux de mémoire* ('sites of memory') as shells on the shore (Nora 1989, 12). To stay with his image, they are moments of history torn from the flow of time as the sea of living memory leaves them in the sand of the collective consciousness of a social group. Although this poetic image may be criticised by scholars who believe that things can only be scientifically accurate if they are presented as lifelessly as possible, it has always seemed to me that no better description of the *lieux de mémoire* has ever been given. It is precisely this 'standing outside of time' that makes these shells worth collecting. They may refer to the past, but they are crucial to understanding the present of a social group. In the present case study, I examine one of these shells that I came across almost by accident in a remote mountainous area of Iran.

To the west of the modern town of Jiroft in the Kerman Province lies a highland plain usually referred to as Dašt-e Esfandaqeh. At its far western end is a small village called Bāġ-e Borǧ.[42] The reader may already be familiar with the village, as it was the location of another case study in which animal husbandry (*Dām-Dārī*) was described as a ResourceCulture in this specific context (see Chapter 6.1). Nevertheless, this case study is also self-contained and discusses a crucial aspect of this ResourceCulture in more detail. The villagers belong to one lineage called Eskandery and see themselves as a *ṭāyefa*, a social unit based on kinship. At the core of their understanding of being a group is a belief in a common ancestry. This descent is expressed in the person of Karbalāʾ Bārānī, a mythical ancestor who arrived in the area some 350 to 400 years ago. The remembrance of Karbalāʾ Bārānī as a common forefather carries superempirical[43] traits and is highly constitutive of the group's collective identity. When I first came to the village to conduct fieldwork in the summer of 2018, I wondered how to make sense of this strange obsession with a person who died a long time ago and yet is still very much alive in the village.

The following discussion attempts to make sense of this phenomenon by understanding Karbalāʾ Bārānī as a site of memory in the sense of Nora (1989). In doing so, theories of collective memory and identity will be discussed through this particular case study. However, as will become clear in the course of the discussion, established theories of cultural memory and identity are not sufficient to fully comprehend the significance of Karbalāʾ Bārānī in its specific context. Therefore, I intend to expand Nora's concept of sites of memory by additionally describing Karbalāʾ Bārānī as a social resource, analysed through the heuristic tool of ResourceComplexes, that serves to establish social cohesion. From this perspective, Karbalāʾ Bārānī can be seen as a resource that was, and in a sense still is, crucial to the survival of the group. The contribution thus goes beyond the framework of a mere case study by showing that the dynamic understanding of resources has the potential to improve established theories of collective memory and identity and to lead to a more comprehensive picture. This theoretical background is briefly described in the following section.

Theoretical Background

The present case study rests on two main pillars in terms of theoretical framework. One is Nora's concept of sites of memory (1), the other is the ResourceComplexes perspective (2). The following section will briefly describe them, starting with the former.

1) Nora's concept of *lieux de mémoire* ('sites of memory'), like all theories of collective memory in the humanities, is based on the double helix of cultural memory studies established by Maurice Halbwachs ([1925] 1966; [1939] 1967). This double helix is Halbwachs's basic insight that memory is not, as one might intuitively assume, individual and located in individual persons, but collective and located in communities. Of course, Halbwachs did not

42 All names of places, persons and events are transcribed according to the transliteration guidelines of the DMG the first time they are mentioned and subsequently referred to in a Latinised form for better readability. Exceptions to this rule are specific terms for which no adequate Latinised form seems possible without causing potential misunderstanding.

43 I use the term 'superempirical' to avoid the problematic term 'supernatural', following Beckmann et al. (2017).

say anything other than that individual persons are the basic bearers of memory, but what he understood was that the way something is remembered is inevitably shaped by the socio-cultural frameworks of the communities in which these persons live. The fact that today we tend to take this insight for granted, and are tempted to add that every perception is already shaped by these frames, only shows how much Halbwachs's idea has become a paradigm.

One of the most famous and widely discussed intellectual offspring of Halbwachs's concept is Nora's idea of *Les lieux de mémoire*, the sites of memory (Nora 1989). According to Nora, in 'modern' societies there is no longer a collectively shared memory capable of providing orientation and belonging to all members of a community, although he seems to believe that such a thing used to exist. He concludes that '[w]e speak so much of memory because there is so little of it left' (Nora 1989, 7). The collective memory has been replaced by history, which, according to Nora, has a very different character. He points out that '[m]emory is life, borne by living societies' and '[...] remains in permanent evolution [...]' (Nora 1989, 8). History, on the other hand, '[...] is the reconstruction [...] of what is no longer' (Nora 1989, 8). Nora must be criticised for seeming to ignore that even history in the sense of historiography (*Geschichtsschreibung*) is always only a product of its temporal and spatial context (Graves/Rechniewski 2010, 3). Nora's texts sometimes give the impression that history could be 'the past, as it really happened', which is not the case. Nevertheless, his distinction is useful. History and historiography in modern states that follow European/North American models are indeed, as Nora noted, concerned with 'a representation of the past' (Nora 1989, 8). They are reflexive in nature, and although this representation should not be confused with 'truth', they give the impression of being based on objective facts that are taken as truth within certain discursive formations.[44] As such, they fall within the realm of what Michel Foucault subsumed as 'Governmentality' (Foucault 2005) and serve to establish and legitimise power relations. Memory, however, serves a different purpose. It is affective, highly flexible and tells us who we are and what our place is in the universe. In other words, it is crucial to our identity. That is why we cannot let go of it, even though we have a historiography that also tells us about the past. But the past is not what memory is really about. Unlike history, memory is about the present. There are human groups without history, but there never has been and never will be a human group without memory.[45] And so we cling to it, even though historiography tries to occupy its space.

Nora has also been criticised in other respects, notably for his strong nationalist focus, which does not take into account conflicting and divergent memories within a society (Neumann 2003, 56). In this respect, his work even seems to be a step backwards compared to Halbwachs, who more strongly emphasised the plurality of collective memories (plural!) within a single society.[46] Moreover, Nora's idea of the emergence of sites of memory is

44 A simple example might serve well to illustrate this: history books tell us that the Americas were discovered by Spanish conquistadors at the end of the 15th century. This is hardly the truth, since people had been living there since the end of the last glacial period and had settled almost everywhere on the continent, so that there was nothing left to discover except in a few extremely remote areas. Nevertheless, this 'fact' is taken as truth in the European historiography.

45 A reference to Claude Lévi-Strauss's *opus magnum* 'La pensée sauvage' is in order here (Lévi-Strauss [1962] 1968). On the basis of extensive cross-cultural comparisons of ethnographic material from all over the world, Lévi-Strauss distinguished between societies with history ('hot societies') and those without ('cold societies'). Cold societies, however, are not devoid of memory. It is quite the opposite; they very actively work on collective memory in the form of myths and descent.

46 In 'Les cadres sociaux de la mémoire' Halbwachs originally identified only three types of social groups as constitutive of different collective memories within a broader society: family, religious groups and social classes. In 'La mémoire

questionable, as it seems to suggest that they only emerge when memory is threatened by history, which tries to take its place, which is also untrue. All groups that understand themselves as such have common sites of memory. Having made all these criticisms, one might ask why anyone should bother to use the concept of sites of memory at all. The answer, in my opinion, is that Nora got one fundamental thing right that deserves to be followed: that groups constitute their collective memory, and hence their identity, through particular sites where beliefs, emotions and conceptions of being and belonging merge and crystallise. It could even be argued that it is the existence of such sites, understood in a similar way by all members of a given community, that makes them a group. Without them, they would be a bunch of strangers in the strict sense of the word: they would know each other, but they would remain strange to each other. When these sites are identified and understood, they can be the key to understanding a community under study.

2) A heuristic tool that can be of great help in this quest for understanding is the perspective of ResourceComplexes (Teuber/Schweizer 2020). This approach has already been explained in this volume, so it is only worth repeating its essence here: it is based on the idea that resources do not exist in isolation, but in combinations of different elements that are needed to serve a purpose. Although we tend to identify a particular resource as the centre of the complex, it is in fact the realisation of that purpose that constitutes the complex and keeps it going. What that purpose is depends essentially on what a community believes is worth striving for, which means that from the perspective of ResourceComplexes it is possible to see how networks of resources realise certain socially embedded values. ResourceComplexes help to understand how such networks combine different elements in a rational, goal-oriented way to realise these values.

Regional Context

As the regional context has already been described in the case study on animal husbandry (*Dām-Dārī*) in the same volume (see Chapter 6.1), it will not be repeated here. The reader should only be reminded of some basic facts that seem to be crucial for the argumentation.

Kerman Province is the largest of Iran's 31 provinces and is located in the south-east of the country. It is bordered to the east by the vast Dašt-e Lut, a salt desert that is said to be one of the hottest and driest places on the planet, and to the west by patches of wasteland (notably Kavir-e Namak-e Sirjān, which locals refer to as Kafa-ye Qaṭru) (Borjian 2017). The climate is mostly arid and the region suffers from water scarcity, which partly explains why the population density is low and the total population does not even reach the three million mark (2011 census, SCI 2014). The urbanisation rate is well below the national average (58 % to 68 %, Zanjani/Nejatian 2017). Key to understanding Kerman is the interplay between areas of warm climate (locally referred to as *Garmsīr*) and areas of cold climate (*Sardsīr*). The two distinct microclimates result in fundamentally different flora and fauna, and the local population has always sought to combine the resources of these two habitats in one way or another. The Eskanderies, for instance, followed a biannual migration pattern for centuries, spending half the year in the lowlands and the other half in the highlands. The area where they used to spend their summers in the *Sardsīr* lies at the western end of a highland plain called Dašt-e Esfandaqeh. Today it is home to a small mountain village called Bagh-e Borj.

collective', however, he broadened this perspective to emphasise a plurality of these groups (Damir-Geilsdorf 2008, 45–46).

Bagh-e Borj and the *ṭāyefa* Eskandery

The Persian term Bagh-e Borj can be translated literally as 'The Garden of the Tower'. A local chronicle I found in Jiroft mentions that the highest peak of the aforementioned mountain range was once called *Borǧ* ('tower'), which explains the name of the village (Ibrāhīmī/Ibrāhīmī-Pūr 2017, 95). Following a common pattern in Iran, the village is divided into two parts, most commonly referred to as 'the upper part' (*bālā*) and 'the lower part'(*pāyīn*). According to the 2006 census, the population of the two parts of the village was 118 and 48 respectively (SCI 2006). The village's infrastructure is very basic, consisting of a mosque and a primary school housed in the same building. There are no shops or stores in the village, and the nearest (larger) villages with such facilities are about 25 km away. As there is no asphalt road connecting Bagh-e Borj to these (or other) places, the passage is occasionally impassable in winter due to heavy snowfall.

All the inhabitants of the village belong to the *ṭāyefa* Eskandery. *ṭāyefa* is an indigenous term that refers to a kinship-based social unit and has already been described in the case of the Eskanderies (see Chapter 6.1). Another term of importance for the village community is *ʿašīra/ʿašāyer*. The phrase *mā ʿašāyer hastšīm*, which literally translates as 'We are nomads!', was a topos in the village when I was doing fieldwork there. This puzzled me, too, because the Eskanderies are sedentary and their main source of income is stone quarrying. Until 70 years ago, however, they followed a bi-seasonal migration pattern from *Garmsīr* to *Sardsīr* (and vice versa) and their livelihoods depended on animal husbandry. Bagh-e Borj was their *Sardsīr*, where they spent the summers when the climate was moderate in the mountains and hot in the lowlands. It was not a village in the classical sense of the word, since the Eskanderies lived in tents, and there were only a few permanent structures that could not be moved. One of these, however, will be important for the following discussion.

The Eskanderies follow an endogamous marriage pattern and only intermarry among themselves. This strict tendency towards endogamy is legitimised by a lively reference to a common ancestor: no descendant of a mythical ancestor named Karbalāʾ Bārānī should marry anyone who is not. The following section deals with this ancestor and his importance to the village community.

Karbalāʾ Bārānī

In order to give an appropriate idea of the importance of Karbalāʾ Bārānī to the villagers, it is most useful to let one of them speak for himself (admittedly through my translation from Farsi into English) before I offer any further interpretations:

'Someone (*fardī*) with the name Karbalāʾ Bārānī came from another place (*ǧā-ye dīgeh*), Sir Karbalāʾ Bārānī was a faithful and religious (*muʾmen wa moʿtaqed*) man. That is why people, from 300 years ago until now, still (*hanūz*) they have a special worship (*ehterām-e ḫāṣṣ*) [for Karbalāʾ Bārānī] and they go to his tomb, they make a pilgrimage [to his tomb]. People who don't have children, for example, go there and say a prayer, of course in the name of God (*beh esm-e ḫodāwand*). And then they have children, thanks to God! [...] Or people who are sick, they also go there. Our ancestor (*ǧad-e mā*), everyone in Bāġ-e Borǧ, everyone who lives here, are all descendants (*nawādegān*) of the same Sir Karbalāʾ Bārānī. Seven generations (*haft nasl*), that means that I am the seventh generation, and all of them, they all come from one person and that is Karbalāʾ Bārānī! [...] The people of Bāġ-e Borǧ are good-hearted (*sāde del*; literally: "simple-hearted") with a positive spirit (*bā tafakkurāt-e mos̱bat*), just as their father (*pedar-e ānhā*), their ancestor (*ǧad-e ānhā*), Sir Karbalāʾ Bārānī, was a faithful (*muʾmen*) and righteous (*dorost*) man.'

This relatively short passage from a much longer interview reveals some important peculiarities that stand out. The first is that all the villagers are descended from the same ancestor, who commands special respect or even worship (*ehterām*, the Persian term, can mean both). Second, the villagers share certain (positive) traits that stem from their common ancestor, such as faithfulness and righteousness, traits that are highly valued in Iran. Third, the person of Karbalāʾ Bārānī seems to be associated with the sphere of the superempirical. This is evidenced by the fact that the villagers do not simply 'visit' his tomb (which lies a few kilometres outside the upper part of the village), but make a pilgrimage (*ziyarāt*) there. Sick or childless villagers make this *ziyarāt* to find a cure for their problems. Therefore, they pray at his tomb. The villager quoted above also directly emphasised that it is not Karbalāʾ Bārānī who is prayed to, but that the prayer is directed to God and is made in God's name (*beh esm-e ḫodāwand*). This is self-evident, since the villagers are (Shia) Muslims, and to pray to anyone other than God would be *širk*, the acceptance of other deities alongside God, which is a major sin in Islam. Since the villagers of Bagh-e Borj, as in most rural areas of Iran, tend to be quite religious, my interviewer wanted to avoid giving the impression that they would do such a thing.

Karbalāʾ Bārānī and his tomb rather bear the hallmarks of an *emāmzādeh*. The Persian term *emāmzādeh* has two interrelated meanings: it refers to a holy person in Shia Islam, usually a descendant of one of the holy Imams, and to the tomb of such a person. These tombs are usually built as shrines and are places of (general) devotion. While some of them are large and famous even outside Iran, such as the shrine of Fāṭima al-Maʿṣūma in Qom, most of them are rather small and local and are scattered in great numbers throughout the country.[47] They are 'sites where divine favor and blessing occur' (ʿAbbās Qomī, quoted by Algar 1998) and have the ability to cure diseases or grant wishes when pilgrims visit and pray there. Karbalāʾ Bārānī and his tomb seem to serve a similar purpose. They provide a point of intersection where different spheres collide: the past and the present, the empirical and the superempirical. Although Karbalāʾ Bārānī himself is not a deity or a saint, through him the boundaries of these spheres can be transcended. He serves as a mediator between these different spheres.

Karbalāʾ Bārānī as a Site of Memory

As outlined above, memory is not really concerned with 'truth' or 'facts' but its primary aim is to establish commonality (Giesen 1999, 44). The example of Karbalāʾ Bārānī illustrates this assumption in a most striking way. What is surprising is that the centrality of Karbalāʾ Bārānī stands in relatively sharp contrast to what the villagers actually know for certain about their ancestor. For instance, there is not even a consensus in the village as to where Karbalāʾ Bārānī came from before arriving in the area. Some say Teheran, others mention Hamadān.[48] This lack of known facts about a person of such vital importance to the community seems, at first glance, to be a contradiction in terms. But it is important to remember that memory is

47 The precise number is not known. In the 'Encyclopædia Iranica', Varjāvand mentions estimates ranging from 138 to 1,059. However, he suggests that even the higher estimates are probably too low (Varjāvand 1998).

48 This sounds plausible since the dialect spoken in Bagh-e Borj does indeed have some similarities with the dialect spoken in Hamadān.

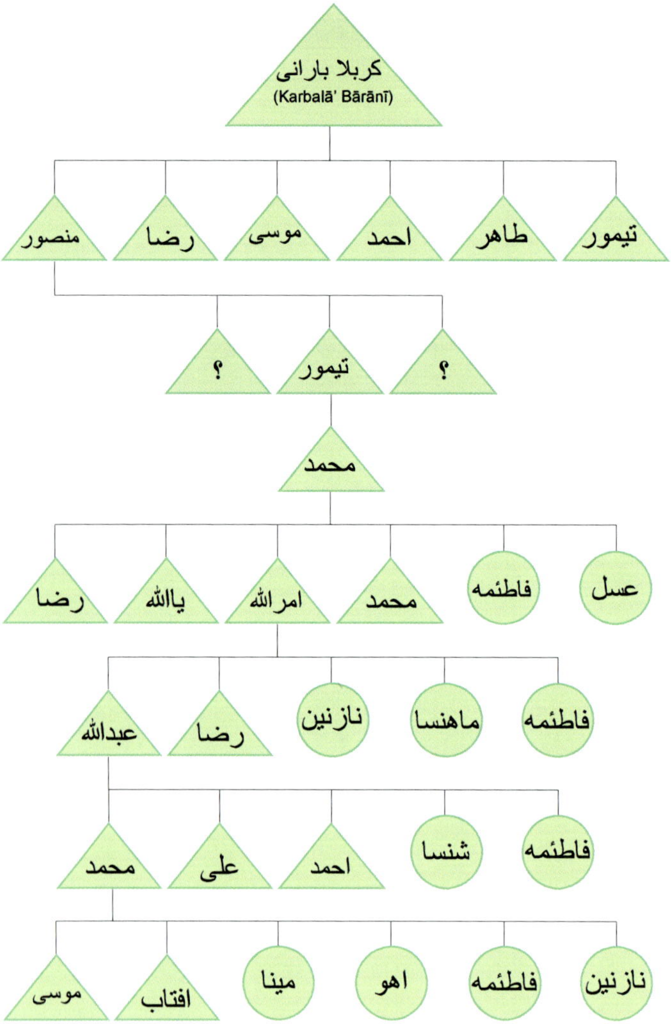

Fig. 52 A genealogy of a villager that leads back to the mythical common ancestor Karbalā' Bārānī. The illustration is not complete: occasionally the names of ancestors were unknown (marked with '؟' in the diagram) and female ancestors occur 'only' from the fourth generation after Karbalā' on. Note also that it is anonymised by changing most names (Data: Wulf Frauen; Graphic design: Richard Szydlak).

about affectivity, not factuality. This makes it a source of collective identity, because all sites of memory have one thing in common: people associate something with them that they consider crucial to their identity. And this is exactly the case with Karbalā' Bārānī. Beyond 'facts' and 'details' there is really only one fundamental aspect that is essential to the villagers: that Karbalā' Bārānī is the ancestor of all the villagers, and that all the villagers are equally descended from him. And this belief is hard to shake: villagers can trace their respective lineage all the way back to Karbalā' Bārānī. The following example illustrates this (see fig. 52).

The villager who gave this genealogy could do so even by heart, without consulting any notes or sources other than his memory. Other villagers could do the same from their respective positions in the family tree.[49] Of course, villager A can only do this from his point

49 The metaphor of the tree seems an excellent one here, as some villagers described their *ṭāyefa* as *mes̱l-e deraḫt* – 'like a tree'.

of view and not from the point of view of villager B. But all these lines approach each other, eventually overlapping and crystallising in the person of Karbalā' Bārānī. Karbalā' Bārānī is thus a site of memory that constitutes an identity *ex negativo* in the sense that it has an exclusionary character: whoever is not a descendant of Karbalā' Bārānī is not a member of the *ṭāyefa* Eskandery.

Karbalā' Bārānī as a Resource to Establish Social Cohesion

The genealogies mentioned in the previous section can be understood as a specific kind of objectification of memory. According to Aleida Assmann, such objectifications are crucial for groups, because otherwise memory would slowly fade away (Assmann 2007b, 55).[50] In addition to the genealogies, there are other forms of objectification of memory in Bagh-e Borj regarding the case of Karbalā' Bārānī. The most obvious one has already been briefly mentioned: his tomb. Already Halbwachs realised that memory has a tendency to spatialisation (Assmann 2005, 79). The example proves this assumption right, since the intangible site of memory Karbalā' Bārānī becomes a spatial entity with his tomb. In fact, intangible sites of memory (almost) always find tangible materialisations, just as tangible sites of memory (almost) always contain immaterial aspects.[51] On closer inspection, therefore, the site of memory of Karbalā' Bārānī is not a single person who died a long time ago, but an accumulation of tangible and intangible elements and social practices. As such, it can be understood in terms of ResourceComplexes as a network of various interdependent elements. A simplified diagram is depicted in fig. 53.

The elements of the ResourceComplex shown are all connected to memory. This is hardly surprising, since Karbalā' Bārānī is understood here as a site of memory. But they are also interrelated and interdependent: the memory of Karbalā' Bārānī is spatially manifested in his tomb, and this spatial manifestation is necessary for the performance of social practices related to him (such as *ziyarāt*). Ultimately, all these elements work together to carry out the reconstructive act of remembering. It can be seen that memory is something that is socially constructed and would disappear if this work were not to be carried out any longer.

One might now ask what is the added value of interpreting this site of memory as a ResourceComplex, apart from the fact that it nicely illustrates the social construction of memory? The answer is that a ResourceComplex perspective, as mentioned above, makes it possible to identify networks that enable groups to realise socially embedded values. These values are in turn the result of the historical processes that created them. The question now is, what are these values and what historical processes have contributed to their creation?

---

50 Assmann draws on Hannah Arendt, who in the early 1980s noted that such objectifications are essential for groups to make memory concrete (Arendt 1981, 87).

51 Let me illustrate this claim with an example from Etienne François's and Hagen Schulze's edited volume on German sites of memory: Johann Wolfgang Goethe is a German poet that passed away centuries ago, but there is a Goethe House in Frankfurt where the *Volksdichter*, as he is often referred to in German, was supposedly born. In fact, the house in which he was actually born was destroyed in an air raid during Second World War, and the Goethe House merely stands in a similar location. Goethe's poems are, of course, immaterial, but they are printed in tangible books. If these poems were not read out of these books and discussed in schools, for instance, then both the poems and the books would be irrelevant. The site of memory 'Goethe' is therefore not only a person who died long ago, but an accumulation of tangible and intangible elements and social practices. To a certain extent, this applies to all sites of memory.

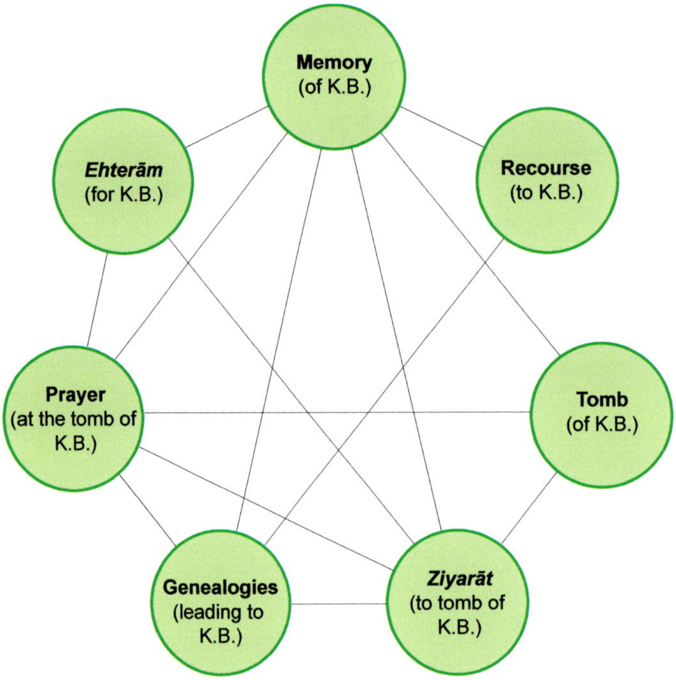

Fig. 53 ResourceComplex of Karbalā' Bārānī (Graphic: Wulf Frauen; Graphic design: Richard Szydlak).

In terms of values, it can be said that the various tangible and intangible elements of the resource network associated with Karbalā basically serve to establish and strengthen social cohesion and *communitas*, the spirit of being a community. Unity, a strong family bond and a connection to the land that hosted the group's common ancestors are the underlying values that are realised through this network. To understand the fundamental significance of these values for the *ṭāyefa* Eskandery, it is helpful to consider the group's past: until 70 years ago, the Eskanderies were pastoral nomads who followed a biannual migration cycle and lived in tents. At that time, the combination of autonomous, self-sufficient households with a very high degree of mobility was a potential threat to the community – in the event of a dispute, individual households could theoretically have split from the group at any time.[52] The resource of Karbalā' Bārānī was crucial to the survival of the group at that time: it legitimised the strict system of endogamous marriage patterns that the villagers refer to as *dayn-dār*[53] and brought all Eskanderies back to the tomb of Karbalā' Bārānī every year, where they had their summer pastures. In this context, Karbalā' Bārānī can be understood as part of the wider ResourceCulture around animal husbandry that I have described elsewhere (see Chapter 6.1).

Concluding Remarks

This case study has shown that Pierre Nora's concept of *lieux de mémoire*, despite all its shortcomings, can be used to describe social phenomena in a meaningful way. These phenomena

52 Already Fredrik Barth, in his classic 'Nomads of South Persia', remarked that this combination was potentially lethal for pastoral-nomadic groups (Barth 1961).

53 Which theoretically, literally translated, should mean 'faithfulness'. In this context, I understand it to mean 'remaining faithful to the lineage'. No descendant of Karbalā' Bārānī should marry someone who is not.

can be found all over the world and in all communities that consider themselves to be such. One of Nora's major shortcomings was that he attributed *lieux de mémoire* only to 'modern' nation-states in which memory is 'threatened' by history. This is wrong and unnecessarily restrictive in two ways. The case study has also shown that the ResourceComplex perspective can be used to expand Nora's concept in a meaningful way. It allows us to understand the underlying logics that led to the genesis of a particular site of memory. More importantly, it makes visible the intrinsic values and meanings that were hidden within a site of memory. Nora understood sites of memory, as the word 'site' implies, in a very one-dimensional way, whereas in reality they are complex multi-dimensional networks that actors use to socially express values that are important to them.

The specific site of memory, Karbalā' Bārānī, is linked in its genesis to the very functional aspect of holding together a group that would otherwise have slowly disintegrated. This was essential for survival in the mountains of Kerman. Today, this aspect is less important, since the Eskanderies have changed their *modus vivendi*. Nevertheless, the memory of Sir Karbalā' Bārānī is very vivid in the village. This is not a contradiction, because regardless of its genesis, the site of memory Karbalā' Bārānī still serves its most important function: it tells them about their place in the universe, where they are and who they are. Finally, Nora's image of the *lieux de mémoire* as 'shells on the shore when the sea of living memory has receded' (Nora 1989, 12), beautiful though it is, must be rejected. This suggests that the sites of memory are isolated. But they are not. They are all part of the humanly woven 'web of significance' (Geertz 1973) that we call culture and that we need to make sense of the world around us. In other words, these shells were not left behind by the sea, it is the connection between them that makes up the ocean.[54]

6.9 *Heroa* of City Founders as Resources of Civic Identity and Cultural Memory. Beyond Colonisation, Decolonisation and Romanisation

Beat Schweizer

Introduction. Founders of Cities, Worshipping of Heroes and *Heroa* as Sites of Cultural Memory

The founders of cities in the archaic and classical periods of Greek history were ascribed a special status for civic identity and cultural memory. The literary and epigraphic sources called them *oikist* – οἰκιστής – and equated them with heroes. This meant not only a kind of retrospective recognition or revaluation of a particular person, but above all an incorporation into the urban cults and the city's festival calendar.

Herodotus (6, 38) wrote about the Athenian Miltiades, tyrant of the Thracian Chersonese in the 6th century BC and considered by the Chersonesitai as their founder, that they sacrificed to him 'as it is the norm for a founder' – θύουσι ὡς νόμος οἰκιστῇ – (translation of Ekroth 2002, 199; cf. Malkin 1987, 190; Greco 2021, 299). Herodotus's label *nomos* was intended as a reference to customary, quasi-legal social or political practices. According to Thucydides (5, 11, 1), Brasidas, a Spartan commander in the Peloponnesian War who died

54 In the case of Bagh-e Borj, since these various aspects are related to animal husbandry, they can be broadly understood as the ResourceCulture of animal husbandry (*Dām-Dārī*) described earlier (see Chapter 6.1).

in the battle of Amphipolis in the region of Thrace in the late 5th century BC, was buried facing the *agora*. It was offered to him ὡς ἥρῳ, as a hero, and at the same time the city was dedicated to him ὡς οἰκιστῇ, as a founder. And in one of Pindar's Odes it is stated that Battus, founder of Cyrene in the 7th century BC, was worshipped as a hero – ἥρως δ' ἔπειτα λαοσεβής – (Pyth. 5, 94–95) and that his tomb was built in the area of the *agora* (Pyth. 5, 93).

To be worshipped as a founder, to become a hero, was therefore ritually linked to acts of sacrifice and spatially bound to tombs and sacred precincts in the centre of the city. The latter, called *heroa*, were the material points of reference for the construction of cultural memory and the identity of cities and their history. Continuities and changes can be traced far beyond the respective founding periods, sometimes over centuries, in politically, ethnically and governmentally different power and rule relations.

Amphipolis, Cyrene and – from an archaeological perspective – also Poseidonia in the Gulf of Salerno can be discussed as examples of the connection between immaterial and material dimensions of cultural networks or assemblages centred around these *heroa* as parts of sacred landscapes, as sites of offerings and regularly held processions with various kinds of sacrifices and games or with poetry and poetic performances commemorating founding myths. Insofar as the construction of meaning and value, of knowledge of the past and notions of sociality or identity was deeply intertwined with presentations of texts, images and things, as well as with ritual bodily practices and emotions, *heroa* were media of ongoing reconstructions of the past. They served as resources for civic identities. Processes of constructing cultural memory as well as processes of cultural forgetting determined the emergence and decline of a ResourceCulture, very specifically in each city, but also in the wider setting of comparable cultural contexts.

War and Identity Politics. Burying an Army Commander, Forgetting the Founder and Making a New Hero in Amphipolis

In 437/436 BC, during the growing political tensions between Athens and Sparta for control of Greece, Amphipolis was founded in the northern Aegean as an Athenian *apoikia* by Hagnon, the son of Nicias, sent as a founder (οἰκιστής) (Thuc. 4, 102, 1–4). The new city, near the mouth of the Strymon River at a site formerly known as *Ennea Hodoi* (Nine Ways), was established to control a hinterland rich in gold, silver and timber, as well as the trade routes from west to east along the Aegean coast or north along the Strymon far into Thrace. According to Thucydides, the foundation was named Amphipolis by Hagnon, because it 'was conspicuous both seaward and landward' (Thuc. 4, 102, 4), and Amphipolis 'was useful [to Athens] for the importation of timber for ship-building and for the revenue it produced' (Thuc. 4, 108, 1, translations of Thucydides by Smith 1919–1923). Thucydides, whose work is the main source for the events, himself held rights to gold mines in this region (Thuc. 4, 105, 1).

Athenian control of Amphipolis was lost shortly after its foundation in the first phase of the Peloponnesian War, when after 424 BC a Spartan army led by Brasidas and reinforced by regional allies attempted to drive the Athenians out of the region. During this campaign, Amphipolis was besieged but politically persuaded by Brasidas to switch sides without a fight. A small Athenian fleet, stationed at nearby Thasos and led by Thucydides himself, arrived too late to intervene. According to Thucydides, Brasidas 'made haste to seize the city if possible before he should come [...]' (Thuc. 4, 105, 1). In any case, Thucydides was banished from Athens for 20 years.

In 422 BC, the Athenians, led by Cleon, attempted to retake Amphipolis, but lost the battle in which not only Cleon was killed but also Brasidas was fatally wounded. Thucydides (5, 11, 1) described the effects on Amphipolis as follows:

'After this all the allies gave Brasidas a public burial (δημοσίᾳ ἔθαψαν) in the city at a spot facing what is now the market-place [ἀγορά], following his body in full armour. And the Amphipolitans fenced in his monument [μνημεῖον] and have ever since made offerings to him as a hero (ὡς ἥρῳ), giving honours and instituting games and yearly sacrifices. They also adopted him as founder of the colony (καὶ τὴν ἀποικίαν ὡς οἰκιστῇ προσέθεσαν), pulling down the edifices of Hagnon (Ἁγνώνεια οἰκοδομήματα) and obliterating whatever was likely, if left standing, to be a reminder [μνημόσυνον] of his settlement [...].'

It has often been assumed that the reference to the 'edifices of Hagnon', translated accordingly as 'Hagnoneia' (Malkin 1987, 228) and 'cult buildings of Hagnon' (Hornblower 1996, 452), indicate that Hagnon himself was already being worshipped as a hero.[55] However, Thucydides wrote of 'Hagnoneian buildings' (cf. Leschhorn 1984, 152; Mari 2012, 328) and at the end of the passage (5, 11, 1) of honours: 'Hagnon, in consequence of their hostile action against Athens, would not in like manner as before receive their honours'. The honours for the living Hagnon must have been different from those for the dead Brasidas (Mari 2012, 350), to whom both games and '*thysiai* sacrifices' involving the sharing of meat were given. The 'offerings to him as a hero' were literally '*entemnein* sacrifices' or 'cutting of the throats of victims'-sacrifices (cf. Ekroth 2002, 184–186), which were also associated with war dead (Ekroth 2002, 135–136; 2007, 107). Furthermore, a 2nd century AD passage by Polyaenus tells of a local Thracian hero Rhesus, son of Clio and the river god Strymon, whose alleged bones were transferred from the Troas, fulfilling the prescription of a founding oracle (Mari 2010, 411; 2012, 335 note 24): 'for it is not decreed, until you find and bring back from Troy the remains of Rhesus and bury him piously in his ancestral land'. Therefore: Hagnon 'buried Rhesus' bones by the river. Digging a ditch, he fortified the place' (Polyaenus VI 53, translation by Krentz/Wheeler 1994, 617–618). Thus, a mythical hero and, in fact, the transfer of bones as well as a reburial could be part of a foundation and of the appropriation of land (Figueira 2008, 454; cf. Malkin 1987, 81–84). Obviously, this cult, inaugurated by Hagnon, was preserved despite all the changes that took place, even after the Macedonian conquest of Amphipolis in the middle of the 4th century BC (Mari 2010, 412).[56]

What Thucydides described as the treatment of the dead Brasidas was initially a public funeral ceremony and the burial of a fallen military leader within the city. Then, first, the location was emphasised at a place for communal meetings – 'at a spot facing what is now the market-place' (according to Smith 1919–1923) or 'in front of what is now the Agora'

---

55 Cf. Hornblower 1996, 452–455 with discussion, following mainly Malkin 1987, 228–232 on Hagnoneia; therefore Simonton 2018, 6 note 17: 'Scholars are generally agreed that the Hagnonian buildings were cult structures dedicated to Hagnon's worship as a hero during his lifetime'.

56 '[...] un interessante fossile dell'apoikia del 437, e uno dei più duraturi: ma se l'heroon fondato da Agnone sopravvisse alla damnatio memoriae del suo fondatore fu proprio perché esso non richiamava il legame originario e speciale della città con Atene, ma semmai la volontà di conciliazione e (forse) la compresenza di elementi etnici diversi' (Mari 2010, 412).

(Hornblower 1996, 449; Simonton 2018, 5)[57] – and second, the sacred precinct, which was linked to the establishment of offerings to Brasidas and annual sacrifices, and thus a hero cult. Third, and implicitly tied to the tomb in the city and heroic honours, Brasidas was ascribed the status of a founder of the colony,[58] while buildings related to Hagnon, the actual founder, were demolished in order to destroy the memory of him and his deeds.

Although the events described by Thucydides must have left material traces, not much can be said from an archaeological point of view. The location of the *agora* has been inferred from literary sources rather than archaeological evidence. However, during excavations near the museum, within but close to the ancient city walls, a cist grave with a cremation burial in a silver larnax crowned with a gold diadem was discovered (Lazaridis 1997, 83 fig. 42). A votive deposit was found nearby and the grave was said to have been architecturally framed (Koukouli-Chrysanthaki 2002, 66–72, 61 fig. 2).[59] This grave was assigned to Brasidas by the excavator – 'a working hypothesis' (Koukouli-Chrysanthaki 2002, 69; 2011, 415; also Lazaridis 1997, 21; von Mangoldt 2013, 17–19 No. A 1), but this was viewed rather sceptically in archaeological and historical research (cf. Schörner 2007, 23 note 150; Mari 2010, 409).

The story of Amphipolis, of Hagnon and Brasidas (Mari 2012) has been well discussed in relation to the conflict between Athens and Sparta in the early years of the Peloponnesian War, as well as to Thucydides's high valuation of Brasidas in relation to an event of contemporary history, in which he himself was involved (most recently Plant 2023). The passage quoted above has been part of historical discourses on the establishment of hero cults and their rituals (Malkin 1987, 228–232; Ekroth 2002, 184–186; 208–209) and also on Hagnon or Brasidas – the latter was called *soter*/saviour (Leschhorn 1984, 154–155) – as possible forerunners of the heroisation of living people in Hellenistic times (Hoffmann 2000; Currie 2005, 158–200). Whatever the case, making Brasidas a founder was seen as 'a significant religious act on the part of the city', a 'symbolic refounding of the city' through which the 'loyalty of the city' was 'transferred on a religious level to an enemy of Athens' (Plant 2023, 26) and was also associated with a political shift in Amphipolis from democracy to oligarchy (cf. Simonton 2018).

More recently, the issue of cultural memory and identity, or the destruction of cultural memory, has been raised (Mackowiak 2018; Simonton 2018; Mauersberg 2022). For Amphipolis, then, it was not just a matter of changing alliances, from Athens to Sparta, but of changing identities.[60] In doing so: 'Manipulation of public commemoration did not simply reflect a pre-existing change in power, it was itself a constitutive element of that change' (Simonton 2018, 13). Cultural memory was actively and performatively reshaped through

---

57 The burial must have taken place 422 BC, Thucydides died about 20 years later. Cf. 'Thucydides demonstrates personal knowledge when he slips beyond the narrative time to a later (narratively speaking) present' (Plant 2023, 25) against 'The relocation of the agora perhaps followed the demolition of the Ἁγνώνεια, which must have stood in the city's first agora' (Koukouli-Chrysanthaki 2011, 415).

58 Based on this passage, Malkin 1987, 229 defined four points as 'characteristics of oikist cults': public funeral, tomb and sacred enclosure inside the city, continuing hero cult and annual 'honours'. Using the examples of Miltiades and Brasidas, Proietti 2014, 202–204 argued against a direct link between hero cult and annual honours, relating the latter more to their status as war dead.

59 '[...] foundations of a two-roomed temple-like building covering a cist grave' (Koukouli-Chrysanthaki 2011, 415), 'the funerary heroon was demolished' in the middle of the 4th century BC (Koukouli-Chrysanthaki 2011, 420).

60 'Détruire le culte préparé pour Hagnon consistait à assumer une décision identitaire: celle du non-retour à Athènes [...] revient fondamentalement à effacer une mémoire du passé historique de la communauté' (Mackowiak 2018, 322).

the establishment of a new origin (Mauersberg 2022, 47). Materially, the monument in its place became 'a "focal point" of group action' (Simonton 2018, 18[61]), and its location in a sacred space in the centre of the city served as a context of social structuring and symbolic valuation.[62] 'Commemorative festivals' meant the 'concentration of citizens in the city-center', and this 'was not merely a symbol of "people-power": it was itself an element of democracy in action' (Simonton 2018, 16).

Thus, the heroisation of Brasidas involved the building of a tomb in a sacred space that served as a monument (μνημεῖον) and as a crucial tangible element of civic and political identity, as well as the destruction of Hagnon's immaterial memory and material memorial (μνημόσυνον; Hornblower 1996, 452: 'solid memorials'). The founding was therefore seen less as an important event of the past and more as an event of the present. 'Obliterating whatever was likely, if left standing, to be a reminder of his [Hagnon's] settlement' (Thuc. 5, 11, 1), the destruction of buildings or building inscriptions, thus the elimination of Hagnon's 'inscription' within the urban space, was obviously an act of violence in connection with a change of rule or a new power constellation. The intended obliteration or erasure of Hagnon – ἀφανίζω literally means 'to make unseen' or 'to make disappear' the material memorial – can be understood as forgetting in the form of a condemnation of memory, in the sense of Connerton's 'repressive erasure' (Connerton 2008, 60–61) or Assmann's 'repressive forgetting' (Assmann 2016, 49–53). Furthermore, Connerton spoke not only of a form of 'forgetting, which is constitutive in the formation of a new identity', but also of 'prescriptive forgetting' in the sense of believing it to be in the interests of all parties or groups in a society (Connerton 2008, 63–64). Assmann (2016, 57–63) combines both aspects in her term 'constructive forgetting'.

At the moment when Hagnon or material traces of him in the cityscape were made unseen or made disappear as resources of identity, the new monument to Brasidas could have unfolded its effect as a resource of new identity with full force. From this perspective, however, Thucydides's account of events was problematic in that the historian preserved the memory of Hagnon and thus a counter-narrative to the newly constructed cultural memory. For, according to Thucydides, the citizens of Amphipolis acted as they did because 'they courted the alliance of the Lacedaemonians through fear of the Athenians, thinking Brasidas to have been their saviour (σωτήρ), whereas Hagnon, in consequence of their hostile action against Athens, would not in like manner as before receive their honours either with benefit to themselves or with pleasure to himself' (Thuc. 5, 11, 1).

Civil War and Political Identities. The Construction, Reconstruction and Abandonment of the Founder's Tomb in Cyrene

According to written sources, Cyrene was founded in 631 BC about 10 km behind the Mediterranean coast of eastern Libya, on a ridge of the fertile Jebel Akhdar uplands near modern Shahat. The founding story, including details of earlier attempts on an island and at another site on the mainland, as well as of Delphic oracles (Malkin 1987, 60–89) linked

61 Simonton referred here, however, to statues of tyrant killers, another kind of heroes, in the *agorai* of Athens and Erythrae.

62 '[...] le sacré, comme le rite, est indissociable de la vie sociale et de sa perpétuation; il établit les "social basic acts" du type de ceux qui permettent au groupe humain de formaliser un événement social et de le traduire en haut degré symbolique' (Mackowiak 2018, 324).

either to Battus, the founder, or to the king of Thera, the place of origin of the settlers, has been intensively researched, especially because of the interrelationship between 'Greek colonisation' in archaic times and the – decisive – role of Delphi (Malkin 1987, 60–69; Murray 1993, 117–123; Bernstein 2004, 171–222; Osborne 1996, 8–17) or because of the figure of Battus (Leschhorn 1984, 60–72; Giangiulio 1981; Dougherty 1993).

The historical source material on the foundation and history of Cyrene is relatively good as it shows great diversity.[63] Basic information was provided by Herodotus (cf. Corcella 2007, 669–694) in the context of his description of Egypt or Libya, the 'Libyan Campaign Logos' (Giangiulio 2011, 705), in which he also reported two versions of the founding story, that told by the citizens of Thera and that of the Cyrenians themselves, and then a smaller part on which both agree (Herodotus 4, 150,1–158,2).[64] The Theran story is more concerned with the 'prehistory' of the *apoikia*, the Delphic advices, a seven-year drought, explorations of the route and destination, selection by lot in families with more than one son, and a first settlement on Platea, an island off the Libyan coast. The Cyrenian story focuses on Battus, the founder, and his family background, including folktale motifs, then the first attempts at Platea with a thwarted attempt to return. Both stories merge with the move from the island to the mainland, from Platea first to Aziris and then to Cyrene (Murray 1993, 117–118; Osborne 1996, 10–12), a place with a good water supply: 'where there was a hole in the sky', according to Herodotus (4, 158).

In addition to Herodotus's records of oral histories there is a 4th century BC stone stele inscribed with a decree granting civil rights to the Theraians in Cyrene, found reused in the Byzantine Bath, but probably originally set up in the sanctuary of Apollo in Cyrene. For historical anchoring, the alleged original agreement on the establishment of an *apoikia* was added, the so-called 'oath of the settlers' (Murray 1993, 119–120), which spoke of selection, which could not be avoided on pain of punishment, but also of the right to return after five years in case of failure. Furthermore, partly congruent, partly divergent information was given in Pindar's Fourth and Fifth 'Pythian Odes' about the victory of Arcesilaos IV, the eighth and last king of the Battiad dynasty in Cyrene, in the Pythian Games at Delphi in 462 BC.[65]

The discourses of historical research have focused on the comparison of the two versions of the founding history and the question of the authenticity of the 4th century BC inscription, always with a view to the 'correct' reconstruction of the history of Cyrene – stripped of the legends of lore (Leschhorn 1984, 62–63). However, 'once we appreciate the factors which shape these stories, we can see that it is in vain to seek historical truth from either account' (Osborne 1996, 12)[66]. Another, more presentist, direction has recently been taken with the comparison of the Theraian version with Pindar's Fourth, both of which have elements of a much more distant myth-historical past linking Thera to Sparta, but which were appropriate to the times (Thomas 2018; cf. Miller 1997, 96–97). A similar approach has emphasised the differences in time and context between the sources. Pindar wrote poems for the winners at Olympia and Delphi. His Fourth and Fifth 'Pythian' Epinicia were dedicated

63 According to Calame 1990, 280 'the narrative of the founding of the great Greek colony of Libya exists for us solely in the widest variety of literary forms.' Not all, especially of the later sources, are mentioned here.

64 'These two tales seem to embody different "local knowledges" […]' (Neer/Kurke 2019, 165).

65 Cf. the semiotic or historic overviews: Calame 1990, 281–300 on Pindar's Fourth 'Pythian', 309–313 on Pindar's Fifth 'Pythian', 313–319 on Herodotus; Miller 1997, 96–102 on Pindar's Fourth 'Pythian', 102–110 on Herodotus 4, 150–158, 207–214 on Pindar's Fifth 'Pythian' and Battus as founder.

66 The opposing position was taken by Malkin 2003.

to the last king of the Battiad dynasty and were probably performed at the *Karneia*, an important festival of Apollo in Cyrene.[67] There was also a strong reference to Battus the Founder, also as founder of the shrines (Pindar V 89). Herodotus wrote history at a time when Cyrene had an oligarchic or democratic constitution.[68] Since his narrative of Cyrenean history ends with the incorporation of Egypt and the Cyrenaica into the Persian Empire, he is relatively silent about the last two kings who ruled during the Persian Wars (Laronde 1990). However, Arcesilaus III and his mother Pheretime in particular were described in an 'ambience of terror and cruelty' (Giangiulio 2011, 708). Finally, the 4th century BC inscription of the 'oath of the settlers' presents a narrative of equality or equal rights between the original settlers and between earlier and later settlers. This indicates the social and political situation of the 4th rather than the 7th century BC (Osborne 1996, 12–15), but there was still talk of Battus as leader and king.

For archaeological topographical research, one piece of information from Pindar's Fifth 'Pythian' has been of particular interest, according to which Battus the Founder was worshipped as a hero (5, 94–95) and his tomb was located on the edge of the *agora* (Pyth. 5, 93; cf. Greco 2021, 306).[69] The *agora* of Cyrene, located north of an important paved road used for processions in honour of Apollo (Pyth. 5, 90–93), is a site that has been archaeologically studied for many years. Apart from a few special monuments, the buildings on the sides of the square have been gradually published since 1965 (Stucchi 1965; Bacchielli 1981; Bacchielli/Stucchi 1983; Purcaro 2001). Building on this, slightly divergent attempts at an overall view have been presented by Stucchi (1967, 47–93), Ensoli (2000), Giudice (2006) and, finally, Scott (2013, 14–44), who tried to reconcile the architectural and political developments. All in all, the development of the square, which began in the late 7th century BC, tended in the first centuries of its existence to have small official or religious buildings on the west and east sides and small halls on the north side. Major changes, with larger hall buildings, altars and monuments, occurred in the Hellenistic and Roman periods, for the later especially in the Augustan period and, after destructions during a revolt, in the Hadrianic period and later.

According to Scott's summary, construction on the *agora* began at the end of the 7th century BC. *Oikoi* on the south-east and south-west corners, associated with a local deity (Stucchi 1965, 33–58)[70] and Apollo Archegetes, already defined the extension of the *agora* (Scott 2013, 17–18).[71] In the early 6th century BC a tumulus was erected on the east side over a secondary cremation. The ashes of a deceased person and soil[72] had been poured into a shallow pit and then covered with a stone packing. An overlying layer of ash indicates

67 Calame 1990, 278 connected the Fifth with the *Karneia*, the Fourth with a banquet in the king's palace. More generally, Neer and Kurke (2019, 161–162) set the objective 'not to integrate the narratives and thematic emphases of Pindar's Cyrenaean odes into a single synthetic account'.

68 Cf. Proietti 2012, 195: 'La plasticità della tradizione battiade si manifesta anzi in più fasi proprio in rapporto alla sua delficizzazione, funzionale in una prima fase alla legittimazione della monarchia battiade ma poi scomparsa nella tradizione post-battiade, appunto de-delficizzata, recepita da Erodoto'.

69 Neer and Kurke (2019, 162) label the Fifth 'Pythian' 'as the single Pindaric poem most obsessed with topography and the mapping out of civic space'.

70 Contrary to Stucchi and the scholars following him, Parisi Presicce 2007 linked the *oikos* of the south-east corner with the *heroon* and thus related both structures to the cult of Battus.

71 Cf. Neer/Kurke 2019, 174: 'This juxtaposition – kings on one side, Apollo and the *prytaneis* on the other – maps onto the two traditions of Cyrene's foundation, with the human founder, Battos, to the east and the divine founder, Apollo, to the west.'

72 The ashes of the deceased were mixed with earth that obviously did not come from Cyrene: Kreutz 2016, 36 with literature. Did the dead person die in a foreign country?

sacrificial activities above the stone setting before the grave was covered by a mound 6.20 m in diameter. To the north-west of the mound was a small rectangular platform, probably used for rituals at the monument. Thus, on the basis of Pindar's tradition, the ensemble was identified as the tomb of Battus (Stucchi 1965, 58–65). In the second half of the 5th century BC, after the end of the Battiad dynasty, the tumulus was destroyed, but another tomb was built – slightly offset to the east – in the form of the contemporary cist tombs with large stone slabs under a new tumulus comparable in size to the old one (Stucchi 1965, 111–114; Scott 2013, 24). However, this tomb was interpreted as a cenotaph.[73] In the late 4th century BC, the tomb underwent another intervention, when it was incorporated into a rectangular walled area and covered with a roof (Stucchi 1965, 139–142; Scott 2013, 26).[74] It remained in this state until it was destroyed during the Diaspora revolt in 115 AD.[75]

Thus Battus the Founder and his tomb were, for some 750 years, both immaterial and material resources of identities. As Pindar's Odes attest, they were the focus of dynastic identities until the middle of the 5th century BC and the last king of the Battiad dynasty. Despite all the struggles between dynasts and oligarchs or citizens and between members of the dynastic family (Neer/Kurke 2019, 163–165; Delp 2022, 293–297, 302–307), the kings justified their claim to power with reference to Battus the Founder or Apollo (Miller 1997, 96), who was said to have sent Battus to found Cyrene. But beyond this dynastic correlation, both Battus and his tomb were resources of Cyrenean cultural memory.[76] The cult continued not only under monarchical, but also under republican or democratic rule in Cyrene, as part of the Persian and Ptolemaic empires, and finally as one of the capitals of a Roman province, until a symbolic re-founding by Hadrian.[77]

There are interruptions in the 5th and 4th centuries BC and, at the very end, in the 2nd century AD. The *agora* was rebuilt in the late 5th century BC. The original tomb was replaced by a cenotaph in an architecturally modern form, but still under a tumulus.[78] Archaeological research has linked this change to the fighting in the civil war. Bacchielli suggested a destruction of the earlier tumulus by a part of the democrats and the building of the new one as part of a later aristocratic consolidation (Bacchielli 1985; cf. Ensoli 2000, 65), while Kreutz (2016, 40) related this reference to the *oikist* with democratic rule.[79] However, both interpretations work on the assumption of a destruction of the monument as a resource of cultural memory, which seemed to have been appropriated by the dynasty, but which could also be attributed to a broader civic basis. At the end of the 4th century BC, the tomb was made visible within a walled enclosure. This is a sign of the continuation of the Greek *polis* in the Hellenistic period (cf. Giudice 2020). In the 2nd century AD the tomb 'was destroyed

73 Stucchi 1965, 111–114; Kreutz 2016, 38–39. No human remains, but also no grave goods were found.

74 '[…] the tomb of Battus was also monumentally elaborated and differentiated' (Scott 2013, 28). For a short description of these three phases see also Greco 2021, 307.

75 '[…] il venerato Heroon di Batto dovette cadere vittima di fanatismo, e non fu più reparato' (Stucchi 1965, 241); 'Sul lato orientale della piazza l'Heroon di Batto, vittima illustre del fanatismo giudaico, non fu più ricostruita […]' (Ensoli 2000, 77); 'eliminazione dell'Heroon' (Giudice 2008, 15).

76 Malkin 2003, 157: 'But even supposing that the kingship in Cyrene had become unpopular this does not imply that people began despising their founder and first king'; Malkin 2003, 169: 'So we see that Battos really did not become such an embarrassment in democratic, forth-century Cyrene.'

77 Chevrollier 2016, 53–56 discussed the use of the name Battus in the Roman period as evidence of distinction in an aristocratic discourse.

78 Scott 2013, 25 on the redesign in a contemporary form: 'thus denying him a perceptibly historical presence'.

79 According to Giangiulio 2001, 135–136, Herodotus's historical variants are also unthinkable before the democratic form of rule.

in the revolt, seems to have been left unrepaired' (Scott 2013, 38).[80] According to Scott (2013, 41), under Marcus Aurelius 'the tomb of Battus, left visibly in ruins since the Jewish revolt, was now completely covered over [...] with the construction of a new stoa, [...] The agora's Greek heritage, apart from the temple of Apollo, had all but disappeared.' Giudice's reconstruction instead highlighted Hadrian as the new founder of Cyrene. Symptomatic of this would be the rebuilding of the temple of Apollo Archegetes and buildings for public meetings, but not the tomb of the original founder (Giudice 2008, 20).[81] The construction of a hall on top could therefore be interpreted as deliberate destruction. Or perhaps a *heroon* no longer had any significance as an identity-forming resource. In Assmann's terms, this would most likely be automatic forgetting (Assmann 2016, 30–36). Social forgetting was then combined with material disposal.

Changing Ethnicities and (Dis-)Continuities of Identity. Cultural Memory and Cultural Obsolescence in Relation to the *Heroon* of Poseidonia

According to Strabo (5, 4, 13), Poseidonia was founded on the coastal plain of the Gulf of Salerno by settlers from Sybaris, after a first attempt on a promontory to the south (cf. Greco 2014, 20–24).[82] The earliest archaeological finds of necropoleis and settlement point to a date just before 600 BC (Greco 2014, 27). Between the Archaic and Hellenistic periods, the settlement, first Poseidonia, later Paestum, underwent a change of political orders and ethnicities from a Greek city, then to a Lucanian influx and finally to a Latin colony under Roman rule (Mele 1996a; 1996b; Gualtieri 2013). From the beginning, Greek Poseidonia was planned as a large city with a strict separation of residential and public or sacred spaces (e.g. Mertens 2006, 164–169 with 166 fig. 287; Longo 2012, fig. 449–450; Schweizer 2020, 52 fig. 1; 2022, 64 fig. 1). For the latter, large areas in the city centre were kept free from the beginning.[83] A conquest by the Lucanians reported by Strabo (6, 1, 3) was archaeologically correlated with a change especially in burial customs at the end of the 5th century BC (e.g. Mele 1996a, 18; criticised by Gualtieri 2013, 376–377; Nowak 2014, 41–50). In 273 BC, however, the city was re-established as the Latin *colonia* Paestum (cf. Torelli 1999a; 1999b).

The monumentalisation of Poseidonia began with large sacred buildings. South of the *agora*, the temple known as the 'Basilica' and attributed to Hera (Mertens 2006, 139–148) was built between 560–520 BC (Zuchtriegel 2022, 53), the so-called temple of Neptune (Mertens 2006, 283–295), tentatively attributed to Apollo (cf. Zuchtriegel 2022, 53), between 500 and 470–460 BC with modifications in design (Mertens 2019; Zuchtriegel 2022, 90), possibly on the site of an earlier temple (Mertens 2006, 339). The temple of Athena in the north (Mertens 2006, 222–227) was constructed around 500 BC replacing a building

80 '[...] là dove fino alla rivolta giudaica si trovava l'Heroon di Batto e, dopo, erano ancora visibili le sue rovine' (Stucchi 1965, 255). Cf. Chevrollier 2019, 196; for Chevrollier 2019, 198–199 the ruined state of the Temple of Zeus, which had also been destroyed, was intended to preserve the memory of the events of 115/6–117: 'la visibilité des ruines renforçait le processus de construction d'une mémoire et d'une identité civiques dans la Cyrène impériale'.

81 'Emblematica, a tal proposito, è la scelta di obliterare [...] la tomba di Batto' (Giudice 2008, 20). Cf. Laronde 1988, 1038: '[...] cet hérôon fut abandonné lors de la révolte juive de 115–117, ce qui paraît surprenant quand on sait le soin avec lequel Cyrène releva alors ses sanctuaires; [...].'

82 Cf. Hall 2008, 391, 404 on further sources and different founding stories. In this section, parts of Schweizer 2020; 2022 were updated. Overlaps could not be avoided.

83 According to Longo (2012, 335), the street network and the spacing of the agora were defined around 520/510 BC. See Mertens 2006, 164 for earlier discussions of different orientations of sacred buildings and street networks.

from the early 6th century BC (Zuchtriegel 2022, 40). When Paestum was refounded as a Latin colony in 273 BC, these sacred spaces continued to be used for sacred purposes (Torelli 1999a, 43–54), while large parts of the former Greek *agora* were fundamentally redesigned (Longo 2012; 2014). In the southern part, a *forum* was established, with *comitium*, *curia*, amphitheatre and baths, as well as several buildings for the cult of the emperor and the gods (Torelli 1999a, 20–27; 1999b, 46–49). This 'agora inferiore', with remains of older buildings that cannot be interpreted with certainty, but with remains of two *stoai* from the 4th century BC, may also have served more commercial purposes in Greek times.[84]

In the northern part of the Greek *agora*, the 'agora superiore', there were originally buildings with political and religious functions:[85] the *ekklesiasterion* from around 470 BC, supplemented in the 4th century BC by an altar with an Oscan inscription (Mertens 2006, 337–338), the small temple 'T', possibly built for Zeus in the 4th century BC, and the *heroon*. In Roman times the *ekklesiasterion*, the former meeting place of the Greek community, was abandoned and used as a quarry. It has been suggested that this part of the former *agora* was then largely covered by private buildings due to the loss of political functions (Longo 2012, 335).

For the *heroon*, however, continuities and changes between the Late Archaic and Roman periods have been identified (Mertens 2006, 166–167, figs. 288–289; Greco 2014).[86] Originally it was built as a monumental cist grave in a depression cut into the natural rock and covered by a tumulus (Greco 2014, 32, figs. 28–29). To the east of it, bowls and *lekythoi* have been found that are contemporary with the *heroon's* objects. These clay vessels, including those with graffiti with the letter *my* and therefore associated with a founder Megyllos (Ficuciello 2014b, 47 figs. 41; 48; Greco 2014, 17 figs. 18; 43), were interpreted as the remains of the foundation sacrifice (Ficuciello 2014b, 45–50 for the excavation, 46–47 for the deposit). When the Latin colony of Paestum was established, the tomb was repaired and provided with a new roof, but it was now surrounded by a walled compound.[87] However, the things found inside the tomb during excavations in the 1950s (cf. Ficuciello 2018 for the history of the research) date back to the 6th century BC.

The 'grave goods' consisted of eight bronze vessels on the sides of the chamber, iron spits on a stone table in the centre and an Attic painted amphora (Greco 2014, 8–15 figs. 7–16, 50–57 figs. 42–47). The iron spits represented animal sacrifice, the consumption of the divided meat. The bronze vessels were filled with honey, a substance that has been associated with the process of heroisation (Greco 2014, 40). Correspondingly, the images on the painted amphora depicting the introduction of Hercules to Olympus and Dionysus with Hermes can be seen as metaphors for heroisation.[88] Thus, a grave in a familiar form, but monumentalised in the centre of the city, and 'usual' things in an unusual combination and function

84 Summary of the findings in Longo 2012, 335–336.

85 Longo 2012, fig. 450 with phase plans of the 5th and 4th centuries BC; see Longo 2014, 255–256; 2012, 335 also on deposits from the Archaic period that have become known through recent excavations.

86 On the state of research also Ficuciello 2014a; 2018, 203–205; Zuchtriegel 2022, 59–64. Hall 2008, 411 is critical of the overall interpretation as *heroon*, while Rausch 2000 is completely different. Cf. the discussion of Schörner 2007, 152–167; Greco 2021, 313–315. Cf. Ficuciello 2018, 205–211 on new investigations east of the Peribolos.

87 Ficuciello 2014b, 48–49 on the most recent finds within the *recinto* from the early 3rd century BC, which corresponds to the time of the colony's founding in 273 BC.

88 Represented are two of the few mythological figures that can be associated with an *apotheosis*. See Lyons 2010, esp. 80–82.

show how material was used to create a central monument of urban or settler identity and at the same time a central place of civic cult. As a symbol of the community and a resource for citizens, the tomb was already a fictional grave at the time of its construction.[89]

The further cultural embedding of this monument of religious and civic identity over time can be reconstructed by comparing the images and furnishings of the tombs in the necropoleis. In the so-called Tomba del Tuffatore a banquet in Greek style was painted (e.g. Zuchtriegel 2018), in other tombs of the 5th century BC equipment and vessels represented the realm of athletics (e.g. Cipriani 1989). In contrast, in tombs of the late 5th century BC bronze helmets, breastplates and belts were discovered, like those of the warriors depicted in 4th century BC grave paintings from Poseidonia. These weapons and warriors were associated with a kind of 'decolonisation' of Poseidonia, at least with an integration of Lucanians (cf. Nowak 2014, 41–50).[90] At the same time, however, the city centre, the temples and the *agora* were characterised by a continuous use or further development with the construction of buildings that were related with political and religious functions of a *polis* (cf. Longo 2012, fig. 450; 2014, 255–256; Svoboda-Baas 2019, 19–54). The clearest evidence of this change is the 4th century BC altar with an Oscan inscription in the *ekklesiasterion* (Mertens 2006, 337–338). Furthermore, a literary text with a core dating back to the late 4th century BC and directly related to the *heroon* (e.g. Ficuciello 2014b, 49)[91] criticises the inhabitants of Poseidonia for having completely abandoned their Greek language and customs:

'[…] Aristoxenus, in his book entitled Promiscuous Banquets, says: "We act in a manner similar to the people of Paestum who dwell in the Tyrrhenian Gulf; for it happened to them, though they were originally Greeks, to have become at last completely barbarized, becoming Tyrrhenians or Romans, and to have changed their language, and all the rest of their national habits. But one Greek festival they do celebrate even to the present day, in which they meet and recollect all their ancient names and customs, and bewail their loss to one another, and then, when they have wept for them, they go home. […]" ' (Athenaeus, The Deipnosophists XIV 632; translation by Yonge 1854).

Finally, with the establishment of the colony of Paestum, the *heroon* was incorporated in a completely different social and cultural context. Nevertheless, the *heroon* was preserved, albeit architecturally redesigned, and subjected to a kind of historical conservation. Even the foot of the Attic amphora was repaired with lead (Greco 2014, 39 fig. 36 a).[92] Longo called the monument the 'memoria di Poseidonia' (Longo 2014, 256). Greco, however, suggested that although the sacred space had been preserved, the original discourses for it had probably been lost (2014, 35).[93] In any case, in contrast to the usually assumed breaks in urban development and history between the Greek, Lucanian and Latin/Roman phases (most recently Humm 2018), continuities of use and gradual redesign must be seen for Poseidonia/Paes-

89 According to Greco 2014, 43 and also Mertens 2006, 166 a cenotaph.

90 On the 'Lucanian' phase of Poseidonia and Paestum cf. Crawford 2006, 64–67; Zuchtriegel 2022, 93–126 with regard to images and architectural forms.

91 Torelli 1999b, 52, 77–78 connects the passage with the Heraion on the Sele. For interpretations of the text, see the overview of Gualtieri 2013, 380–382.

92 For Greco (2014, 39), however, the repair was carried out in order to deposit the image of the introduction of Hercules to Olympus. Thus, for him, the built space, the core of the *heroon*, remained untouched since the late 6th century BC (Greco 2014, 9).

93 Ficuciello 2014a, 250 on opposing (?) opinions: as well as the *ekklesiasterion* 'fu eliminato anche il sacello', but 'fu anche restaurato'.

tum (cf. Crawford 2006; Gualtieri 2013). The best evidence of this is the *heroon*, which has existed in very different cultural contexts over a period of at least 300 years. The use of the past, including the bronze vessels and the iron implements as well as the Attic amphora, as a resource of identity, as early as the late 6th century BC and then into the 4th century BC, and fundamentally changed in the early 3rd century BC, characterised the construction and maintenance of cultural memory. However, as the short text of Aristoxenus shows, cultural memory is accompanied by a kind of cultural obsolescence[94] or forgetting.

Heroa as Resources of ResourceCultures

The historical sources and monuments discussed here attest to the particular importance of the *heroa* of city founders for discourses and negotiations of civic identities in relation to the past. They were the real or fictional tombs of the city's founders. They were part of sacred areas, landmarks for offerings and processions, and points of reference for the preservation or conflict of political power and rule. They were sites of memory or monuments of cultural memory, because cultural and social identities were constructed through recourse to the past with specific media, spaces and things, within a specific epistemic setting.[95] It was a matter of 'functional memory' or 'inhabited memory' with group relevance, selectivity, value relations, future orientation and identity building in the sense of Aleida and Jan Assmann (1994, 123; engl. Erll 2011, 36 table II 3).

Heroa of city founders could be seen as prime examples of cultural memory in the sense of Aleida and Jan Assmann, for 'What counts for cultural memory is not factual but remembered history. One might even say that cultural memory transforms factual into remembered history, thus turning it into myth. Myth is foundational history that is narrated in order to illuminate the present from the standpoint of its origins' (Assmann 2011, 37–38). This notion was based on a conceptualisation of myth as a foundational, legitimising and world-modeling narrative ('fundierende, legitimierende und weltmodellierende Erzählung': Assmann/Assmann 1998, 180, 185–195) relevant to religious studies and ethnology. This, in turn, was particularly influenced by Bronislaw Malinowski with his insistence on ('charter') myth as an active cultural force of ethnicities – for Malinowski, tribes – strongly linked to 'their ritual acts, their moral deeds, their social organization, and even their practical activities' (Malinowski 1926, 11).[96]

Transferred to the field of early Greek history: 'The forms of the remembered past entail myth and history without any distinction between them. The past that is fixed and internalized as foundational history is myth, regardless of whether it is fact or fiction' (Assmann 2011, 59). In the examples discussed here, Brasidas was seen as a founder, as was Rhesus, or Battus, as was Apollo. Megyllos was compared to Hercules. In Assmann's words: 'Through

94 The term has been adopted from Connerton 2008, 66–67, who, however, applied it to products of modern consumer society. Cf. Assmann's 'automatical forgetting' (2016, 30–36).

95 Cf. Malkin 2003, 156: 'one also needs to take into account non-narrative, social and religious modalities of collective memory'; Malkin 2003, 164: 'We need to look at such non-oral *lieux de mémoire*, limiting and fossilizing the elements of tradition.'

96 'Myth fulfils in primitive culture an indispensable function: it expresses, enhances, and codifies belief; it safeguards and enforces morality; it vouches for the efficiency of ritual and contains practical rules for the guidance of man. Myth is thus a vital ingredient of human civilization; it is not an idle tale, but a hard-worked active force; it is not an intellectual explanation or an artistic imagery, but a pragmatic charter of primitive faith and moral wisdom' (Malinowski 1926, 23).

memory, history becomes myth. This does not make it unreal – on the contrary, this is what makes it real, in the sense that it becomes a lasting, normative, and formative power' (Assmann 2011, 38).

However, the materiality of the *heroa* opened up the possibility of being incorporated into changing state, political and ethnic contexts. Obviously, they are signs of an ongoing reconstruction of the past. Despite their importance, the founders could be – easily? – replaced, damned or forgotten, the monuments could be destroyed and even forgotten, or relocated, re-erected, but also redesigned in a contemporary form.

Discourses on the founding of cities in the context of the so-called Great Greek Colonisation have regularly focused on push and pull factors, 'resources' in the sense of favourable or unfavourable conditions of raw materials and 'natural' environments and the resulting social crises. These points have already been made in the historical sources discussed here. Wood, timber, gold and control of territories and trade routes were mentioned as 'resources' of war, as well as drought and expulsion as opposed to a land with a good water supply in a place with 'a hole in the sky'.

Conceptualising resources as means of constructing, maintaining and transforming social relations and units, as well as identities, also allows us to see *heroa* as resources in the sense of networks or bundles of tangible and intangible elements. Material dimensions included sacred orders and bodily practices such as sacrifices and processions, presentations of texts, images, things and monuments in spaces set apart. Intangible dimensions were related to meaning and value, such as knowledge of the past, notions of sociality or identity, religious imaginaries, ritual practices and imaginaries of urban and extra-urban landscapes. The *heroa* of city founders were seen as part of the 'resource production' specifically associated with these foundations (Itgenshorst 2021, 104),[97] whether as part of an initial institutional act or as part of the construction of cultural memory.

ResourceScapes were constituted and centred on *heroa*, sustained by regularly held feasts with various kinds of sacrifices or games, or by poetry and poetic performances commemorating the foundation myth. They 'can be analysed as ResourceComplexes [Teuber/Schweizer 2020], if seen as a network of distinctive purposeful linked elements, or as ResourceAssemblages, if the focus is more on relational contingent connections' (Schweizer 2021, 310) through time, on historicity. This refers not only to the ongoing recontextualisations of the monuments and their respective mythical narratives within continuous changes in politics, populations and forms of rule, nor to climatic or man-made ruptures. Changes in materiality were also important, leading to repairs, redesigns and new constructions.

The examples analysed therefore testify to a ResourceCulture that has *heroa*, monuments in the centre of the city, as its core. For research, this offers the opportunity to remove the monuments from their usual historical and archaeological narratives, in which they have been assigned to the historical time frame of their construction and, respectively, to the research field of colonisation. Looking at the long histories of these monuments as media of cultural memory and reconstruction of the past reveals continuities and changes that go beyond the usual periodisations of colonisation, decolonisation and Romanisation.

97 Itgenshorst 2021 with a differentiation between orientation on existing immaterial resources ('Ressourcenorientierung') and the production of new resources ('Ressourcenproduktion'). Both are resources of residence in terms of the differentiation between resources of residence and resources of mobility according to Delp 2022, 63–69.

On the other hand, *heroa* were focal points of the cultural memory of ResourceCultures, which could be part of interstate or intrastate struggles and wars, could be deliberately destroyed and rebuilt. Cultural memory then interacted closely with forms of forgetting such as repressive erasure and repressive forgetting, or a form of forgetting that was constitutive of the formation of a new identity, in the sense of prescriptive forgetting. However, forms of automatic forgetting or cultural obsolescence ultimately point to the end of *heroa* as resources and thus of the respective ResourceCultures. Cultural memory and forgetting are thus two sides of the same coin (cf. Assmann 2016).

6.10 Heritage as a Resource for Environmental Activism. A Case Study from the Garhwal Himalayas

Lokesh Ohri and Karin Polit

'You and I, Arjuna
Have lived many lives.
I remember them all:
You do not remember'

Krishna in the 'Bhagvad Gita' (cited in Uttaranchal Tourism 2002, 9)

Introduction: Heritage as a Resource

Garhwal, located in the mountainous state of Uttarakhand in northern India, is one of the regions referred to as *devbhoomi*, the land of the gods (Ohri 2019a). The landscape is ecologically fragile. The relatively young mountain range is regularly shaken by earthquakes of varying magnitude and is characterised by very steep slopes, thin soils and susceptibility to erosion, exacerbated by heavy monsoon rains. The region shows strong interdependencies between human and non-human actors.

Today, this fragile system is under threat as various human interests enter the region, some of whom include more-than-human actors in their visions for the future. While state actors and private stakeholders see the mountains as a place that provides ample resources for hydropower, irrigation, timber and tourism, the communities that have lived in and with the landscape for generations see it as a source of life and survival in a very different sense. These mountains also house the realms of their gods and sustain them both spiritually and materially.

This case study takes a closer look at these 'zones of awkward engagement' (Tsing 2004) between the various actors who live, work or travel in the region and those responsible for policy and its implementation to develop a deeper understanding of the interplay between environmental activism and heritage. Arguing that heritage can serve as a resource for environmental activism and for persuading state and private actors to agree on more sustainable planning for the future development of the region, we use the perspective of ResourceComplexes and the ResourceCultures concept to analyse the ongoing negotiation processes in the state of Uttarakhand.

In order to scrutinise this interplay of different interests and viewpoints, the ResourceComplexes perspective is useful to think analytically about heritage as both a resource and a network of resources. Cultural heritage as a resource forms the basis for how modern communities deal with their past, which creates a particular present and is thus always embedded in specific social and cultural frameworks. At the same time, however, cultural heritage is integrated into complex processes of cultural negotiation (see Hardenberg et al. 2017; Chapter 2), in which different resources are brought together in a network of 'cultural heritage'. Examining intangible and tangible cultural heritage from the perspective of ResourceComplexes helps us to understand, in general, how people in modern societies find commonalities in their coexistence and consolidate them in resources organised as networks around cultural heritage. Analysing how cultural heritage is created, negotiated, used and changed gives us clues as to how the past is made usable for the present. The interplay of people, institutions, practices, objects, past political processes and media of knowledge transfer creates cultural heritage (Brosius/Polit 2011). Once established, cultural heritage becomes a resource. We base this idea on the assumption that something can become cultural heritage if it anchors a past through value creation processes that can be used to construct identities in the present. In this case study, we also examine the power relations that frame both the production and the distribution, use or rejection of cultural heritage as a resource. Of particular interest is how the resource 'cultural heritage' is produced and then used to underline different narratives and interests that underpin different positions. For example, we examine the extent to which cultural heritage as a resource is used as a currency in conservation negotiation processes, and how it can be both divisive and exacerbating conflict, but also unifying and reconciling, for example by emphasising an inclusive, shared cultural heritage resource. According to our research, cultural heritage can function as a resource that enables people, even in extreme situations, to actively create their own spaces as a kind of counter-public (Warner 2002). We build on the basic assumption made in this book (see e.g. Chapter 2) that resources do not exist *a priori* outside of societies, but that they emerge in their specific forms through value-creation processes and are therefore dynamic and can also be used strategically. Seen from to the ResourceComplexes perspective, the network of resources around cultural heritage includes both material and immaterial elements, namely all objects, practices or places that are considered valuable at a certain time and under certain conditions, in a certain society, by certain groups, i.e. as anchors to make the past usable for identity constructions in the present (Polit 2011). For example, old Hindu temples, statues of a Hindu goddess carved in stone, a cave that became a temple of a Hindu god or performances such as the 'Ramman', a very unique local enactment of the Ramayana Epos in the Garhwal Himalayas, which at certain points in time come to be identified as national and specific cultural and intangible heritage through value-creation processes. These attributions are always dynamic and embedded in power relations, and therefore rarely consensual but rather contested. The very negotiations about what does and does not belong to the cultural heritage of a society, a religious community or a region, provide information about power and knowledge hierarchies that are highly informative for understanding cultural patterns of interpretation, social and political transformation processes and processes of coming to terms with the past (Polit 2012a).

The Production of Heritage

In this section we examine how cultural heritage is created and produced, used and evaluated as a resource within cultural patterns of interpretation. We show how cultural heritage, as a network of resources in the sense of a ResourceComplex, changes in the context of socio-economic and political transformation processes, and the role that cultural heritage plays as a resource for activist claims (environmental protection) and for negotiating similarities with more-than-human actors in the Garhwal Himalayan landscape. We show how different actors use cultural heritage as a resource for meta-narratives of dominant groups as well as for counter-narratives and historiographies of marginalised groups. We argue that the impacts of global climate change, global policies and national infrastructure projects threaten the intangible cultural heritage of gods, mountains and people alike. Counter-movements must therefore be understood as joint actions of these human and non-human actors, in which Himalayan people protect their relationship with the gods and mountains against outsiders' efforts to annex them for their own purposes. The data we present here are based on anthropological fieldwork and experiences in Garhwal over the last 23 (Polit) and 35 (Ohri) years. Before completing his doctorate in social and cultural anthropology at the South Asia Institute in Heidelberg, Lokesh Ohri helped set up a local NGO called REACH (Rural Entrepreneurship for Arts and Cultural Heritage), which still organises an annual heritage festival called *Virasat* (heritage) in the state capital Dehra Dun. We analyse grey literature such as brochures and pamphlets, oral texts recorded in the Garhwali language, unstructured and semi-structured interviews, informal conversations and information from years of participant observation.

The Land of the Gods

In ancient texts, academic writings and modern travel brochures, the Garhwal Himalayas in the northern Indian state of Uttarakhand are often referred to as *devbhoomi*, the land of the gods (Linkenbach 2007; Ohri 2019b; Polit 2012b). The people who live here experience their landscapes as divine landscapes, landscapes that they share with deities (Ohri 2019b). Most of the local deities, such as Narsingh or Dhari Devi, serve the communities as so-called *kshetrapals* or guardians of the landscape, its people, wildlife and resources, as well as *bhumiyals*, lords of the land (Niebuhr 2023).

While the region remains marginalised in terms of economic development, basic infrastructure and political recognition (Yadav et al. 2018), the landscape, ritual traditions and architecture are unique and recognised nationally as important cultural heritage sites. The Nanda Devi National Park and the Valley of Flowers are UNESCO World Heritage Sites, and one of the region's ritual performances, a local version of the famous Indian epic 'Ramayana' called 'Ramman', was inscribed on UNESCO's 'Representative List of the Intangible Cultural Heritage of Humanity' in 2009 (UNESCO 2009). While this has not brought significant economic benefits to the community, the award has inspired other communities to seek recognition of their own local traditions through similar programmes. This includes having the region's religious architecture, mostly associated with the Hindu god Shiva, recognised as a national heritage site. Associations with the great epics of the 'Ramayana' and 'Mahabharata' and the high gods of Hinduism are important in both local and national meta-narratives that construct the region as a centre of Hindu heritage. The construction of 'cultural heritage' as a resource is complex and multi-layered. It is evident, for example, in the name used for the region in many Hindu scriptures: Kedarkhand - the land of the god Shiva. The local high-altitude temple of Kedarnath (also called Kedareshvar) is a Shiva temple, and many ritual activities in the region are dedicated to Shiva. The Himalayas are generally considered to be the seat of Shiva, and his wife Uma (also known as Parvati) is worshipped in her various aspects throughout the region (Ohri/Bishnoi 2016). According to local legends, five local temples are associated with the five places where the five Pandava brothers found parts of Shiva's body, who had transformed himself into a bull to avoid a meeting with the Pandavas. It is said that after the great war of Kurukshetra, the Pandava brothers and their common wife Draupadi retreated to the Himalayas to seek Shiva's forgiveness for the sin of fratricide they had all committed during the war. A local myth told in the small town of Guptkashi says that the Pandavas came to Kashi in search of Shiva to atone for their bad *karma* and find salvation. The Lord of the Three Worlds, not wanting to impede the flow of *karma*, even for the sake of righteousness, quickly disappeared into the seclusion of the Himalayas. The Pandavas pursued him. Shiva roamed the Himalayan wilderness incognito in the form of a bull, and at Gupt, or the hidden Kashi, Bhima, the strongest of the Pandava brothers, sensed his presence. Shiva, however, managed to evade them here and at the other five places worshipped as Panch Kedar or the Five Kedars. He did, however, leave parts of his body in the region, which are said to remain visible in the five temples dedicated to him. His hands or arms remain at Tunganath, his face at Rudranath, his navel at Madhmaheshwar and his matted locks at Kalpeshwar. When Bhima finally succeeded in locating the bull in the marshes of Kedarnath, he could only briefly grasp the tail of the transformed Shiva as the latter plunged into the earth to avoid granting audience to sinners and thus salvation. So Kedarnath is Shiva's hump. It is said that the five temples, the Panch Kedar, were built by the Pandava brothers themselves in gratitude for the sight (*darshan*) of Shiva and the forgiveness he eventually granted them. The region of Kedarkhand is therefore associated with Shiva's body and is inherently a land of gods (*devbhoomi*), but is also strongly associated with the 'Mahabharata' and its heroes (Ohri/Bishnoi 2016). Many ritual traditions are associated with this great Indian epic, and many Rajput communities in the region trace their lineage back to the heroes of this great epic (Ohri 2019b). For most people in Garhwal, the 'Mahabharata' and the characters of this epic are more than just myths or characters in a story. The landscape of the region is linked to this epic, most of the local deities have a connection to the story, and thus the 'Mahabharata' is deeply connected to the Garhwali personality (Lipner 1994). For example, in many villages, individual episodes of the 'Mahabharata' are brought into the present by enacting them in the 'Pandav Nritya'. The 'Pandav Nritya' is a ritual performance of the lives of the Pandavas. This ritual practice is also a celebration of the local Rajputs, who associate their ancestors with the Pandavas and consider themselves descendants of the royal family. According to Purohit (1993), the specific episodes were usually performed during the ritual dance of the 'Pandav Nritya'. The ritual practices called 'Pandav Nritya' (the dance of the Pandavas) consisted of the recitation and singing of episodes from the 'Mahabharata' epic in the form of *jagars* (ritual ballads) and the dancing of the Pandavas in the bodies of human mediums in the courtyard of the largest house in the village. For ten days the Pandavas returned to the human realm to join their descendants. While bards and drummers from the oppressed castes sang the ballads, the Kshatriyas (Rajputs), who had been taken over by the Pandavas, danced the episodes. For the villagers, this is interpreted as part of the movement of the Pandavas themselves on earth in the human bodies of their devotees (Purohit 1993). Many inherited social roles in local communities are linked to this relationship with the great epic.

The gods need the people as much as the people need them. The gods belong to the region like the trees and the mountain goats. They make it possible for people to live here. This relationship defines them and is an integral part of their personality. Damage to the young and fragile mountains is therefore much more than the destruction of cultural heritage. It is the land of the gods that is being destroyed, and if the gods no longer live here, the world as it is known will come to an end. This is one of the reasons why catastrophes, often called 'natural catastrophes', such as floods and landslides, are framed differently in Uttarakhand, and why activism to prevent activities that could lead to further disasters is often also about local deities. In such situations, heritage discourse can be a resource for bridging the gaps between local populations and non-human agents on the one hand, and state actors as well as private stakeholders on the other. We will briefly demonstrate this by analysing the example of the largest of the Panch Kedar temples – Kedarnath.

The Re-Construction of Kedarnath

On the night of 16–17 June 2013, over 4,000 people died in a flash flood in the Kedarnath Valley (Niebuhr 2023; Whitmore 2018). So much rain had fallen in the preceding days that the Mandakini River and the Chorabari glacial lake burst their banks almost simultaneously in a flash flood. Within hours, the Kedarnath temple and the entire area around the Mandakini River were buried in mud and water. Hundreds of villages disappeared overnight, and thousands of people were swept away by the water and mud. The famous Shiva temple of Kedarnath itself was not destroyed, as a large boulder was rammed in front of the temple just before the mud flooded in (Niebuhr 2023). The few pilgrims who were inside the temple at the time also survived the disaster unharmed. Soon after the disaster, Indian Prime Minister Narendra Modi took an interest in the development of the temple and attached his name to its reconstruction. He announced his intention to rebuild the temple town and the path and roads leading to it. He was quoted in the media as saying:

'If so much money is to be invested and a huge infrastructure is to be created, then all environmental regulations will be taken into account,' and continued, 'It will be rebuilt in harmony with nature. It will be modern, but its spirit will be the same that the soil of Kedarnath has preserved for centuries.' (Joshi 2017).

Since then, Kedarnath has become a major tourist destination. The pilgrimage, once an act of devotion, has changed its meaning. The Uttarakhand tourism department estimated that around 29 lakh (2.9 million) people visited Kedarnath in 2022. It is a new beginning that serves the new Hindu nationalist idea of modern India, and at the same time brings such masses through the pilgrimage routes that local media reports say that around 300 people died there in 2022. The heritage designation has made Kedarnath a vlogger's paradise. Thanks to the Chardham Mahamarg Pariyojana, or simply the all-weather road that connects the four main pilgrimage sites in the region, travelling to Kedarnath has become much easier. Several helicopter companies operate in the region, dropping off pilgrims from the air several times a day for a quick *darshan* or audience with Shiva. This new reading of religious and cultural heritage, which requires an infrastructure that encroaches on the landscape to such an extent that major disasters occur time and again, is part of a new, different network of heritage resources. Kedarnath has quickly become a heritage site associated with Narendra Modi and a particular brand of modern and political Hinduism. So much so that in July 2024, the chief minister of Uttarakhand, Pushkar Singh Dhami, inaugurated the construction of a replica in New Delhi (Bhat/Talwar 2024). The idea is opposed by the priests and other functionaries of the Kedarnath temple, who claim that Shiva can only be present in the Himalayan temples because of his connection to ancient history, ancient scriptures and the sacred landscape of the Himalayas (Niebuhr 2023). According to several local voices, the balance of the Himalayan landscape, and thus the future of the entire country, depends on the benevolence of the divine energy residing in this region (Niebuhr 2023, 229–231). Restoring this balance, however, requires a more nuanced interpretation and use of the resource 'heritage'. As such, Kedarnath has become a contested heritage that has been used to support certain political agendas, while the same resource is increasingly being used as a resource for environmental activism and opposition to the extractivist and exploitative developments in Kedarnath since 2013.

Heritage as a Resource for Decolonial Environmental Activism

Being part of the resource network of 'cultural heritage in Uttarakhand', which includes the temple of Kedarnath, the river Alaknanda, several other deities, temples, myths, epics and ritual practices, is an embodied, existential experience for local actors. This does not mean that modernity and development are rejected; on the contrary, people want to be better connected to the rest of India. For them, roads and infrastructure mean better access to all the things that connect them to a modern lifestyle, including education and medical care. What they demand is respect for their needs, because their lives and cultures are deeply intertwined with the gods and the landscape, and this includes respect for the Himalayas and all the human and non-human agents that inhabit them. Going back to the local explanations for the flash flood of 2013, it is clear that even then the cause of the disaster was disrespect for the divine beings. The fact that the 1,200-year-old temple had survived the natural disaster seemingly unscathed was seen as a sign of Shiva's power. In the debate over the causes of the catastrophe, scientific and environmental arguments quickly merged with those of the local population, who suspected that their local gods were behind the disaster. In the preceding years, the central government in Delhi had been working with successive governments in the state of Uttarakhand on new dam projects. As a result, the seat of the king of Tehri was submerged, as was a locally important temple to the goddess Dhari Devi, which was relocated in 2012. While environmentalists argued that the ecological balance of the Himalayan region had been destroyed, many locals spoke of the threat to spiritual order posed by the modernist aspirations of governments.

For local people, these modernist aspirations are a continuation of the experiences of more than two centuries of British colonialism when rights to labour, forests, grazing and fishing were taken from the region's farmers and given to local kings. The immense forest wealth was misappropriated for colonial projects (Linkenbach 2007). Independence brought little change for the hill people, who remained part of the larger state of Uttar Pradesh, where they were a minority. The state, which saw large dams as the 'temples of modern India', continued to exploit Himalayan hydropower, timber and labour for the 'greater common good' of the nation (Kumar 2015, 290). The region continued to provide strong men for the Indian army, and its forest wealth was auctioned off to developers. Sacred rivers were dammed to meet the growing needs of an urbanising and industrialising India, with little improvement in local living conditions. After the creation of the state of Uttaranchal (later Uttarakhand) in the early 21st century, people's expectations for participation, recognition and improved livelihoods increased (Pathak 2021). However, the new state, like its predecessors, failed to meet these expectations or give local people control over water, forests, timber, land and wind. The question of sustainable modernisation that benefits both the local population and the ecosystem is more pressing than ever. For our colleagues at local universities and many environmental activists in Garhwal, the controversial development around Kedarnath as a heritage site and the associated construction of the all-weather road is a symbol of the marginalisation of local needs and disregard for divine rights. This construction is necessary to meet the demands of urban populations in the Indian plains for quick access to pilgrimage sites and to the electricity generated in the Himalayas that supports modern life in India's urban centres. However, it does not do justice to local needs or divine rights.

The growing importance of Kedarnath is reflected in the controversial transformation of Kedarnath from a sacred site with local significance to an international tourist site with high political significance. Both are part of different networks of heritage resources that have a direct impact on the shaping of the site through the restoration of the temple. State actors are trying to convince devout Hindus of a vision of a Hindu *Rashtra* (nation) through the creation of infrastructure, which in turn leads to environmental, social and spiritual damage. For centuries, Kedarnath has been characterised by the idea that ascetic practices such as fasting, celibacy and – in the case of Himalayan pilgrimages – arduous journeys lead to greater religious merit (Ohri/Bishnoi 2016). These different readings of the sacred site can be brought together through the resource of cultural heritage.

As Kedarnath has become increasingly important to different interest groups in recent years, it could become a resource as an important heritage site that allows for compromise between different parties. For example, by emphasising the proximity of Shiva to the Kedarnath marshes, a rethinking of the development of the temple region could begin – strengthening the arguments of environmentalists with those of cultural heritage. Emphasising the spiritual significance of the temple in the region and in India could continue to be a resource for political gain, while also leading to a renewed and reconfigured approach to the routes to the temple and the temple itself. We have seen this happen very cautiously in recent years, but with increasing support. In this way, cultural heritage as a resource and as part of resource networks in the sense of ResourceComplexes can play an increasingly important role for environmental activists who at the same time demand respect for local actors and contribute to a sustainable future.

6.11 Memories Are Liquid Stones. Resources and Languages of Memory in a Mobile Society – The Case of the Northern Apennines (8th–5th Centuries BC)

Raffaella Da Vela

Introduction

This case study[98] considers funerary stones as resources capable of supporting and constructing collective cultural memories. In the Northern Apennines during the Iron Age (8th–5th centuries BC), stones were used as markers of rich tombs, both undecorated and decorated with images and – in some cases – with inscriptions. These memorials were associated with personal and social mobility, as evidenced by onomastic details and the official titles included in some inscriptions. Moreover, their images and decoration attest to the intersection of different identities defined by age, gender, status and religion. From the ResourceComplexes perspective, it is possible to reconstruct for many of these stones the meshwork of resources

98 I wish to thank Julien Vogel, Patrick Rieger and Verena Meyer, who acompanied me in the research campaigns in the Apennines, as well as Rubymaya Jaeck-Woodgate for the stylistic editing of the text.

associated with the preservation of memory, the memory of mobility, personal and collective identities, social statements and images.

Also analysed from the ResourceAssemblage perspective, the spatialisation of collective memories in the urban necropoleis led to the formation of funerary landscapes in which different identities conferred by gender, status and religion were intertwined.

In a ResourceCulture of 'stones of memory', the solid materiality of funerary stones is transformed into a fluid means of expression and communication, capable of shaping social identities and negotiating power relations across generations. This interpretative framework invites us to reconsider ancient sources of human mobility in the region as mythological topographies rather than documents of colonisation processes. These topographies narrate the founding of cities and the settling of heroes after long journeys, and represent an attempt to reconstruct the memory of geographical and social mobility.

Materiality and Memory

In his essay 'Matter and Memory', first published in 1896, Henry Bergson explored the relationship between the materiality of images and the immaterial processes of memory associated with their perception and recognition. He proposed the existence of two kinds of individual memories, one capable of recording the images of the experienced past in the form of learned content, and the other, called imaginative memory, which arises from this learning process. This second form of memory emancipates the images from the past and stages them as part of the present (Bergson [1912^2] 2004, 92–93).

Although Bergson's argument must be framed within philosophical and psychological research on memory and cognition, his idea of an imaginative memory acting in the construction of the present is a valid introduction to this contribution. Here, the materiality of funerary stones is seen as a resource for the construction of social memories. The materiality of these stones is no longer limited to the raw materials, mostly of local origin, but also consists of their form, decoration, dimensions, position in the funerary landscape and inscriptions. These funerary stones, in the form of *cippi* (spherical or partially rounded) and stelai (stone slabs), were erected by local families or communities over or near graves. They functioned as memorials to the deceased and their social identities, but also as a means of reinforcing or challenging existing social structures.

Some thirty years after the publication of Bergson's essay, Maurice Halbwachs ([1925] 2022, 52–72, 181–190) challenged Bergson's vision of the relationship between matter and memory, denying that memories could arise from a simple contact of the brain with the material world. Rather, he placed this interaction in a relational or discursive dimension, in which every memory is the result of a process of socialisation: individual memories are socially framed and exist only as a process of communication within society.[99] However, given the persistence of such monuments over centuries within the funerary landscape, their values as media have changed over time. The question of the permanence and impermanence of the social message contained in this interaction between materiality and memory is approached here within the theoretical framework of ResourceCultures.

99 On the main differences between Bergson's and Halbwachs's approach to the materiality of memory, see Voigt 2014.

For local communities in this region and at this time, different resource networks around the funerary stones can be analysed from the perspective of ResourceComplexes as semantic systems aimed at meeting their specific social needs and tailored to their cultural backgrounds. These local clusters were not closed systems, but rather permeable to new ideas and communication strategies, thanks to the mobility of people and objects across the Apennines (Sassatelli 2018, 360; Santocchini Gerg 2021, 64). Innovations were introduced and became co-dependent with local habits and networks. For example, growing urbanisation from the second half of the 6th century BC onwards (Sassatelli 1990, 60–64) affected traditional imagery and introduced new iconographic schemes linked to Greek culture and the urban ideal of citizenship. These dynamics can be analysed using the ResourceAssemblages perspective, by observing how generational memories are linked to the materiality of the stones visible on ancestral graves over several decades. Despite the fluidity of the messages linked to the materiality of the funerary stones in space and time, these media of memory were very persistent and characterised a ResourceCulture of 'stones of memory' in the Northern Apennines over several centuries.

Geography and Chronology of the Funerary Stones in the Northern Apennines

The funerary stones that are now available for archaeological analysis constitute a rich corpus. Given their specific use and destination, these objects had both a practical and a symbolic function, since they marked the position of a specific grave within a necropolis, but also carried religious and social messages. In the Northern Apennines during the Iron Age, these markers are often found on or near the tombs of the social elite, since they are associated with monumental structures and rich burial goods. As the funerary stones have been grouped into several sub-regional clusters according to their morphology, geography and chronology, they will be distinguished in this discussion according to the current classifications (fig. 54). The two main morphological groups are *cippi* and stelai. *Cippi* tend to be rounded at the top or bottom and are further classified according to their main morphological aspect (*round cippi*, *cippi a cipolla* – in the shape of an onion, *cippi a clava* – in the shape of a club, etc., cf. Steingräber 2018, 12). *Cippi* are usually undecorated or only partially decorated. The term stelai is used for flat, discoidal, drop-shaped, square or horseshoe markers. Some classes of funerary stones, such as the '*pietre fiesolane*' (Cappuccini 2009; Amann 2018), could combine both aspects, consisting of a stele surmounted by a *cippus* (*cippi fiesolani*). The chronological designation of the objects refers to the period of production, which usually took place relatively close to the time when they were placed on the graves. However, it is very difficult to determine their period of use in the necropoleis, since not all archaeological contexts have preserved them in their primary disposal layers.

The oldest monuments are isolated. These include the few late Villanovan stelai from Felsina and the surrounding area (protofelsinean stelai: Marchesi 2000, 336), which are also interpreted as markers of the entire burial area (Sassatelli 1988, 204–211). Between the 7th and the first half of the 6th century BC the use of funerary markers was still exceptional, but is attested by the two *cippi* of Rubiera in the north-west of the Apennines (Amann 2008). At this time, rounded *cippi* with plain or decorated bases and club *cippi* were already in use in Pisa and its territory (Maggiani 2018; Sassatelli/Mancuso 2018, 103).[100] Between the 6th

100 *Cippi* and stelai have also been found in Volterra and the surrounding area (Ciampoltrini 1980; 1981), but have been excluded from the present analysis.

century and the beginning of the 5th century BC, the tradition of funerary stones seems to have spread in the upper and middle Arno valley, in the form of '*pietre fiesolane*' (Amann 2018, 63); in Pisa, decorated club *cippi* and square decorated *cippi*, such as the one known as 'cippo della Figuretta', are attested, and a new typology, bulb marble *cippi*, appeared in this city as well as in the north, in Marzabotto (Sassatelli/Mancuso 2018, 104; Maggiani 2018, 83–84). In the middle of the 6th century BC, coinciding with urban growth and the establishment of new settlement patterns, the production of felsinean stelai (*stele felsinee*) began in Bologna. This production showed a significant qualitative and quantitative increase during the 5th century BC and remained in vogue until the beginning of the following century (Morpurgo 2013; 2014; 2015a; 2015b).

Funerary Stones as Resources of Collective Memory

The function of undecorated *cippi* and stelai has alternatively been seen as delimiting the funerary space (Colonna 2015, 14) or preserving the memory of the deceased (Ortalli 2011, 65). However, in a society with a low level of literacy, these are not mutually exclusive, since undecorated *cippi* and square markers could have served as memorials, accompanying oral memory and at the same time as a pragmatic delimitation of the funerary space. The different functions covered different spheres of memorialisation. The oral tradition of the communities could have found a kind of material support in the stones, linking their position, shape and dimensions to the stories of the deceased and her or his family over several generations. Only the first two or three generations may have been able to preserve an oral memory of the deceased and their role in the community and among their descendants. Consequently, families whose memories were lost may have resorted to inventing descent from mythical ancestors in order to construct and maintain social status and power (Palumbo 1997, 46–48; Schweizer 2022). Given their permanency, funerary stones also covered the function of a long-term archive, preserving the location of the grave even after the extinction of genealogy and the loss of communicative memory (Assmann [2016] 2020, 36–37).[101] This function may have been particularly valued in this region, where most of the necropoleis were located near seasonally flooded streams.

This double function of preservation of memory and delimitation of sacred space can be analysed in the tumulus of San Jacopo in Pisa (fig. 54.1). The isolated tumulus, with a diameter of about 30 m, was located to the north of the ancient city, on the main route between Pisa, a river port at the mouth of the Arno, and the Apennines. In the same area other smaller graves covered by tumuli of smaller dimensions have been found (Mandolesi 2020, 53–54).

The peculiarity of this funerary monument, which can be dated to the end of the 8th century BC, is the absence of the body in the grave, which configures this tumulus as a cenotaph, a grave *in absentia* (Floriani/Bruni 2006). In terms of function, then, the whole context is clearly intended to ritually preserve the memory of a real or imaginary person (perhaps a hero or mythical ancestor). An altar was placed on top of the mound. Several undecorated stelai of different typologies and materials delimited the area of the tumulus, where feasts and rituals would also have preserved the collective memory of the community (Colonna 2015, 14).

101 The term was first coined by Jan Assmann to describe a form of collective memory in which knowledge is diffused within a group and communicated in everyday interaction (Assmann 2008).

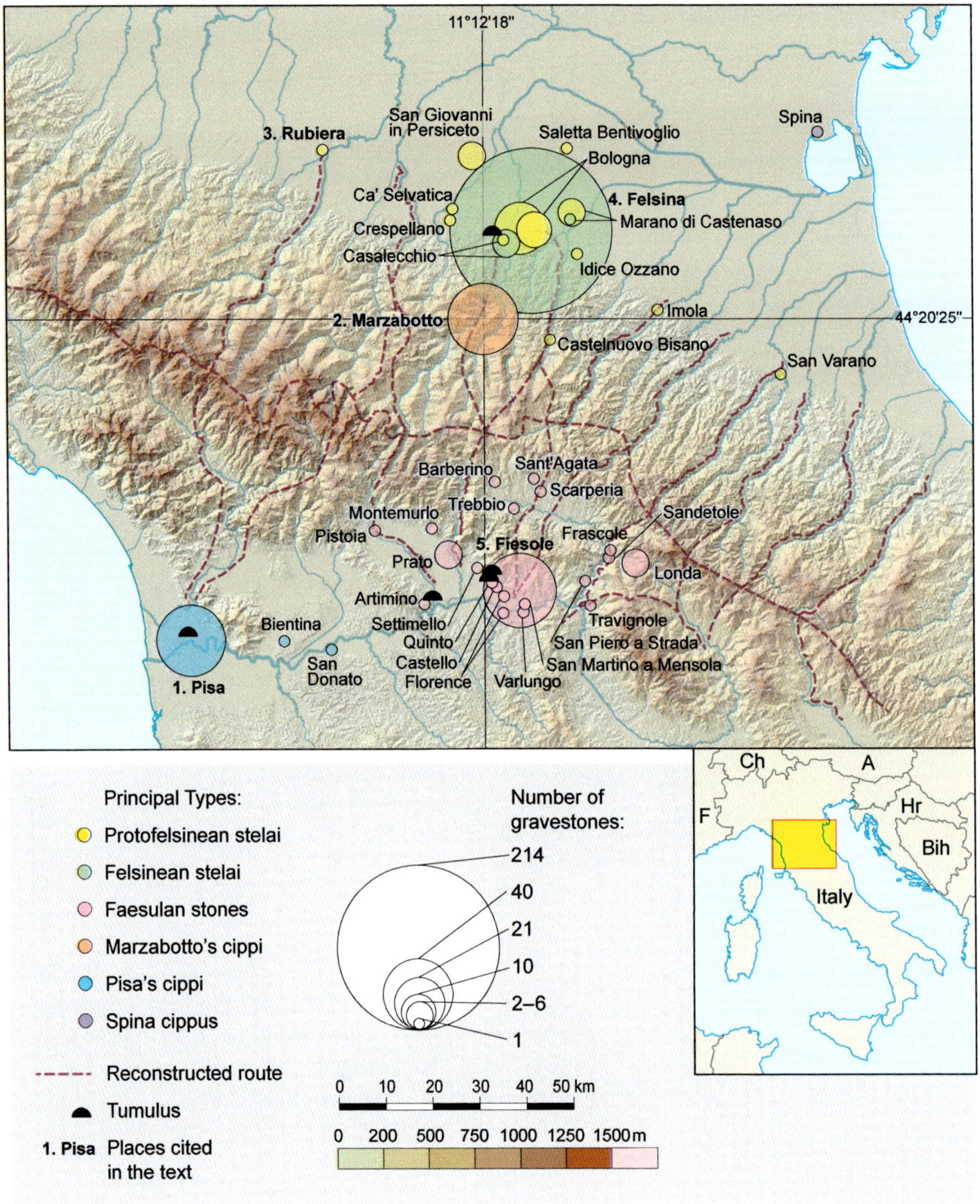

Fig. 54 Places quoted in the text with the quantitative distribution of funerary stones of different types (Data sources: CLMS EEA; Map: Raffaella Da Vela; Graphic design: Richard Szydlak).

The other rare contexts of aniconic funerary stones are not always so clear in their function, in terms of strategies of construction and preservation of cultural memory. The fluvial rounded stones or marmor *cippi* placed atop the individual graves in the necropoleis of Marzabotto from the 6th and 5th centuries BC (fig. 54.2) seem to be linked to an urban ideal of shared habits rather than to any intention of social distinction, thus leading to an isonomic conception of the funerary landscape.

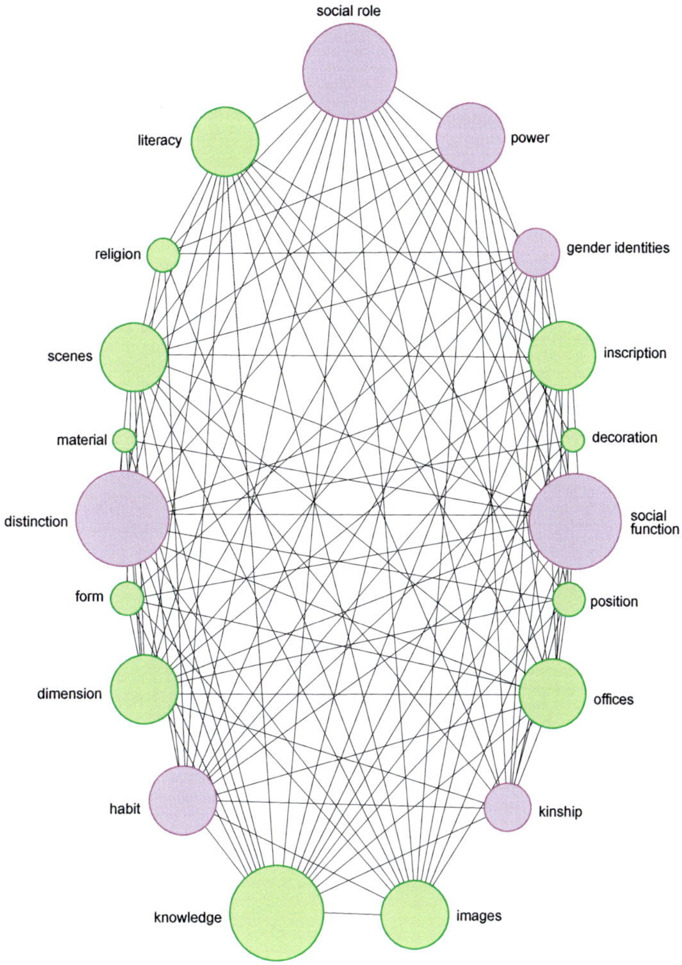

Fig. 55 ResourceComplex (Graphic: Raffaella Da Vela; Graphic design: Richard Szydlak).

Decorated and undecorated funerary stones often coexist within the necropoleis, suggesting that their function in preserving collective memories overlapped. The main difference lies in their value as media of cultural memory (Assmann 1988, 16), since the complexity and degree of institutionalisation of the message conveyed by the images and inscriptions implies a greater variety in the levels of communication at play here.

Decorated and Inscribed Funerary Stones as Resources

The introduction of decorations and inscriptions on funerary stones extends their function as media, suggesting that they were not only resources for the preservation of memory, but also part of a complex semantic network of material and immaterial resources (e.g. type of stone, dimensions, images, messages, inscriptions) and resource-related socio-cultural dynamics (e.g. representation, distinction, habits, literacy). This network can be analysed from the perspective of ResourceComplexes (fig. 55) as will be presented in the following discussion of the 'cippi di Rubiera' (fig. 54.3). These two funerary stones were found in 1983/1984 a few hundred metres from the left bank of a tributary of the Po, the Secchia, which rises in the Apennines and flows northwards (Marchesi 2011, 140–149). The *cippi* were found in fragments, embedded in a disturbed stratigraphy within a cobble quarry. However, given their size and state of preservation, it is possible that they crowned two graves of a neighbouring necropolis, some of whose remains have been found in the same area (Marchesi 2011, 145; Maggiani 2024, 147). The main settlement or group of settlements associated with this necropolis was probably close to the modern Rubiera, at the confluence of the route coming from the Apennines along the Secchia and another important route, later the Roman Via Aemilia.[102] Both funerary stones are made of sandstone, which is not found in the alluvial plains (Ortalli 2011, 58), but rather in the Apennines. The *cippi* both have a cylindrical body with an ogival top. The first 'cippo di Rubiera 1' (Bermond Montanari in Malnati/Bermond Montanari 1989, 1570–1573; 1986, 239–240; Marchesi 2011, 140–144) is preserved at a height of 1.41 m and a diameter of about 0.40 m. Although the surface is corroded, it is possible to reconstruct part of the decoration on three registers. The upper and middle registers are higher and contain figurative friezes with a procession of fantastic animals in an orientalising style. A smaller lower register is occupied by a palmette frieze. The upper part of the monument was decorated with painted motifs, which are now in a poor state of preservation. The fragmentary inscription is carved on two small bands, each above one of the figurative registers (ET^2 Pa 1.1): [a] *mi avileś ⁝ amθuraś ⁝ iṃa ⁝ ame ⁝* [-6/8-]*eiuś ⁝* l-*r* [b] [-?-]-*a* : *al* [ ).[103]

The second stone, 'cippo di Rubiera 2' (Bermond Montanari 1986, 240–243; Marchesi 2011, 145–148) can be reconstructed from two fragments. It was 1.77 m high, and the underground, foundational part of its base was 0.72 m high. The maximum diameter was 0.31 m. It was therefore slightly smaller than the 'cippo di Rubiera 1'. However, the organisation of the decorative friezes is similar: two superimposed friezes with processions of fantastic animals and felines, some of them with prey in their mouths. The animal friezes were framed above and below by two registers decorated with a row of palmettes. The top of this *cippus* was also painted with geometric and vegetal motifs, which have now largely disappeared. Unlike the first *cippus*, the second also has a vertical element that starts at the top and descends towards the cylindrical body, interrupting the horizontal friezes. This element has been convincingly interpreted by Petra Amann (2004, 209–210) as a woman's braid, demonstrating the anthropomorphisation of this stone (Amann 2008). The fragmentary inscription, which occupies the small rim between the figurative friezes, reads: (ET^2 Pa 1.2) *kuvei puleisnai mi iśive miśe*[--]*ḳṣ́*[-15/16-]*ẹṇḳe zilaθ miśalalati amake.*[104] The *cippus* probably belonged to a noblewoman, *kuvei huleisnai*, and was dedicated by a man with an official role in the community (*zilaθ*), a magistrate, a king or a military chief (Amann 2004, 211–213; Maggiani 2024, 150–158).

Taking into account the various discussions on their chronology, new findings in the area and stylistic comparisons, these two *cippi* should be dated between the last quarter of the 7th and the first quarter of the 6th century BC (for a synthesis of current proposals: Marchesi

102 Luigi Malnati in Malnati/Bermond Montanari 1989, 1570; vs Colonna 1988, 35 proposing the settlement in the area of the modern city of Modena; for further discussion: Amann 2004, 203.

103 An alternative reading has been proposed by Adriano Maggiani (2010–2013, 277): *mi avileś ḳamϑuraś : ana : ar* (o *m*?) x[---*v*?]*etuṣ́*. *Lạr*//[---] ϑX[---]*ia an*[---].

104 Only a partial translation is possible: 'I am Kuvei Puleisnai, the *iśive* (funerary monument? Memorial?) […]-enke, magistrate (*zilath*) of Misala had me do'. Alternative reading proposed by Adriano Maggiani (2024, 150): *kuvei huleisnai mi isive mi ś*xxxx[-20?-]*e*(?)*n*(?)*ke zilaθ misalalati amake.*

2011, 142–143). The inscriptions follow different graphic conventions for punctuation and for the shape of the letter θ, indicating different or non-normative writing cultures.

The most probable scenario is that these markers belonged to one or more tombs of a social segment that had the necessary economic, social and cultural capital to have the stones carved, decorated and inscribed. This social segment would have been aware of the decorative motifs of the Orientalizing Period also used in the Etruscan heartland, south of the Apennines, and would have possessed a form of literacy, albeit still with a non-normative orthography.

The analysis of these *cippi* from the perspective of ResourceComplexes allows us to understand their multiple layers of communication and social functions. On the one hand, through the durability of their material, these stones were intended to ensure the perpetuity of the personal memory of the family and the deceased, as suggested by the formal inscription and the anthropomorphisation that made the *cippus* a *simulacrum* of the deceased (or the ancestor), guaranteeing the permanence of her or his social status. On the other hand, they contributed to a collective social memory, as they were erected to be visible to the public. Celebrating Mediterranean family networks and customs, the markers were intended to maintain or construct hierarchies within the local community, but at the same time they served to disseminate images and cultural models, building a collective semantic memory (Papadimitriou 2019). Religious concepts, such as references to a local conception of the afterlife, were also part of this network (for these references, see in particular Krämer 2015 and Roncalli 2015).

Decorated funerary stones from other sub-regions of the Apennines provide further examples of such resource networks, in which the language of images and inscriptions constructed a complex system of memorialisation. In the following centuries, the '*pietre fiesolane*' (Amann 2018) fulfilled similar functions in the preservation, transmission and construction of social memory throughout the Arno Plain. Although their contexts have been lost, these stones clearly offer differentiated degrees of complexity, but the most basic represent a single figure, male or female, and more rarely a wild or fantastic animal standing on two legs, to fill the vertical field of the panel; the decoration of the most articulate and complex '*pietre fiesolane*' was composed in several registers, with feasting and banqueting scenes.

Decorated and inscribed *cippi* usually involve or address three generations: the first is that of the deceased, the second that of the dedicators who erected the monument and the third that of the future generations to whom the monument is addressed (see also Menichetti/Pellegrino 2020, 120–121). Since these *cippi* were embedded in necropoleis, the communication was not only intended for different generations within a family, but for the entire historical generation of their communities.

Urban Necropoleis as ResourceAssemblages of Generational Memories

The social statements and communicative memory entrusted to funerary stones in urban necropoleis were certainly an expression of competition between local families, but also the consequence of political dynamics within the city or settlement. Moreover, their effects extended over several generations, even if they were more immediately comprehensible to the most recent and closest generations. A similar situation is known, for example, within the Athenian necropolis of Kerameikos (Hölscher 2019, 145; Stroszeck 2020, 405–406), and we know that the Athenian lifestyle was a strong model followed in contemporary Felsina.[105] The form of the funerary stones and their iconographic system housed ideological programmes of memory, planned in interaction with the pre-existing monuments. The spatial and temporal dynamics arose as a continuum was created with each new burial, since new generations that placed stones with their own messages had to take into account the older pre-existing stones placed in the same burial space, even if they may have only partially understood their messages (Sassatelli 2015b, 58). In this way, the new monuments transformed the meaning of the old ones, but these older monuments also exercised agency over the new ones. From a ResourceAssemblage perspective, this mutual influence can be analysed as a longitudinal meshwork of resources and meanings embedded in each necropolis (fig. 56). The city of Felsina (fig. 54.4) is an ideal example for the application of this perspective, thanks to the preservation of about 220 funerary stones, most of them in disposal layers close to their original location. Here the first settlement, and later the city, was bounded by two streams, the Aposa to the east and the Ravone to the west (Sassatelli 2015c, 409–411). The necropoleis, in continuous use between the 8th and 4th centuries BC, were located outside the settlement, on the other side of the streams, along the main routes towards the Apennines.[106] In the middle of the 6th century BC, at the same time as the restructuring of the town plan, these roads were monumentalised with a partial cobblestone pavement and lateral ditches for water drainage; the stratigraphy indicates that the areas of the necropoleis were slightly sloping, creating the ideal scenario for the display of *cippi* and stelai along the sepulchral streets (Sassatelli 1988, 212–219). Not only the sides of the streets, but also other areas were reserved for specific groups of tombs (Desantis 2015a, 102; Sassatelli 2015b, 56). During the 5th century BC, the funerary stones (‘*stele felsinee*’) took on peculiar forms and became an expression of civic identity (Sassatelli 2015a, 106; 2015b, 56).

Although many of the necropoleis were excavated in the 19th century, with the consequent loss of their contextual information, recent detailed studies (Sassatelli/Govi 2007; Govi 2014; Morpurgo 2018) have allowed a partial reconstruction of the position of the stelai in the funerary landscapes, as well as their social and chronological stratification. Thinking of these funerary landscapes as ResourceAssemblages has the advantage of allowing us to visualise and analyse three layers of memory in their reciprocal interactions: 1) the intersectional layer; 2) the heterotopic layer; 3) the intergenerational layer.

1) The intersectional layer. Elisabetta Govi (2015, 119; translated by the author) has already given an excellent description of how the stones are mirroring a ‘civic society structured in classes and articulated in genders, which nevertheless leaves space for forms of individual expression’. For example, the topographical arrangement of the graves within the funerary space was clearly determined by social hierarchies. The stelai of larger dimensions were generally placed close to the sepulchral street and were associated with the richest assemblages of grave goods, revealing the habits and lifestyles of the social elite and their Mediterranean cultural and trade networks (Ortalli 2011, 64–65).[107] These funerary stones also bore more com-

105 Giuseppe Sassatelli (in Sassatelli/Cerchiai 2014, 201) suggests that new models within the funerary practices at Felsina during the 5th century BC were introduced from Athens.

106 An exception is the Villanovan necropolis of Piazza Azzarita, which was adjacent to the northern boundary of the settlement (Santocchini Gerg 2015, 38).

107 See, for example, the rich banqueting set found in the monumental tomb 109 of the De Luca necropolis, associated with stele no. 137 of a magistrate: Morpurgo 2015a, 85–86; 2015b, 125–126; for a similar interpretation of the ‘*pietre fiesolane*’: Poggesi 2014, 100. The association of funerary stones with the most prominent graves is also known for the smaller urban centres of the late Archaic and Classical periods, for example in Casalecchio di Reno, Zona A:

Fig. 56 ResourceAssemblage (Graphic: Raffaella Da Vela; Graphic design: Richard Szydlak).

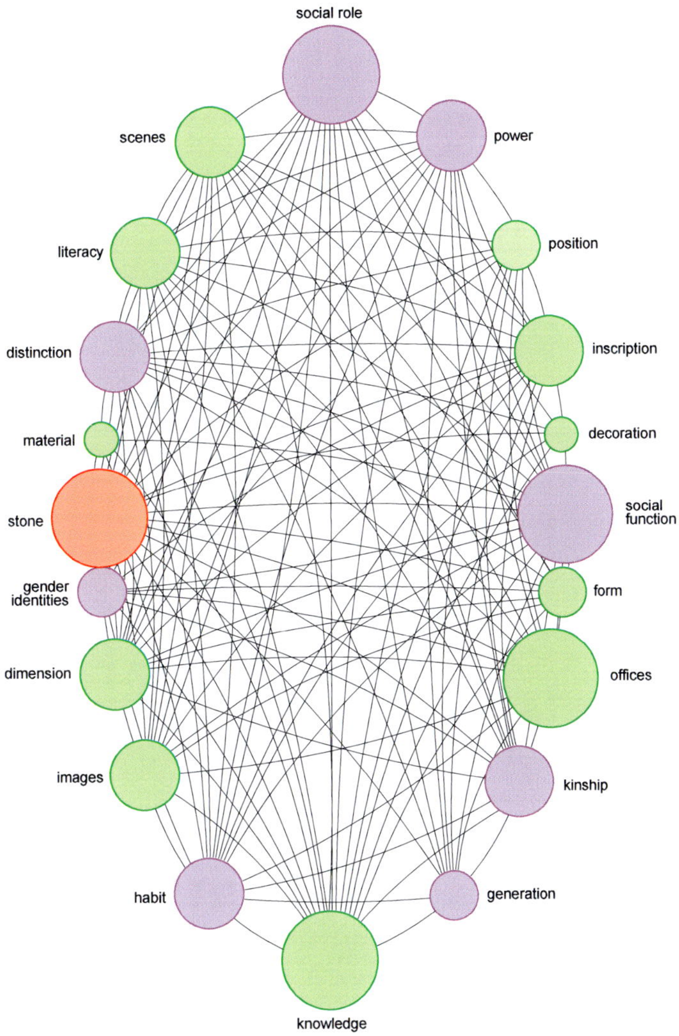

plex iconographies and inscriptions and, in some rare cases, references to the public role of the deceased, represented as a civic magistrate (Sassatelli 1988, 233–235; Govi 2015, 112). These stelai were reserved for men. Women, represented more frequently in the 5th century BC (Govi 2015, 113), were here confined to a more private dimension of memory, related to their role in the household and religious practices (Sassatelli 2015a, 107–108). In the inscriptions, their names are mostly given in the nominative case, while men's names appear in the genitive case, followed by a family name, underlining the collective family memory as the framework of their commemoration. It seems that women had an indirect right to memory, since they did not 'own' their stones as media of the memory, but were represented on them by their social group. This interpretation is confirmed by a few cases in which women's names are followed by the genitive of the husband's name (Sassatelli 1988, 243–244).

Ortalli 2011, 62. However, Giuseppe Sassatelli (in Sassatelli/Cerchiai 2014, 201) invites a certain flexibility in the interpretation of social hierarchies within the horizontal stratigraphy of the western felsinean necropoleis.

There is thus an intersection between the topographical position of the stones and the socio-economic and gender identities of the deceased, with the consequent overlapping of private and public levels of memorialisation. The magistrates' funerary stones represent this intersectionality well. Magistrates carry their symbols of political and religious power, such as *insigna*, sceptres and staffs (Sassatelli 1990, 82), and their inscriptions record their first name, family name and the office they held (Sassatelli 1988, 236–242). Famous examples are the stelai Giardini Margherita no. 10 (by *vel kaikna*) and De Luca no. 137. The latter, dating back to the first half of the 5th century BC (Sassatelli 1988, 246–255; Maggiani 2024, 160), expressed a redundancy of memory, both in images and words, and contributed to the wider dissemination of memory in society, including the analphabetic public. The inscription of no. 137 begins with the expression *mi suθi* ('I am the funerary monument', or, according to Rix [1982, 315]: 'I am the grave'). The funerary stone itself spoke to the onlooker or wayfarer, asserting its own function as a grave marker. The storage of the memorial for public offices in the stone archive of the stelai represents a choice towards the permanent preservation of civic memory. In other cases, the memory of the offices must have been limited to the oral tradition and finally to the staging and ritualisation ('rituelle Inszenierung': Assmann 2007a, 56–59) during the public funeral feast (for these feast practices: Cerchiai 2015, 59; Morpurgo 2015b, 123). This is the case, for example, of the tomb of Arnoaldi 80, marked by a large undecorated spherical *cippus*, in which the ashes of the deceased were placed on a magisterial chair (*sella curulis*), and also of the 'Tomba dello sgabello' in the necropolis of Giardini Regina Margherita, which contains a similar chair (*diphros*) (Morpurgo 2015b, 125).

2) The heterotopic layer. The monumentalisation of the necropoleis, with the consequent projection of urban social structures outside the city, represented a metaphorical extension of the urban space towards the outer territories and the city's networks across the Apennines and overseas, around the Mediterranean. This projection was not only a geographical expansion of the city's cultural and economic space, but also an extension of its imagery, with the assimilation and reworking of Greek iconography within the traditional scenes of Etruscan funerary sculpture (Govi 2015, 111; Sassatelli 2015b, 49–50). The Mediterranean-Greek references were not merely a representation of the socio-economic status and cosmopolitan habits of urban elites, but also a voluntary projection of imported images and models of consumption into the sphere of public memory, resulting in a consequent diffusion of the models themselves (Papadimitriou 2019, 246–247). The presence of these images in the grave goods assemblages and the associated funerary stones can be considered part of what Luca Cerchiai defines as the 'city of images'.[108] Moreover, several stones represented on the one hand the social status and role of the deceased, partly through their Greek way of life, and on the other hand her or his journey into the afterlife (from the 5th century BC onwards with full reference to the Dionysian cult). These assemblages of social life and the afterlife projected personal memories beyond the physical world. This alienation of memory has been noted by Elisabetta Govi (2015, 114) in relation to several scenes of the stelai. The funerary stones can thus be seen as Foucauldian heterotopias in which the communication of memory is projected in a parallel world that not only reflects the real world but also exerts an influence on it (Foucault [1966] 2021, 43–49). In this sense, the metaphor of the mirror

108 Cerchiai 2015, 59. The term is reminiscent of the French expression '*cité des images*', introduced in 1984 as the title of the famous work 'La Cité des images: religion et société en Grèce antique' with contributions by J.-P. Vernant, C. Bérard, F. Lissarrague, C. Bron, F. Frontisi-Ducroux, A. Schnapp and J.-L. Durand, who proposed an anthropological reading of Greek figurative pottery.

seems more appropriate to express this entanglement of social life and the afterlife on the stones (Govi 2015, 117–119), although the formal similarities between the felsinean stelai and the engraved mirrors found in Etruscan graves are still debated (Sassatelli 2015a, 106 vs. Roncalli 2014, 226–230; Pizzirani 2015, 77–78).

3) The generational layer. The concept of historical generations adopted here is based on the sociological category introduced by Karl Mannheim, who defined historical generations not as a concrete social group, but rather as an occasional group within society, bound by contingency in time (Mannheim 1952, 288–292). However, the association of historical generations with the political turn that occurred in the new urban structures in Felsina reinforces the idea of an intrinsic link between the concept of historical generations and that of the elite construction of public meaning, as proposed by Beate Fitze (2009).[109] The intergenerational function of stelai in the construction of memory can be explored archaeologically in terms of those belonging to the same family,[110] but it could also be extended theoretically to the wider civic community. It is not always easy to establish chronological relationships between graves with stones dedicated to the same family. In the necropolis of Giardini Margherita, for example, the inscription on the famous stele no. 10, depicting a magistrate on a ship (a navarch?), suggests a familial relationship with the inscription of a man on the nearby grave of stele no. 15, but a comparison of the epigraphic record and the grave goods shows no chronological gap and the graves probably simply belonged to two brothers (Sassatelli 1988, 243; 1990, 81; 2009, 835).

On the other hand, a good example is the recently excavated Grave 11 in via Saffi, where anthropological analysis has provided additional information. The large pit grave was used for three deceased, an adult (20–24 years old), a young adult (12–14 years old) and a third person, probably a child.[111] They were buried at different times, and the grave was reopened for successive burials, as confirmed by the chronology of the rich grave goods found at the site, which include a banqueting set. A stele with figural relief on both sides and an inscription was placed with the first burial, that of an adult man, *lars atiniu*, while two aniconic grave markers, egg-shaped and troncoconic, were placed on the grave after the last closure, probably after the burial of the child (Desantis 2015a, 111; 2015b, 131–135).[112] The decorat-

109 Fitze 2009, 105–106: 'Historische Generationen verstanden als kollektive Akteure konstituieren sich nur dann, wenn sich altersabhängige Situationsdeutung und Akteursposition (d.h. eine auf den öffentlichen Diskurs ausgerichtete Eliteposition) überschneiden [...] Generationsspezifische, kollektive Akteure konstituieren sich [...] durch die historisch kontingente Überschneidung von altersspezifischen Situationsdeutungen und Prozessen sozialen Wandels im Medium öffentlichen, politischen Diskurses' (translated by the author: 'Historical generations, understood as collective actors, are constituted only when the age-specific situational meaning and their actor position (meaning their role as an elite able to construct public opinion) intersects [...]. Generationally related, collective actors are constituted through the contingent intersection of age-specific and situational attribution of meaning, and the processes of social change in the medium of a public, political discourse').

110 Ortalli 2011, 65 Anm. 34. For the coherence of messages on stelai of the same family, for example in the Polisportivo necropolis, see Sassatelli 2015b, 57. Clusters of family graves in topographical proximity and marked by stelai are also well known in the area, for example in the necropolis of Casalecchio di Reno, Zona A, where the graves of an extended family, located near a paved street, can be dated between the end of the 6th and the end of the 5th century BC (Ortalli 2011, 62).

111 The presence of this third deceased is attested by a unique fragment of bone. As the grave was violated, it is difficult to determine whether the rest of the bones were lost as a result of the violation, or whether the child's bone was residual, and thus there were only two male individuals buried in the grave: Tanganelli et al. 2015, 138.

112 This coincidence between the number of grave markers and the number of people buried could suggest an individual connection with the stones.

ed stele bore two inscriptions made at different times. The first gave the name of the family, *atiniu* (which, for philological reasons, is considered to belong to the traditional elite of the city: Maras 2014, 337–338). This inscription is highlighted epigraphically by the doubling of the lines of the letters, coloured with a black rubrication, and the use of more space between the letters. The second inscription was carved later. Although it has barely survived, Daniele F. Maras has suggested it reading the word *mlach* (noble), which would be a way of reconfirming the social status of the deceased for the following generation (Maras 2014; 2015, 216–217). Considering the imagery of the stele, which contains iconographic elements appropriate to the time of the first burial in the second quarter of the 5th century BC (such as the Greek *chiton* and references to the Dionysian cult), as well as archaising elements such as the Italic chariot, the Etruscan *daimon* and the charioteer's hairstyle, Paola Desantis emphasises this interpretation, proposing the reconstruction of genealogies and the projection of family memories that take place in an invented past (Desantis 2015b, 132–135). If her interpretation is correct, the mixing of objects from different periods and the retrospective projection of collective memories by parts of successive generations could be a process similar to that which led to the formation of the Homeric poems, which grew out of different oral traditions.

If stelai can be seen as heterotopias, then the necropoleis in which they were embedded can be seen as heterochronias, in which real and imagined layers of time are intertwined and affect each other (Foucault [1966] 2021, 45–46). Unfinished or partially decorated stelai are quite common in Felsina (Sassatelli 1988, 233) and may indicate that these monuments could be reworked, integrated, remodeled and finished after being placed on the graves, thus becoming fluid in their materiality as they could physically change over time and allow for the manipulation of family memories.

Memories Are Liquid Stones: Material Permanence and Message Impermanence in a Mobile Society

The heterochronic space of interaction within the necropoleis affects the materiality of the stelai. Although their raw material was stone and thus solid, their materiality as resources in the construction of memory became fluid as they were moved, reworked and manipulated, as shown in the above-mentioned felsinean stele of Grave 11 on via Saffi. This fluid materiality was used throughout the Iron Age to shape social identities and negotiate power relations within the societies of the Northern Apennines.

The ResourceCulture of 'stones of memory' was very long-lasting and widespread in the region. This ResourceCulture can be seen in the networks of resources (approached from the perspective of ResourceComplexes) constituted by the funerary stones and their complex of socio-cultural dynamics linked to the preservation, construction and transmission of memory. In addition to such networks (and seen from the perspective of ResourceAssemblages), the necropoleis in which the stones were placed also showed a series of fluid interactions in which memories were reinvented to form an intergenerational dialogue. This dialogue was a process embedded in the historical and cultural framework of the region, characterised by increasing urbanisation and various forms of mobility. Both mobility and urbanisation consistently affect the preservation, transmission and invention of memories. While the framing function of urbanisation was reflected in the stones as media of memory, the framing function of mobility was embedded in the strategies of communicative memory. Urbanisation, for example, framed the memory of public offices by celebrating civic magistrates on their funerary stones. In this case, it could be useful to confront the different systems of memory created around the office of the *zilath* at Rubiera in the 7th century BC (Colonna 1988, 25), probably a small proto-urban complex based on clan structures, and the office of the same name – probably different – on stele no. 137 at Felsina, Arnoaldi necropolis, dating from the end of the 5th century BC (for the evolution of the term: Maggiani 1996, 101–109). In the former case, the *zilath* office was associated with a single body-like *cippus*, which represented the family genealogy and was probably related to a military power, as well as the name of the city in which he was serving (Amann 2004, 211). The later *zilath* memorialised in stele no. 137, however, was part of a social interaction with three other figures; he carried the symbol of his office and the city is not mentioned, as the scene is contextualised within the extended memory of the city itself (Hölscher 2019, 145). Instead, the framing function of mobility was embedded in the strategies of representation: in Rubiera, recurring Mediterranean images and a variable use of writing conventions; in Felsina, representing the journey to the afterlife with references to the Greek world, but also projecting the collective cultural memory along the major communication routes outside the city. It is therefore possible to conclude that urbanisation framed the political dimension of memory, while mobility contributed to the construction of knowledge orders.

A point of contact between these two frameworks can be found in the stories of the journeys of Greek and Etruscan heroes along the Po Valley and especially towards Bologna, as reported in Greek and Roman literary sources (Harari 2010, 52). These '*Meistererzählungen*', reflected in the Greek pottery produced on demand for the Etruscan market, as well as in the few mythical scenes on the felsinean stelai (Maggiani 1997; Cerchiai 1999; Sassatelli 2009, 836), preserved the memory of mobility and cultural contacts across the region, but at the same time constructed memories and identities for the new urban structures. Considering these stories within the ResourceCulture of 'stones of memory' allows us to read them as part of the construction of memory linked to the growing importance of the city, rather than as documentary sources of the so-called 'Etruscan colonisation' of the Po Valley. Although the later literary sources, given their etic perspective (Zamboni 2018, 237), are further biased by their own historical context, they still provide an echo of these mythical topographies constructed within a coherent system of reconstructed genealogies and celebrated civic structures.

Image and Map Credits

Fig. 6 Photo: Molgreen, CC BY-SA 3.0, <https://creativecommons.org/licenses/by-sa/3.0>, via Wikimedia Commons; <https://commons.wikimedia.org/wiki/File:20131002_xl_wiki_5203.JPG>

Fig. 10 Photo: Jonathan Martz, CC BY-SA 3.0, <https://creativecommons.org/licenses/by-sa/3.0>, via Wikimedia Commons; <https://commons.wikimedia.org/wiki/File:Strasbourg_Cathedral_(cropped).jpg>

Fig. 12 Map: 'Kleine Idrisikarte' of 1192 published 1926 by Konrad Miller; Digitisation: Universitäts- und Landesbibliothek Sachsen-Anhalt (ULB), CC BY 4.0, <https://creativecommons.org/licenses/by/4.0/>; <http://dx.doi.org/10.25673/59240>, cropped

Fig. 13 Photo: Wolfgang Staudt from Saarbrücken, Germany, CC BY 2.0, <https://creativecommons.org/licenses/by/2.0>, via Wikimedia Commons; <https://commons.wikimedia.org/wiki/File:Memorial_to_the_Murdered_Jews_of_Europe_(566295389).jpg>, cropped

Fig. 14 Photo: © Museum of Cultural History, University of Oslo, Norway/ Olaf Væring, CC BY-SA 4.0, <https://creativecommons.org/licenses/by-sa/4.0/>; cropped

Fig. 21; Fig. 40; Fig. 41; Fig. 45; Fig. 46
ESA COP 30; European Space Agency. Copernicus Global Digital Elevation Model. Distributed by OpenTopography. <https://doi.org/10.5069/G9028PQB>. Produced using Copernicus WorldDEM-30 © DLR e.V. 2010–2014 and © Airbus Defence and Space GmbH 2014–2018 provided under COPERNICUS by the European Union and ESA; all rights reserved.

Fig. 25; Fig. 29; Fig. 36a; Fig. 47; Fig. 48a; Fig. 48b
CGIAR-CSI SRTM, <http://srtm.csi.cgiar.org/srtmdata/>

Fig. 34 Open Topography – Global Bathymetry and Topography at 15 Arc Sec: SRTM15+ V2.5.5. Distributed by OpenTopography, <https://doi.org/10.5069/G92R3PT9>

Fig. 49 Goethe- und Schiller-Archiv (GSA); Geoinformation of pathways and waterways from Instituto Nacional de Estadística Paraguay, Digital Mapping 2012

Fig. 54 Eu_dem_v11_E40N20 (Copernicus.eu 2021), <https://land.copernicus.eu/en/products/products-that-are-no-longer-disseminated-on-the-clms-website>. This publication has been prepared using European Union's Copernicus Land Monitoring Service information, <https://doi.org/10.2909/393359a7-7ebd-4a52-80ac-1a18d5f3db9c>; European Environment Agency – Europe coastline shapefile, <https://www.eea.europa.eu/ds_resolveuid/06227e40310045408ac8be0d469e1189>

Bibliography

Abbott 1855

K. Abbott, Geographical Notes. Taken during a Journey in Persia in 1849 and 1850. Joseph Ratzinger Gesammelte Schriften 25, 1855, 1–78.

Adam/Vonderau 2014

J. Adam/A. Vonderau (eds.), Formationen des Politischen. Anthropologie politischer Felder (Bielefeld 2014).

Adey 2017

P. Adey, Mobility (London 2017).

Ahlrichs et al. 2018

J. J. Ahlrichs/K. Riehle/N. Sultanalieva, The Production of Liminal Places. An Interdisciplinary Account. Ethnographisch-Archäologische Zeitschrift 56.1–2 (2015), 2018, 205–242. <https://www.waxmann.com/artikelART102402>.

Alarcón García et al. 2020

E. Alarcón García/A. Moreno Onorato/M. Díaz-Zorita Bonilla/L. Arboledas Martínez, ¿Quiénes eran? La cueva de Biniadrís de Baix Vell (Alaior, Menorca, Islas Baleares). Revista de Menorca 99, 2020, 113–134.

Algar 1998

H. Algar, EMĀMZĀDA i. Function and devotional practice, last updated 2011, <https://iranicaonline.org/articles/emamzada-i> (last access: 01.08.2023).

Alizadeh 2010

A. Alizadeh, The Rise of the Highland Elamite State. Enclosed or Enclosing Nomadism? Current Anthropology 51, 2010, 353–383.

Altamirano García/Alarcón García 2018

M. Altamirano García/E. Alarcón García, Bone Tools for the Deceased. Approaches to the Worked Osseous Assemblage from the Bronze Age Funerary Cave of Biniadrís (Menorca, Spain). Quaternary International 472.A, 2018, 108–114.

Amann 2004

P. Amann, Die etruskischen 'Cippen von Rubiera' aus der südlichen Poebene. Neue Vorschläge und Versuch einer Einordnung. In: H. Heftner/K. Tomaschitz/G. Dobesch (eds.), Ad fontes! Festschrift für Gerhard Dobesch zum fünfundsechzigsten Geburtstag am 15. September 2004 (Wien 2004) 203–217.

Amann 2008

P. Amann, Intorno al cippo II di Rubiera. In: G. M. Della Fina (ed.), La colonizzazione etrusca in Italia, Annali della Fondazione per il Museo Claudio Faina 15, 2008, 247–272.

Amann 2018

P. Amann, 'Le pietre fiesolane'. Repertorio iconografico e strutture sociali. In: S. Steingräber (ed.), Cippi, Stele, Statue-Stele e Semata. Testimonianze in Etruria, nel mondo italico e in Magna Grecia dalla prima Età del Ferro all'Ellenismo. Proceedings Sutri 24th–25th April 2015. Mousai 8 (Pisa 2018) 63–79.

Amores Carredano/Fernández Cantos 2000
F. Amores Carredano/A. Fernández Cantos, La necrópolis de la Cruz del Negro. In: P. Bazán (ed.), Argantonio Rey de Tartessos (Sevilla 2000) 156–163.

Anand 2017
N. Anand, Hydraulic City. Water and the Infrastructures of Citizenship in Mumbai (Durham 2017).

Andermann/Bünz 2019
K. Andermann/E. Bünz (eds.), Kirchenvogtei und adlige Herrschaftsbildung im europäischen Mittelalter. Vorträge und Forschungen 86 (Ostfildern 2019).

Andrina 1992
C. D. Andrina, Das Familienbuch der Herren von Eptingen. Kommentar und Transkription. Quellen und Forschungen zur Geschichte und Landeskunde des Kantons Baselland 41 (Liestal 1992).

Anglada et al. 2017
M. Anglada/A. Ferrer/D. Ramis, Actividad humana en el litoral de Menorca durante la prehistoria. In: L. Gómez-Pujol/G. X. Pons (eds.), Geomorfología litoral de Menorca. Dinámica evolución y prácticas de gestión. Monografies de la Sociedad d´Història Natural de les Illes Balears (Palma de Mallorca 2017) 213–223.

Ansari-Renani et al. 2013
H. Ansari-Renani/B. Rischkowsky/J. P. Mueller/S. Momen/S. Moradi, Nomadic Pastoralism in Southern Iran. Pastoralism. Research, Policy and Practice 3.11, 2013, 1–25.

Apphun 1994
H. Apphun, Johann Siebmachers Wappenbuch von 1605. Die bibliophilen Taschenbücher 693 (Dortmund 1994).

Apter 2005
A. Apter, The Pan-African Nation. Oil and the Spectacle of Culture in Nigeria (Chicago 2005).

Aranda/Esquivel 2007
G. Aranda/J. A. Esquivel, Poder y prestigio en las sociedades de la Cultura de El Argar. El consumo comunal de bóvidos y ovicápridos en los rituales de enterramiento. Trabajos de Prehistoria 64.2, 2007, 95–118.

Arboledas Martínez et al. 2022
L. Arboledas Martínez/A. Moreno Onorato/F. Contreras Cortés, Exploitation of Copper Mining Resources during the Bronze Age in Eastern Sierra Morena. In: M. Bartelheim/F. Contreras Cortés/R. Hardenberg (eds.), Landscapes and Resources in the Bronze Age of Southern Spain. RessourcenKulturen 17 (Tübingen 2022) 179–200.

Archivo Histórico Provincial de Santa Cruz, Her 1-23; 1–27; Her–1 3–1.

Archivo Municipal de La Laguna, A–IX, 7 & 13; C–V; C–VIII.

Arendt 1981
H. Arendt, Vita activa oder Vom tätigen Leben (München 1981).

Argyrou/Agapiou 2022
A. Argyrou/A. Agapiou, A Review of Artificial Intelligence and Remote Sensing for Archaeological Research. Remote Sensing 14.23, 2022, 6000. DOI: <https://doi.org/10.3390/rs14236000>.

Arnold/Greenfield 2006
E. Arnold/H. Greenfield, The Origins of Transhumant Pastorialism in Temperate South Eastern Europe. A Zooarchaeological Perspective from the Central Balkans. BAR International Series 1538 (Oxford 2006).

Arteaga/Roos 1992

O. Arteaga/A. M. Roos, El Proyecto Geoarqueológico de las Marismas del Guadalquivir. Perspectivas arqueológicas de la campaña de 1992. Anuario Arqueológico de Andalucía 1992.2, 1992, 329–339.

Arteaga et al. 2016

O. Arteaga/D. Barragán Mallofret/A. M. Roos/H. D. Schulz, Primicia cartográfica del río Guadalquivir hace 6500 años. Revista Atlántica-Mediterránea 18, 2016, 139–161.

Ash 2002

M. G. Ash, Wissenschaft und Politik als Ressourcen füreinander. In: R. vom Bruch/B. Kaderas (eds.), Wissenschaften und Wissenschaftspolitik. Bestandsaufnahmen zu Formationen, Brüchen und Kontinuitäten im Deutschland des 20. Jahrhunderts (Stuttgart 2002) 32–51.

Assmann 1988

J. Assmann, Kollektives Gedächnis und kulturelle Identität. In: J. Assmann/T. Hölscher (eds.), Kultur und Gedächtnis (Frankfurt am Main 1988) 9–19.

Assmann 1992

J. Assmann, Das kulturelle Gedächtnis. Schrift, Erinnerung und politische Identität in frühen Hochkulturen (München 1992).

Assmann 1999

A. Assmann, Erinnerungsräume. Formen und Wandlungen des kulturellen Gedächtnisses (München 1999).

Assmann 2005

J. Assmann, Das kollektive Gedächtnis zwischen Körper und Schrift. Zur Gedächtnistheorie von Maurice Halbwachs. In: H. Krapoth/D. Laborde (eds.), Erinnerung und Gesellschaft. Mémoire et Société. Hommage à Maurice Halbwachs (Wiesbaden 2005) 65–83.

Assmann 2006

A. Assmann, Der lange Schatten der Vergangenheit. Erinnerungskultur und Geschichtspolitik (München 2006).

Assmann [1997] 2007a

J. Assmann, Das kulturelle Gedächtnis (München [1997] 2007[6]).

Assmann 2007b

A. Assmann, Geschichte im Gedächtnis. Von der individuellen Erfahrung zur öffentlichen Inszenierung (München 2007).

Assmann 2008

J. Assmann, Communicative and Cultural Memory. In: A. Erll/A. Nünning (eds.), Cultural Memory Studies. An International and Interdiciplinary Handbook (Berlin 2008) 109–118.

Assmann 2011

J. Assmann, Cultural Memory and Early Civilization. Writing, Remembrance, and Political Imagination (Cambridge 2011).

Assmann 2016

A. Assmann, Formen des Vergessens (Göttingen 2016).

Assmann [2016] 2020

A. Assmann, Formen des Vergessens. Historische Geisteswissenschaften. Frankfurter Vorträge 9 (Göttingen [2016][5] 2020).

Assmann/Assmann 1994

A. Assmann/J. Assmann, Das Gestern im Heute. Medien und soziales Gedächtnis. In: K. Merten/S. J. Schmidt/S. Weischenberg (eds.), Die Wirklichkeit der Medien. Eine Einführung in die Kommunikationswissenschaft (Opladen 1994) 114–140.

Assmann/Assmann 1998

A. Assmann/J. Assmann, Mythos. In: H. Cancik/B. Gladigow (eds.), Handbuch religionswissenschaftlicher Grundbegriffe. Volume 4: Kultbild – Rolle (Stuttgart 1998) 179–200.

Aubet 2001

M. E. Aubet, The Phoenicians and the West (Cambridge 2001).

Aubet Semmler 1975

M. E. Aubet Semmler, La necrópolis de Setefilla en Lora del Río, Sevilla (Barcelona 1975).

Awinia 2020

C. S. Awinia, The Sociology of Intra-African Pastoralist Migration. The Case of Tanzania. Frontiers in Sociology 5, 2020, 518797. DOI: <https://doi.org/10.3389/fsoc.2020.518797>.

Bacchielli 1981

L. Bacchielli, L'Agorà di Cirene II 1. L'area settentrionale del lato ovest della platea inferiore. Monografie di Archeologia Libica 15 (Rome 1981).

Bacchielli 1985

L. Bacchielli, Modelli politici e modelli architettonici a Cirene durante il governo democratico. In: G. Barker/J. Lloyd/J. M. Reynolds (eds.), Cyrenaica in Antiquity. BAR International Series 236 (Oxford 1985) 1–14.

Bacchielli/Stucchi 1983

L. Bacchielli/S. Stucchi, L'agorà di Cirene II 4. Il lato sud della platea inferiore e il lato nord della terrazza superiore. Monografie di Archeologia Libica 17 (Rome 1983).

Badcock/Johnston 2009

A. Badcock/R. Johnston, Placemaking through Protest. An Archaeology of the Lees Cross and Endcliffe Protest Camp, Derbyshire, England. Journal of the World Archaeological Congress 5.2, 2009, 306–322.

Bagwell 2009

M. Bagwell, After the Storm, Destruction and Reconstruction. The Potential for an Archaeology of Hurricane Katrina. Archaeologies. Journal of the World Archaeological Congress 5.2, 2009, 280–292.

Balaguer et al. 2017

P. Balaguer/F. X. Roig-Munar/J. A. Martín-Prieto/D. I. Carreras, El litoral de Menorca. Marco fisiográfico y tipos de costa. In: L. Gómez-Pujol/G. X. Pons (eds.), Geomorfología litoral de Menorca. Dinámica evolución y prácticas de gestión. Monografies de la Sociedad d´Història Natural de les Illes Balears (Palma de Mallorca 2017) 13–28.

Baldacchino 2006

G. Baldacchino, Islands. Islands Studies Journal 1.1, 2006, 3–18.

Banks et al. 2008

W. E. Banks/F. d'Errico/A. Townsend Peterson/M. Kageyama/A. Sima/M. F. Sánchez-Goñi, Neanderthal Extinction by Competitive Exclusion. PLoS One 3.12, 2008, e3972. DOI: <https://doi.org/10.1371/journal.pone.0003972>.

Barabási 2002

A.-L. Barabási, Linked. The New Science of Networks (Cambridge 2002).

Barbier 2011

E. Barbier, Scarcity and Frontiers. How Economies Have Developed through Nature. Resource Exploitation (Cambridge 2011).

Barjamovic 2008
G. Barjamovic, The Geography of Trade. Assyrian Colonies in Anatolia c. 1975–1725 BC and the Study of Early Interregional Networks of Exchange. In: J. G. Dercksen (ed.), Anatolia and the Jazira during the Old Assyrian Period. Old Assyrian Archives 3 (Leiden 2008) 87–100.

Barjamovic 2010
G. Barjamovic, Sites, Routes and Historical Geography in Central Anatolia. In: I. Singer (ed.), Ipamati kistamati pari tumatimis. Luwian and Hittite Studies Presented to J. David Hawkins on the Occasion of his 70th Birthday. Monograph Series of the Institute of Archaeology of Tel Aviv University 28 (Tel Aviv 2010) 10–25.

Barjamovic 2011
G. Barjamovic, A Historical Geography of Anatolia in the Old Assyrian Colony Period. CNI Publications 38 (Copenhagen 2011).

Barjamovic 2018
G. Barjamovic, Interlocking Commercial Networks and the Infrastructure of Trade in Western Asia during the Bronze Age. In: K. Kristiansen/T. Lindkvist/J. Myrdal (eds.), Trade and Civilisation. Economic Networks and Cultural Ties from Prehistory to the Early Modern Era (Cambridge 2018) 113–142.

Barragán Romano/Zagalsky 2023
R. Barragán Romano/P. C. Zagalsky (eds.), Potosí in the Global Silver Age (16th—19th Centuries). Studies in Global Social History 49/3. Studies in the Social History of the Global South 49/3 (Leiden 2023).

Barriendos 2005
M. Barriendos, Climate and Culture in Spain. Religious Responses to Extreme Climatic Events in the Hispanic Kingdoms (16th–19th centuries). In: W. Behringer/H. Lehmann/C. Pfister (eds.), Kulturelle Konsequenzen der "Kleinen Eiszeit" = Cultural Consequences of the "Little Ice Age" (Göttingen 2005) 380–414.

Bartelheim 2007
M. Bartelheim, Die Rolle der Metallurgie in vorgeschichtlichen Gesellschaften. Sozioökonomische und kulturhistorische Aspekte der Ressourcennutzung. Ein Vergleich zwischen Andalusien, Zypern und dem Nordalpenraum/The Role of Metallurgy in Prehistoric Societies. Socioeconomic and Cultural Aspects of the Use of Resources. A Comparison between Andalusia, Cyprus and the North Alpine Area. Forschungen zur Archäometrie und Altertumswissenschaft 2 (Rahden 2007).

Bartelheim 2022
M. Bartelheim, Societies and Resources in the Bronze Age of Southern Iberia. In: M. Bartelheim/F. Contreras Cortés/R. Hardenberg (eds.), Landscapes and Resources in the Bronze Age of Southern Spain. RessourcenKulturen 17 (Tübingen 2022) 11–32.

Bartelheim 2023
M. Bartelheim, Rekonstruktion prähistorischer Gesellschaftsstrukturen – Konzeptionelle Betrachtungen zur Frühbronzezeit Mittel- und Südwesteuropas. In: K. Massy/C. Horn (eds.), Hierarchische Strukturen in der Bronzezeit. Forschungsstand, Methoden, Modelle und Neu-Evaluierung. Beiträge zur Ur- und Frühgeschichte Mitteleuropas 106 (Langenweissbach 2023) 109–126.

Bartelheim/Montero Ruíz 2009
M. Bartelheim/I. Montero Ruíz, Many Ores and Little Water – The Use of Resources during the Bronze Age on the Iberian Peninsula. In: M. Bartelheim/H. Stäuble (eds.), The Economic Foundations of the European Bronze Age (Rahden 2009) 5–21.

Bartelheim et al. 2015
M. Bartelheim/R. Hardenberg/T. Knopf/A. Scholz/J. Staecker, 'ResourceCultures'. A Concept for Investigating the Use of Resources in Different Societies. In: A. Danielisová/M. Fernández-Götz (eds.), Persistent Economic Ways of Living. Production, Distribution, and Consumption in Late Prehistory and Early History (Budapest 2015) 39–50.

Bartelheim et al. 2017
M. Bartelheim/R. Hardenberg/J. Staecker 2017, The ‚Resource Turn'. A Socio-Cultural Perspective on Resources. In: A. K. Scholz/M. Bartelheim/R. Hardenberg/J. Staecker (eds.), ResourceCultures. Sociocultural Dynamics and the Use of Resources. Theories, Methods, Perspectives. RessourcenKulturen 5 (Tübingen 2017) 13–24.

Bartelheim et al. 2021a
M. Bartelheim/R. Hardenberg/T. Scholten, Ressourcen – RessourcenKomplexe – RessourcenGefüge – RessourcenKulturen. In: T. Schade/B. Schweizer/S. Teuber/R. Da Vela/W. Frauen/M. Karami/D. K. Ojha/K. Schmidt/R. Sieler/M. Toplak (eds.), Exploring Resources. On Cultural, Spatial and Temporal Dimensions of ResourceCultures. RessourcenKulturen 13 (Tübingen 2021) 9–22.

Bartelheim et al. 2021b
M. Bartelheim/L. García Sanjuán/R. Hardenberg, Human-Made Environments. The Development of Landscapes as ResourceAssemblages. An Introduction. In: M. Bartelheim/L. García Sanjuán/R. Hardenberg (eds.), Human-Made Environments. The Development of Landscapes as Resource-Assemblages. RessourcenKulturen 15 (Tübingen 2021) 7–22.

Bartelheim et al. 2021c
M. Bartelheim/D. Chala-Aldana/M. Díaz-Zorita Bonilla, The Known Unknowns. Full Bronze Age Settlement and Landscape Use in the Lower and Middle Guadalquivir Valley. In: M. Bartelheim/L. García Sanjuán/R. Hardenberg (eds.), Human-made Environments. The Development of Landscapes as ResourceAssemblages. RessourcenKulturen 15 (Tübingen 2021) 139–162.

Bartelheim et al. 2022
M. Bartelheim/D. Chala-Aldana/M. Díaz-Zorita Bonilla/R. Hardenberg/M. Melles, Landscape Use and Transhumance in Sierra Morena through the Ages. In: M. Bartelheim/F. Contreras Cortés/R. Hardenberg (eds.), Landscapes and Resources in the Bronze Age of Southern Spain. Ressourcen-Kulturen 17 (Tübingen 2022) 109–134.

Barth 1961
F. Barth, Nomads of South Persia. The Basseri Tribe of the Khamseh Confederacy (Oslo 1961).

Barth 2002
F. Barth, An Anthropology of Knowledge. Current Anthropology 1, 2002, 1–18.

Bar-Yosef 1998
O. Bar-Yosef, On the Nature of Transitions. The Middle to Upper Palaeolithic and the Neolithic Revolution. Cambridge Archaeological Journal 8.2, 1998, 141–163.

Bar-Yosef 2002
O. Bar-Yosef, The Upper Paleolithic Revolution. Annual Review of Anthropology 31, 2002, 363–393.

Bastian et al. 2017
M. Bastian/O. Jones/N. Moore/E. Roe (eds.), Participatory Research in More-than-Human Worlds (London 2017).

Baten et al. 2019
J. Baten/R. Steckel/C. S. Larsen/C. Roberts, Multidimensional Patterns of European Health, Work, and Violence over the Past Two Millennia. In: R. Steckel/C. S. Larsen/C. A. Roberts/J. Baten (eds.),

The Backbone of Europe. Health, Diet, Work and Violence over Two Millennia (Cambridge 2019) 381–396.

Batet Company 2006
C. Batet Company, L'aigua conquerida. Hidraulisme feudal en terres de conquesta. Alguns exemples de la Catalunya Nova i de Mallorca (Valencia 2006).

Baumann/Gingrich 2004
G. Baumann/A. Gingrich, Grammars of Identity/Alterity. A structural Approach (New York 2004).

Beardsley et al. 1955
R. K. Beardsley/P. Holder/A. D. Krieger/B. J. Meggers/J. B. Rinaldo/P. Kutsche, Functional and Evolutionary Implications of Community Patterning. Memoirs of the Society for American Archaeology 11.11, 1955, 130–158.

Beaudry/Parno 2013
M. C. Beaudry/T. G. Parno, Archaeologies of Mobility and Movement (New York 2013).

Beckmann et al. 2017
H. Beckmann/L. Priester-Lasch/J. Widmaier, Networks, Practices, and Tangible Media. Outlining a Comparative Approach to Studying the Ritual Enactment of the 'Superempirical'. In: A. Scholz/M. Bartelheim/R. Hardenberg/J. Staecker (eds.), ResourceCultures. Sociocultural Dynamics and the Use of Resources. Theories, Methods, Perspectives. RessourcenKulturen 5 (Tübingen 2017) 23–49.

Behrens et al. 2018
T. Behrens/K. Schmidt/R. A. Viscarra Rossel/P. Gries/T. Scholten/R. A. MacMillan, Spatial Modelling with Euclidean Distance Fields and Machine Learning. European Journal of Soil Science 69.5, 2018, 757–770. DOI: <https://doi.org/10.1111/ejss.12687>.

Behrens et al. 2019
T. Behrens/R. A. Viscarra Rossel/R. Kerry/R. MacMillan/K. Schmidt/J. Lee/T. Scholten/A-X. Zhu, The Relevant Range of Scales for Multi-Scale Contextual Spatial Modelling. Scientific Reports 9, 2019, 14800. DOI: <https://doi.org/10.1038/s41598-019-51395-3>.

Behringer 2018
W. Behringer, Tambora und das Jahr ohne Sommer. Wie ein Vulkan die Welt in die Krise stürzte (München 2018).

Bellat et al. 2022
M. Bellat/B. Glissmann/T. Rentschler/K. Schmidt/P. Sconzo/P. Pfälzner/T. Scholten, Unraveling Archaeological Settlement, Landscape, and Resource Use Patterns with Machine Learning in Kurdistan (Iraq). In: EGU General Assembly Conference Abstracts, May 2022, EGU22-6021. DOI: <https://doi.org/10.5194/egusphere-egu22-6021.

Bellwood 2022
P. Bellwood, First Farmers. The Origins of Agricultural Societies (Malden 2022).

Bender 1993
B. Bender (ed.), Landscape. Politics and Perspectives. Explorations in Anthropology (Providence 1993).

Bender 1998
B. Bender, Stonehenge. Making Space, Materializing Culture (Oxford 1998).

Bender/Taves 2012
C. Bender/A. Taves (eds.), What Matters? Ethnographies of Value in a Not so Secular Age (New York 2012).

Bendix et al. 2007
R. Bendix/D. Hemme/M. Tauschek, Vorwort. In: R. Bendix/D. Hemme/M. Tauschek (eds.), Prädikat Heritage. Wertschöpfungen aus kulturellen Ressourcen (Berlin 2007) 7–17.

Benítez de Lugo/Mejías Moreno 2022
L. Benítez de Lugo/M. Mejías Moreno, Climatic Crisis, Socio-Cultural Dynamics and Landscape Monumentalisation during the Bronze Age of La Mancha. The Motilla Culture as an Adaptation to the Changes of the End of the 3rd mill. calBC. In: M. Bartelheim/F. Contreras Cortés/R. Hardenberg (eds.), Landscapes and Resources in the Bronze Age of Southern Spain. RessourcenKulturen 17 (Tübingen 2022) 165–178.

Bennett 2010
J. Bennett, Vibrant Matter. A Political Ecology of Things (Durham 2010).

Berger et al. 2010
P. Berger/R. Hardenberg/E. Kattner/M. Prager, Introduction. The Anthropology of Values. In: P. Berger/R. Hardenberg/E. Kattner/M. Prager (eds.), The Anthropology of Values. Essays in Honour of Georg Pfeffer (Delhi 2010) Xv–xxxiii.

Bergmann 1994
G. J. Bergmann, Auslandsdeutsche in Paraguay, Brasilien, Argentinien (Bad Münstereifel 1994).

Bergson [1912[2]] 2004
H. Bergson, Matter and Memory [translated engl. N. M. Paul/W. S. Palmer New York 1912][2] (New York 2004).

Bermond Montanari 1986
G. Bermond Montanari, I cippi di Rubiera. Studi Etruschi 64, 1986, 239–244.

Bernáldez Sánchez et al. 2010
E. Bernáldez Sánchez/E. García Viñas/E. Ontiveros Ortega/A. Gómez Morón/A. Ocaña García de Veas, Del mar al basurero. Una historia de costumbres. In: M. L. de la Bandera Romero/E. Ferrer Albelda (eds.), El Carambolo. 50 años de un tesoro (Sevilla 2010) 345–385.

Bernáldez Sánchez et al. 2013
E. Bernáldez Sánchez/E. García Viñas/M. Gamero Esteban/F. Amores Carredano/A. Ocaña García de Veas, Knucklebones and Other Animal Deposits in the 'Cruz del Negro' Necropolis. Possible Phoenician Funerary Rituals in SW Spain? Anthropozoologica 48.2, 2013, 323–340.

Bernbeck 2008
R. Bernbeck, An Archaeology of Multi-Sited Communities. In: H. Barnard/W. Wendrich (eds.), The Archaeology of Mobility. Old World and New World Nomadism (Los Angeles 2008) 43–77.

Bernstein 2004
F. Bernstein, Konflikt und Migration. Studien zu griechischen Fluchtbewegungen im Zeitalter der sogenannten Großen Kolonisation. Mainzer Althistorische Studien 5 (St. Katharinen 2004).

Bharati/Johnston 2023
A. Bharati/W. Johnston, "monasticism". Encyclopedia Britannica, last updated 17.12.2023, <https://www.britannica.com/topic/monasticism> (last access: 18.12.2023).

Bhat/Talwar 2024
S. Bhat/G. Talwar, Why Priests are Opposed to Kedarnath Temple Replica in Delhi. Times of India, last updated: 17.07.2024, <https://timesofindia.indiatimes.com/city/delhi/controversy-over-construction-of-kedarnath-mandir-replica-in-delhi/articleshow/111793039.cms> (last access: 23.07.2024).

Bickler 2021
S. H. Bickler, Machine Learning Arrives in Archaeology. Advances in Archaeological Practice 9, 2021, 186–191.

Binford 1980
L. R. Binford, Willow Smoke and Dog's Tails. Hunter-Gatherer Settlement Systems and Archaeological Site Formation. American Antiquity 45.1, 1980, 4–20.

Bishop 2006
C. M. Bishop, Pattern Recognition and Machine Learning (New York 2006).

Blau 1964
P. M. Blau, Exchange and Power in Social Life (New York 1964).

Blázquez Martínez 1983
J. M. Blázquez Martínez, Gerión y otros mitos griegos de Occidente. Gerión 1, 1983, 21–38.

Bobeck 1952
H. Bobeck, Beiträge zur klima-ökologischen Gliederung Irans. Erdkunde 6, 1952, 65–84.

Bockmann et al. 2018
R. Bockmann/H. Ben Romdhane/F. Schön/I. Fumado Ortega, The Roman Circus and Southwestern City Quarter of Carthage. First Results of a New International Research Project. Libyan Studies 49, 2018, 177–186.

Boessneck 1969
J. Boessneck, Die Knochenfunde vom Cerro del Real bei Galera (Prov. Granada). Studien über frühe Tierknochenfunde von der Iberischen Halbinsel 1 (München 1969).

Bolus 2009
M. Bolus, Die Neandertaler. Verbreitung und Expansion einer europäischen Menschenform. In: D. Bandini/U. Kronauer (eds.), Früchte vom Baum des Wissens. Eine Festschrift der wissenschaftlichen Mitarbeiter. 100 Jahre Heidelberger Akademie der Wissenschaften (Heidelberg 2009) 21–33.

Borenstein et al. 2006
E. Borenstein/J. Kendal/M. W. Feldman, Cultural Niche Construction in a Metapopulation. Theoretical Population Biology 70, 2006, 92–104.

Borjian 2017
H. Borjian, Kerman, last updated 2017, <http://www.iranicaonline.org/articles/kerman-geography> (last access: 08.03.2022).

Boserup 1965
E. Boserup, The Conditions of Agricultural Growth. The Economics of Agrarian Change Under Population Pressure (New York 1965).

Bosinski et al. 1986
G. Bosinski/K. Kröger/J. Schäfer/E. Turner, Altsteinzeitliche Siedlungsplätze auf den Osteifel-Vulkanen. Jahrbuch des Römisch-Germanischen Zentralmuseums 33, 1986, 97–130.

Bouchaud/Mézard 2000
J. P. Bouchaud/M. Mézard, Wealth Condensation in a Simple Model of Economy. Physica A. 282, 2000, 536–545.

Bouma 2014
J. Bouma, Soil Science Contributions towards Sustainable Development Goals and their Implementation. Linking Soil Functions with Ecosystem Services. Journal of Plant Nutrition and Soil Science 177.2, 2014, 111–120.

Bourdieu/Waquant 1992
P. Bourdieu/L. J. D. Waquant, An Invitation to Reflexive Sociology (Chicago 1992).

Bradburd 1980
D. Bradburd, Never Give a Shepherd an Even Break. Class and Labor Among the Kornachi. American Ethnologist 7, 1980, 603–620.

Bradley 2017
I. Bradley, Water. A Spiritual History (London 2017).

von Brandt 1958
F. von Brandt, Werkzeug des Historikers (Stuttgart 1958).

Branton 2009
N. Branton, Landscape Approaches in Historical Archaeology. The Archaeology of Places. In: T. Majewski/D. Gaimster (eds.), International Handbook of Historical Archaeology (New York 2009).

Brentjes 1972
B. Brentjes, Zur ökonomischen Funktion des Rindes in den Kulturen des Alten Orients. Klio 54, 1972, 9–44. DOI: <https://doi.org/10.1524/klio.1972.54.54.9>.

Bridge 2009
G. Bridge, Material Worlds. Natural Resources, Resource Geography and the Material Economy. Geography Compass 3.3, 2009, 1217–1244.

Brosius/Polit 2011
C. Brosius/K. Polit (eds.), Ritual, Heritage and Identity. The Politics of Culture and Performance in a Globalised World (New Delhi 2011).

Buchholz/Karageorghis 1973
H.-G. Buchholz/V. Karageorghis, Prehistoric Greece and Cyprus. An Archaeological Handbook (London 1973).

Burini et al. 2021
F. Burini/N. Cortesi/G. Psaila, From Data to Rhizomes. Applying a Geographical Concept to Understand the Mobility of Tourists from Geo-Located Tweets. Informatics 8.1, 2021, 1. DOI: <https://doi.org/10.3390/informatics8010001>.

Burkert 1972
W. Burkert, Homo Necans. Interpretationen altgriechischer Opferriten und Mythen (Berlin 1972).

Burkhardt 2018
S. Burkhardt, Am Wasser gebaut. Die Klöster und die aquatischen Welten des Mittelalters. In: M. Krätschmer/K. Thode/C. Vossler-Wolf (eds.), Klöster und ihre Ressourcen. Räume und Reformen monastischer Gemeinschaften im Mittelalter. RessourcenKulturen 7 (Tübingen 2018) 127–137.

Burt 1992
R. Burt, Structural Holes. The Social Structure of Competition (Cambridge 1992).

Byrne et al. 1992
D. E. Byrne/L. R. Sykes/D. M. Davis, Great Thrust Earthquakes and Aseismic Slip along the Plate Boundary of the Makran Subduction Zone. Journal of Geophysical Research 97.B1, 1992, 449–478.

Calame 1990
C. Calame, Narrating the Foundation of a City. The Symbolic Birth of Cyrene. In: L. Edmunds (ed.), Approaches to Greek Myth (Baltimore 1990) 270–341.

Calcara 1851
P. Calcara, Descrizione dell'Isola di Linosa (Palermo 1851).

Cappuccini 2009

L. Cappuccini, Le "Pietre fiesolane". In: L. Cappuccini/C. Ducci/S. Gori/L. Paoli (eds.), Museo Archeologico del Mugello e della Val di Sieve. Museum Catalogue (Montevarchi 2009) 83–93.

Carrasco/Fernández 2020

I. Carrasco/A. Fernández, El «Vaso de los toros» de Écija (Sevilla). Un nuevo ejemplo de cerámica figurada zoomorfa orientalizante. Boletín del Museo Arqueológico Nacional 39, 2020, 27–52.

Carrier 2021

J. G. Carrier, Economic Anthropology (Newcastle upon Tyne 2021).

Castro 1998

E. V. Castro, Cosmological Deixis and Amerindian Perspectivism. The Journal of the Royal Anthropological Institute 4, 1998, 469–488.

Celestino Pérez 2001

S. Celestino Pérez, Los santuarios de Cancho Roano. Del Indigenismo al Orientalismo Arquitectónico. In: D. Ruiz Mata/S. Celestino Pérez (eds.), Arquitectura oriental y orientalizante en la Península Ibérica (Madrid 2001) 17–56.

Cerchiai 1999

L. Cerchiai, La rappresentazione di Teseo sulle stele felsinee. In: Le mythe grec dans l'Italie antique. Fonction et image, Proceedings École française de Rome, Rome, 14th–16th November 1996. Publications de l'École française de Rome 253 (Rome 1999) 353–365, <https://www.persee.fr/doc/efr_0223-5099_1999_act_253_1_5430> (last access: 09.07.2023).

Cerchiai 2015

L. Cerchiai, Le stele come sistema semantico. Ancora una riflessione. In: G. Sassatelli/L. Cerchiai, Riflessioni sull'immaginario delle stele tra committenza e artigiani. In: E. Govi (ed.), Studi sulle stele etrusche di Bologna tra V e IV sec. a.C. (Rome 2015) 58–60.

Chala-Aldana 2022

D. Chala-Aldana, Beyond Culture Areas. Ceramic Typologies as Indicators of Spatial Interactions between the Middle and Low Guadalquivir, Southeast and Southwest Iberia. In: M. Bartelheim/F. Contreras Cortés/R. Hardenberg (eds.), Landscapes and Resources in the Bronze Age of Southern Spain. RessourcenKulturen 17 (Tübingen 2022) 33–88.

Chala-Aldana 2024

D. Chala-Aldana, The Full Bronze Age in the Middle and Low Guadalquivir Valley. A Landscape of Resources. RessourcenKulturen 27 (Tübingen 2024).

Chang/Koster 1986

C. Chang/H. A. Koster, Beyond Bones. Toward an Archaeology of Pastoralism. In: M. L. Schiffer (ed.), Advances in Archaeological Method and Theory 9 (New York 1986) 97–148.

Character et al. 2024

L. Character/T. Beach/T. Inomata/T. G. Garrison/S. Luzzadder-Beach/J. D. Baldwin/R. Cambranes/F. Pinzón/J. L. Ranchos, Broadscale Deep Learning Model for Archaeological Feature Detection Across the Maya Area. Journal of Archaeological Science. DOI: <http://dx.doi.org/10.2139/ssrn.4744069>.

Cherry 1988

J. Cherry, Pastoralism and the Role of Animals in the Pre- and Protohistoric Economies of the Aegean. In: C. R. Whittaker (ed.), Pastoral Economies in Classical Antiquity. The Cambridge Philological Society, Supplementary Volume 14 (Cambridge 1988) 6–34.

Chevrollier 2016
F. Chevrollier, Conscience gentilice et processus d'aristocratisation chez les notables de Cyrénaïque à l'époque impériale. In: V. Michel (ed.), De Leptis Magna à Derna, de la Tripolitaine à la Cyrénaique. Travaux récents sur la Libye antique. Études libyennes 3 (Paris 2016) 39–66.

Chevrollier 2019
F. Chevrollier, Pour une archéologie de la rébellion. Bilan matériel et bilan humain de la révolte juive de Cyrène à la fin du règne de Trajan. In: E. Nantet (ed.), Les Juifs et le pouvoir politique dans l'Antiquité gréco-romaine (Rennes 2019) 189–207.

Christophilopoulou 2022
A. Christophilopoulou, Ancient Migration or Ancient Mobility? Perspectives from Cyprus. In: G. Bourogiannis (ed.), Beyond Cyprus. Investigating Cypriot Connectivity in the Mediterranean from the Late Bronze Age to the End of the Classical Period. AURA Suppl. 9, 2022, 273–286.

Ciampoltrini 1980
G. Ciampoltrini, I cippi funerari della media e della bassa Valdera, Prospettiva. Rivista di storia dell'arte antica e moderna 21, 1980, 77–81.

Ciampoltrini 1981
G. Ciampoltrini, Segnacoli funerari tardoarcaici di Pisa. Studi Etruschi 49, 1981, 31–39.

Cipriani 1989
M. Cipriani, Morire a Poseidonia nel V secolo. Qualche riflessione a proposito della necropoli meridionale. Dialoghi di Archeologia 7.2, 1989, 71–91.

Clark/Haswell 1964
C. Clark/M. R. Haswell, The Economics of Subsistence Agriculture (London 1964).

Clausing-Lage 2019
V. Clausing-Lage, Aratos Phainomena. Astronomische und meteorologische Phänomene als Bilder der Umwelt (Diss. Eberhard Karls Universität Tübingen 2019). DOI: <http://dx.doi.org/10.15496/publikation-35511>.

Cline 2014
E. H. Cline, 1177 B.C. The Year Civilization Collapsed (Princeton 2014).

Codex Sangallensis
Codex Sangallensis (Cod. Sang. 1084), St. Gallen, Stiftsbibliothek. Wappenbuch des St. Galler Abtes Ulrich Rösch, <https://www.e-codices.unifr.ch/de/list/one/csg/1084> (last access: 30.07.2024).

Collantes 2009
F. Collantes, The Demise of European Mountain Pastoralism. Spain 1500–2000. Nomadic Peoples 13.2, 2009, 124–145.

Colonna 1988
G. Colonna, Il lessico istituzionale etrusco e la formazione della città (specialmente in Emilia-Romagna). In: La formazione della città in Emilia-Romagna, Proceedings Bologna/Marzabotto 1985 (Bologna 1988) 15–36.

Colonna 2015
G. Colonna, I tumuli d'Etruria. In: G. M. Della Fina (ed.), Delimitazione dello spazio funerario in Italia dalla protostoria all'età arcaica. Recinti, circoli, tumuli, Annali della Fondazione per io Museo Claudio Faina 22 (Rome 2015) 6–27.

Conard 2002
N. J. Conard, Der Stand der altsteinzeitlichen Forschung im Achtal der Schwäbischen Alb. Mitteilungen der Gesellschaft für Urgeschichte 11, 2002, 65–77.

Conard/Bolus 2006

N. J. Conard/M. Bolus, The Swabian Aurignacian and its Place in European Prehistory. In: O. Bar-Yosef/J. Zilhão (eds.), Towards a Definition of the Aurignacian. Proceedings of the Symposium Held in Lisbon, Portugal, June 25–30, 2002 (Lisbon 2006) 211–239.

Conard/Bolus 2008

N. J. Conard/M. Bolus, Radiocarbon Dating the Late Middle Paleolithic and the Aurignacian of the Swabian Jura. Journal of Human Evolution 55.5, 2008, 886–897. DOI: <https://doi.org/10.1016/j.jhevol.2008.08.006>.

Connell 2011

J. Connell, A New Inequality? Privatisation, Urban Bias, Migration and Medical Tourism. Asia Pacific Viewpoint 52.3, 2011, 260–271. DOI: <https://doi.org/10.1111/j.1467-8373.2011.01454.x>.

Connerton 2008

P. Connerton, Seven Types of Forgetting. Memory Studies 1.1, 2008, 59–71.

Conrad/Hardenberg 2020

R. Conrad/R. Hardenberg, Religious Speech as Resource. A Research Report. International Journal of Practical Theology 24.1, 2020, 165–195.

Corcella 2007

A. Corcella, Commentary on Herodotus Book IV. In: D. Asheri/A. Lloyd/A. Corcella, A Commentary on Herodotus, Books I–IV (Oxford 2007) 543–721.

Cordell et al. 2009

D. Cordell/J.-O. Drangert/S. White, The Story of Phosphorus. Global Food Security and Food for Thought. Global Environmental Change 19.2, 2009, 292–305.

Cortés-Sánchez et al. 2019

M. Cortés-Sánchez/F. J. Jiménez-Espejo/M. D. Simón-Vallejo/C. Stringer/M. C. Lozano Francisco/A. García-Alix/J. L. Vera Peláez/C. P. Odriozola/J. A. Riquelme-Cantal/R. P. Giráldez/A. M. González/N. Ohkouchi/A. Morales-Muñiz, An Early Aurignacian Arrival in Southwestern Europe. Nature Ecology & Evolution 3, 2019, 207–212. DOI: <https://doi.org/10.1038/s41559-018-0753-6>.

Costanza et al. 1997

R. Costanza/R. Arge/R. de Groot/S. Farber/M. Grasso/B. Hannon/K. Limburg/S. Naeem/R. V. Oneill/J. Paruelo/R. G. Raskin/P. Sutton/M. van den Belt, The Value of the World's Ecosystem Services and Natural Capital. Nature 387, 1997, 253–260.

Costello/Svensson 2018

E. Costello/E. Svensson (eds.), Historical Archaeologies of Transhumance across Europe. Themes in Contemporary Archaeology 6 (New York 2018).

Craik 1943

K. J. W. Craik, The Nature of Explanation (Cambridge 1943).

Crawford 2006

M. Crawford, From Poseidonia to Paestum via the Lucanians. In: G. Bradley/J.-P. Wilson (eds.), Greek and Roman Colonization. Origins, Ideologies and Interactions (Swansea 2006) 59–72.

Cresswell 2006

T. Cresswell, On the Move. Mobility in the Modern Western World (New York 2006).

Cresswell 2010

T. Cresswell, Towards a Politics of Mobility. Environment and Planning D. Society and Space 28.1, 2010, 17–31.

Crumley/Marquardt 1990
C. L. Crumley/W. H. Marquardt, Landscape. A Unifying Concept in Regional Analysis. In: K. M. S. Allen/ S. W. Green/E. B. W. Zubrow (eds.), Interpreting Space. GIS and Archaeology (London 1990) 73–79.

Crusius 1596
M. Crusius, Annales suevici sive Chronica rerum gestarum antiquissimae et inclytae Suevicae gentis 3 (Frankfurt am Main 1596).

Cruz García 2004
T. Cruz García, Ensayos sobre economía canaria. Agua y agricultura (Las Palmas de Gran Canaria 2004).

Cubas et al. 2019
M. Cubas/R. Peyroteo-Stjerna/M. Fontanals-Coll/L. Llorente-Rodríguez/A. Lucquin/O. E. Craig/ A. C. Colonese, Long-Term Dietary Change in Atlantic and Mediterranean Iberia with the Introduction of Agriculture. A Stable Isotope Perspective. Archaeological and Anthropological Sciences 11, 2019, 3825–3836.

Currie 2005
B. Currie, Pindar and the Cult of Heroes (Oxford 2005).

Currie 2016
A. Currie, Ethnographic Analogy, the Comparative Method, and Archaeological Special Pleading. Studies in History and Philosophy of Science Part A, Volume 55, 2016, 84–94.

Cziesla 1990
E. Cziesla, Siedlungsdynamik auf steinzeitlichen Fundplätzen. Methodische Aspekte zur Analyse latenter Strukturen (Bonn 1990).

Dahm 2016
V. Dahm, Der Obersalzberg als historischer Ort und als Stätte historisch-politischer Bildung. In: V. Dahm/A. A. Feiber/H. Mehringer/H. Möller (eds.), Die tödliche Utopie. Bilder, Texte, Dokumente, Daten zum Dritten Reich (München 2016) 17–27.

Daily 1997
G. C. Daily, Introduction. What Are Ecosystem Services. In: G. C. Daily (ed.) Nature's Services. Societal Dependence on Natural Ecosystems (Washington 1997).

Dales 1977
G. Dales, Shifting Trade Patterns Between the Iranian Plateau and the Indus Valley in the Third Millenium B.C. In: J. Deshayes (ed.), Le plateau iranien et l'Asie centrale des origines à la conquête islamique. Leurs relations à la lumière des documents archéologiques. Actes du Colloque international 567 du CNRS, Paris, 22–24 mars 1976 (Paris 1977) 67–78.

Daly/Nymoen 2008
A. Daly/P. Nymoen, The Bøle Ship, Skien, Norway. Research History, Dendrochronology and Provenance. International Journal of Nautical Archaeology 37.1, 2008, 153–170.

Damir-Geilsdorf 2008
S. Damir-Geilsdorf, Die nakba erinnern (Wiesbaden 2008).

Danwerth et al. 2018
O. Danwerth/T. Dittmer/S. Hardjana/D. Hausmann/N. Perreaux/L. Richter/C. Scheidler/F. Steinfeld/D. Weidgenannt, Resources in a Social World. In: I. Amelung/H. Leppin/C. A. Müller (eds.), Discourses of Weakness and Resource Regimes. Trajectories of a New Research Program. Discourses of Weakness and Resource Regimes 1 (Frankfurt am Main 2018) 223–253.

Dautel 2021
K. Dautel, The Spatial Ambiguity of the Island. Marion Poschmann's Pine Islands (2017). In: M. Heitekemper-Yates/T. Schwarz (eds.), Pacific Insularity. Imaginary Geography of Insular Spaces in the Pacific (Tokyo 2021) 97–109.

Da Vela 2022
R. Da Vela, The Memory of the Little Things. Mobility and Encounters as Constitutive Elements of Memoryscapes in the Iron Age Apennines. In: A. Gaucci/F. Iacono/F. Zaina (eds.), Mobility and Memory in the Archaeology of the Middle Sea. Ocnus 30, 2022, 183–195.

Da Vela 2023
R. Da Vela, Networks of Cult Practices as Resources of Cohesion in Transhumant Societies of the Apennines (6th to 1st Centuries BC). In: R. Da Vela/M. Franceschini/F. Mazzilli, Networks as Resources for Ancient Communities, RessourcenKulturen 22 (Tübingen 2023) 185–206.

Da Vela et al. 2023
R. Da Vela/M. Franceschini/F. Mazzilli (eds.), Networks as Resources for Ancient Communities. RessourcenKulturen 22 (Tübingen 2023).

David/Thomas 2008
B. David/J. Thomas, Landscape Archaeology. Introduction. In: B. David/J. Thomas (eds.), Handbook of Landscape Archaeology. World Archaeological Congress Research Handbooks in Archaeology (Adington 2008) 27–43.

David/Thomas 2012
B. David/J. Thomas (eds.), Handbook of Landscape Archaeology. World Archaeological Congress Research Handbooks in Archaeology 1 (London 2012).

Davidov 2014
V. Davidov, Land, Copper, Flora. Dominant Materialities and the Making of Ecuadorian Resource Environments. Anthropological Quarterly 87.1, 2014, 31–58.

Davies 2014
W. Davies, The Time of the Last Neanderthals. Nature 512, 2014, 260–261.

Dawson 2014
H. Dawson, Mediterranean Voyages. The Archaeology of Island Colonisation and Abandonment. Institute of Archaeology Publication 62 (Walnut Creek 2014).

Dawson 2019
H. Dawson, Island Archaeology. In: C. Smith (ed.), Encyclopedia of Global Archaeology (New York 2019) 1–8.

Dawson et al. 2023
H. Dawson/L. Picornell-Gelabert/M. Calvo-Trias/G. Servera-Vives/A. Valenzuela-Oliver, The "Island Laboratory" Revisited. Integrating Environmental and Sociocultural Approaches. The Journal of Island and Coastal Archaeology 18.4, 2023 <https://doi.org/10.1080/15564894.2023.2235673> (last access: 22.02.2024).

Decker 2009
M. Decker, Plants and Progress. Rethinking the Islamic Agricultural Revolution. Journal of World History 20.2, 2009, 187–206.

De Gruchy/Cunliffe 2020
M. De Gruchy/E. Cunliffe, How the Hollow Ways Got Their Form and Kept Them. 5000 Years of Hollow Ways at Tell al-Hawa. In: D. Lawrence/G. Philip/M. Altaweel (eds.), New Agendas in Remote Sensing and Landscape Archaeology in the Near East. Studies in Honour of Tony J. Wilkinson.

New Agendas in Remote Sensing and Landscape Archaeology in the Near East (Oxford 2020) 124–143.

DeLanda 2006
M. DeLanda, A New Philosophy of Society. Assemblage Theory and Social Complexity (London 2006).

Deleuze/Guattari [1980] 1992
G. Deleuze/F. Guattari, Tausend Plateaus. Kapitalismus und Schizophrenie II (Berlin [1980] 1992).

Delgado Hervás 2008
A. Delgado Hervás, Alimentos, poder e identidad en las comunidades fenicias occidentales. Cuardernos de Prehistoria y Arqueología de la Universidad de Granada 18, 2008, 163–188.

Delgado Ramos 2008
J. M. Delgado Ramos, Sociedad. Gente modos y formas. In: La Cuenca minera. Historia, voz y paisaje (Sevilla 2008) 185–196.

Delile et al. 2019
H. Delile/E. Pleuger/J. Blichert-Toft/J.-P. Goiran/N. Fagel/A. Gadhoum/A. Abichou/I. Ben Jerbania/E. Fentress/A. I. Wilson, Economic Resilience of Carthage during the Punic Wars. Insights from Sediments of the Medjerda Delta around Utica (Tunisia). PNAS 116.20, 2019, 9764–9769.

Delle 1998
J. A. Delle, An Archaeology of Social Space. Analyzing Coffee Plantations in Jamaica's Blue Mountains. Contributions to Global Historical Archaeology (New York 1998).

Delle 1999
J. A. Delle, The Landscapes of Class Negotiation on Coffee Plantations in the Blue Mountains of Jamaica: 1790–1850. Historical Archaeology 33.1, 1999, 136–158.

Delp 2022
D. Delp, Zwischen Ansässigkeit und Mobilität. Die sogenannte Große Kolonisation der Griechen aus migrationstheoretischer Perspektive (Göttingen 2022).

DeMarrais et al. 1996
E. DeMarrais/L. J. Castillo/T. Earle, Ideology, Materialization and Power Strategies. Current Anthropology 37.1, 1996, 15–31.

Dendorfer 2014
J. Dendorfer, Gescheiterte Memoria? Anmerkungen zu den „Hausklöstern" des hochmittelalterlichen Adels. Zeitschrift für Württembergische Landesgeschichte 73, 2014, 17–38.

Dendorfer 2017
J. Dendorfer, Raumwirkungen hochmittelalterlicher Klostergründungen. In: S. Brather/J. Dendorfer (eds.), Grenzen, Räume und Identitäten. Der Oberrhein und seine Nachbarregionen von der Antike bis zum Hochmittelalter. Archäologie und Geschichte. Freiburger Forschungen zum ersten Jahrtausend in Südwestdeutschland 22 (Ostfildern 2017) 533–554.

Depraetere 2008
C. Depraetere, The Challenge of Nissology. A Global Outlook on the World Archipelago. Part I and II. Island Studies Journal 3.1, 2008, 3–36.

Depraetere/Dierksmeier 2023
C. Depraetere/L. Dierksmeier, Island Freshwater Scarcity. Digital Encyclopaedia of Island Studies. Centre for Comparative Studies of the University of Lisbon (2023), <https://insularidades.letras.ulisboa.pt/?s=Scarcity&lang=en> (last access: 15.07.2024).

Desantis 2015a
P. Desantis, Un caso di committenza funeraria a Felsina. La tomba con stele della necropoli di via Saffi. In: E. Govi (ed.), Studi sulle stele etrusche di Bologna tra V e IV sec. a.C. (Rome 2015) 101–125.

Desantis 2015b
P. Desantis, Le nuove scoperte. La tomba con stele della necropoli di via Saffi di Bologna. In: G. Sassatelli/A. Russo Tagliente (eds.), Il viaggio oltre la vita. Gli Etruschi e l'Aldilà tra capolavori e realtà virtuale, Exhibition Catalogue Bologna (Bologna 2015) 131–135.

Descola 2011
P. Descola, Jenseits von Natur und Kultur (Berlin 2011).

De Vincenzo 2013
S. De Vincenzo, Tra Cartagine e Roma. In centri urbani dell'eparchia punica di Sicilia tra VI e I sec. a. C. Topoi. Berlin Studies of the Ancient World 8 (Berlin 2013).

Diamond 2000
J. Diamond, Arm und Reich. Die Schicksale menschlicher Gesellschaften (Frankfurt am Main 2000).

Díaz-Zorita Bonilla et al. 2021
M. Díaz-Zorita Bonilla/E. Subirá/M. Fontanals-Coll/K. J. Knudson/E. Alonzi/K. Bolhofner/B. Morell/G. Remolins/J. Roig/A. Martín/P. González Marcén/J. Plasencia/J. M. Coll/J. Gibaja, Neolithic Networking and Mobility during the 5th and 4th Millenia BC in North-Eastern Iberia. Journal of Archaeological Science 125, 2021, 105272. DOI: <https://doi.org/10.1016/j.jas.2020.105272>.

Díaz-Zorita Bonilla et al. 2022
M. Díaz-Zorita Bonilla/J. Escudero Carrillo/D. Chala-Aldana/M. Bartelheim, Connectivity, Interaction and Mobility during the Bronze Age in SW Spain. In: M. Bartelheim/F. Contreras Cortés/R. Hardenberg (eds.), Landscapes and Resources in the Bronze Age of Southern Spain. RessourcenKulturen 17 (Tübingen 2022) 89–108.

Diekjobst/Hohlstein 2018
A. Diekjobst/M. Hohlstein, Erinnerungskultur und Herrschaftspolitik. Kloster und Welt, In: S. Volkart-Baumann (ed.), Der Thurgau im späten Mittelalter 3–4 (Zürich 2018) 221–229.

Dierksmeier 2020
L. Dierksmeier, Historical Water Scarcity on the Canary Islands, 1500–1800 AD. In: S. Teuber/A. Scholz/T. Scholten/M. Bartelheim (eds.), Waters. Conference Proceedings for 'Waters as a Resource' of the SFB 1070 ResourceCultures and DEGUWA (Deutsche Gesellschaft zur Förderung der Unterwasserarchäologie e.V.), RessourcenKulturen 11 (Tübingen 2020) 39–47.

Dierksmeier 2021
L. Dierksmeier, Introduction. In: L. Dierksmeier/F. Schön/A. Kouremos/A. Condit/V. Palmowski (eds.), European Islands between Isolated and Interconnected Life Worlds. Interdisciplinary Long-Term Perspectives. RessourcenKulturen 16 (Tübingen 2021) 17–30.

Dierksmeier et al. 2021
L. Dierksmeier/F. Schön/A. Kouremenos/A. Condit/V. Palmowski (eds.), European Islands Between Isolated and Interconnected Life Worlds. Interdisciplinary Long-Term Perspectives. RessourcenKulturen 16 (Tübingen 2021).

Digard 1989
J. Digard, Boz, last updated 1989, <https://iranicaonline.org/articles/boz-the-domestic-goat> (last access: 07.07.2021).

Dotsika et al. 2019

E. Dotsika/G. Diamantopoulos/S. Lykoudis/S. Gougoura/E. Kranioti/P. Karalis/D. Michael/E. Samartzidou/E. Palaigeorgiou, Establishment of a Greek Food Database for Palaeodiet Reconstruction. Case Study of Human and Fauna Remains from Neolithic to Late Bronze Age from Greece. Geosciences 9.4, 2019, 165.

Dougherty 1993

C. Dougherty, It's Murder to Found a Colony. In: C. Dougherty/L. Kurke (eds.), Cultural Poetics in Archaic Greece. Cults, Performance, Politics (Cambridge 1993) 178–193.

Douglas 1966

M. Douglas, Purity and Danger. An Analysis of Concepts of Pollution and Taboo (London 1966).

Douglas 1986

M. Douglas, How Institutions Think (New York 1986).

Downey et al. 2020

S. S. Downey/D. Gerkey/S. A. Scaggs, The Milpa Game. A Field Experiment Investigating the Social and Ecological Dynamics of Q'eqchi'Maya Swidden Agriculture. Human Ecology 48, 2020, 423–438. DOI: <https://doi.org/10.1007/s10745-020-00169-x>.

Dudd/Evershed 1998

S. N. Dudd/R. P. Evershed, Direct Demonstration of Milk as an Element of Archaeological Economies. Science 282, 1998, 1478–1481. DOI: <http://doi.org/10.1126/science.282.5393.1478>.

van Duijvenvoorde et al. 2023

J. van Duijvenvoorde/G. Fairclough/S. Lekakis/T. Meier/S. Turner/N. Vuletic, The Importance of Landscape for Identity in the Aftermath of Devastation and Dislocation Past and Present. Journal of European Landscapes 4.1, 2023, 1–9.

Durkheim [1912] 1976

E. Durkheim, The Elementary Forms of Religious Life (London [1912] 1976).

Eck 1998

D. Eck, Darsan. Seeing the Divine Image in India (New York 1998).

Eckart 2012

H. Eckart, Wappen. Die archivalischen Quellen. Mit einer Einführung in die Historischen Hilfswissenschaften (Köln 2012).

Eggert 2007

M. K. H. Eggert, Wirtschaft und Gesellschaft im früheisenzeitlichen Mitteleuropa. Überlegungen zum „Fürstenphänomen". Fundberichte Baden-Württemberg 29, 2007, 255–302.

Ehlers 2002

E. Ehlers, Mountain Nomadism. In: R. Tapper/J. Thompson (eds.), The Nomadic Peoples of Iran (London 2002) 40–48.

Ehlers 2011

E. Ehlers, Nomadism, last updated 2011, <https://iranicaonline.org/articles/nomadism> (last access: 07.07.2021).

Ehrmann 1979

G. Ehrmann, Georg von Ehingen, Reisen nach der Ritterschaft. Edition, Untersuchung, Kommentar. Göppinger Arbeiten zur Germanistik 262 (Göppingen 1979).

Ekroth 2002

G. Ekroth, The Sacrificial Rituals of Greek Hero-Cults in the Archaic to the Early Hellenistic Period. Kernos Supplemente 12 (Liège 2002).

Ekroth 2007
G. Ekroth, Heroes and Hero-Cults. In: D. Ogden (ed.), A Companion to Greek Religion (Malden 2007) 100–114.

Eliade 1959
M. Eliade, The Sacred & The Profane. The Nature of Religion. The Significance of Religious Myth, Symbolism, and Ritual within Life and Culture (New York 1959).

Ellis et al. 2010
E. C. Ellis/K. Klein Goldewijk/S. Siebert/D. Lightman/N. Ramankutty, Anthropogenic Transformation of the Biomes, 1700 to 2000. Global Ecology and Biogeography 19.5, 2010, 589–606.

Ensoli 2000
S. Ensoli, L'Agora. In: N. Bonacasa/S. Ensoli (eds.), Cirene (Milano 2000) 59–80.

Erll 2011
A. Erll, Memory in Culture (Basingstoke 2011).

Escacena Carrasco 2001
J. L. Escacena Carrasco, Fenicios a las puertas de Tartessos. Complutum 12, 2001, 73–96.

Escacena Carrasco et al. 2018
J. L. Escacena Carrasco/Á. Gómez Peña/L. G. Pérez Aguilar (eds.), Caura. Arqueología en el estuario del Guadalquivir. Spal monografías arqueología 26 (Sevilla 2018).

ET[2]
G. Meiser (ed.), Etruskische Texte. Edition Minor (Hamburg 2014).

Evans 1977
J. D. Evans, Island Archaeology in the Mediterranean. Problems and Opportunities. World Archaeology 9.1, 1977, 2–26.

Fagan 2011
B. Fagan, Elixir. A History of Water and Humankind (London 2011).

Farina et al. 2024
M. Farina/A. Lavazza/G. Sartori/W. Pedrycz, Machine Learning in Human Creativity. Status and Perspectives. AI & SOCIETY, 2024, 1–13.

Feldhay/Hertz 2019
R. Feldhay/G. Hertz, Migrating Knowledge and the Mobility of Ideas. Transfers 9.1, 2019, 62–74.

Felicetti et al. 2021
A. Felicetti/M. Paolanti/P. Zingaretti/R. Pierdicca/E. Malinverni, MO.SE. Mosaic Image Segmentation Based on Deep Cascading Learning. Virtual Archaeology Review 12, 2021, 25–38.

Fernández-Armesto 1997
F. Fernández-Armesto, Las islas Canarias después de la conquista. La creación de una sociedad colonial a principios del siglo 16 (Las Palmas de Gran Canaria 1997).

Fernández Flores/Rodríguez Azogue 2010
F. Fernández Flores/A. Rodríguez Azogue, El Carambolo, secuencia cronocultural del yacimiento. Síntesis de las intervenciones 2002–2005. In: M. L. de la Bandera/E. Ferrer Albelda (eds.), El Carambolo, 50 años de un tesoro. Serie Historia y Geografía 165 (Sevilla 2010) 203–270.

Ferrer Albelda et al. 2022
E. Ferrer Albelda/F. Chaves Tristán/M. L. de la Bandera Romero/F. J. García Fernández/M. Oria Segura/E. García Vargas/S. Contreras Rastrojo, Montemolín (Marchena, Sevilla). Las campañas estratigráficas (1980–1981). Spal Monografías Arqueología 45 (Sevilla 2022).

Ferry/Limbert 2008
E. E. Ferry/M. E. Limbert (eds.), Timely Assests. The Politics of Resources and Their Temporalities (Santa Fe 2008).

Ficuciello 2014a
L. Ficuciello, Il sacello ipogeico. In: C. Rescigno/F. Sirano (eds.), Immaginando città. Racconti di fondazioni mitiche, forma e funzioni delle città campane. Catalogo della Mostra tenuta a Santa Maria Capua Vetere nel 2014 (Napoli 2014) 248–253.

Ficuciello 2014b
L. Ficuciello, Lo scavo del 'sacello ipogeico'. In: E. Greco, La 'tomba' del fondatore e le origini di Poseidonia (Paestum 2014) 45–50.

Ficuciello 2018
L. Ficuciello, Il sacello-heroon di Poseidonia-Paestum. Le nuove indagini dell'Università degli Studi di Napoli. Newsletter di Archeologia CISA 9, 2018, 203–223.

Figueira 2008
T. Figueira, Colonisation in the Classical Period. In: G. R. Tsetskhladze (ed.), Greek Colonisation. An Account of Greek Colonies and Other Settlements Overseas 2 (Leiden 2008) 427–523.

Filippucci 2023
P. Filippucci, Landscape. In: F. Stein (ed.), The Open Encyclopedia of Anthropology. Facsimile of the First Edition in The Cambridge Encyclopedia of Anthropology, [2016] 2023. DOI: <http://doi.org/10.29164/16landscape>.

Firth 1973
R. Firth, Symbols. Public and Private (London 1973).

Fischer 2011
A. Fischer, NomadInnen. In: F. Kreff/E. Knoll/A. Gingrich (eds.), Lexikon der Globalisierung (Bielefeld 2011) 304–308.

Fischer 2015
M. Fischer, Images from the Colony Nueva Germania in Paraguay. In: M. Fischer/M. Kraus (eds.), Exploring the Archive. Historical Photography from Latin America. The Collection of the Ethnologisches Museum Berlin (Vienna 2015) 319–339.

Fischer-Treuenfeld 1903
R. Fischer-Treuenfeld, Paraguay in Wort und Bild. Eine Studie über den wirtschaftlichen Fortschritt des Landes. Herausgegeben von L. Rehwinkel, Generalkonsul von Paraguay für das Königreich Preußen (Berlin 1903).

Fitze 2009
B. Fitze, Historische Generationen. Über einen sozialen Mechanismus kulturellen Wandels und kollektiver Kreativität (Bielefeld 2009).

Floriani/Bruni 2006
P. Floriani/S. Bruni, La Tomba del Principe. Il tumulo etrusco di via San Jacopo (Pisa 2006).

Floss 2003
H. Floss, Did They Meet or Not? Observations on Châtelperronian and Aurignacian Settlement Patterns in Eastern France. In: J. Zilhão/F. d'Errico (eds.), The Chronology of the Aurignacian and of the Transitional Technocomplexes. Dating, Stratigraphies, Cultural Implications, Proceedings of Symposium 6.1 of the XIVth Congress of the UISPP (University of Liège, Belgium, September 2–8, 2001) (Lisbon 2003) 273–287.

Floss 2009
H. Floss, Menschen mit Migrationshintergrund. Materialien unterwegs. In: N. J. Conard/ H. Floss/M. Barth/J. Serangeli (eds.), Eiszeit. Kunst und Kultur. Begleitband zur Großen Landesausstellung Baden-Württemberg (Stuttgart 2009) 180–182.

Floss/Wolf 2021
H. Floss/S. Wolf, Elfenbein. Der Werkstoff, der den Menschen zum Menschen macht. In: A. Saviello/ S. Müller-Wolff (eds.), Schrecklich schön. Elefant – Mensch – Elfenbein (München 2021) 30–35.

Floss et al. 2017
H. Floss/C. T. Hoyer/H. Würschem, Le Châtelperronien de Germolles (Grotte de La Verpillière I, commune de Mellecey, Saône-et-Loire, France). Paléo 27, 2017, 149–176.

Flügel et al. 2018
C. Flügel/H. Dolenz/C. Baur/M. Bolder-Boos/I. Ben Jerbania/K. Mansel/K. Schmidt/F. Schön/ H. Töpfer/A. Willmitzer, Carthage, Rue Ibn Chabâat. Le développement urbanistique de la période punique ancienne à la période punique tardive. Rapport préliminaire des fouiles menées par le DAI Rome et l'INP Tunis 2009–2012. Antiquités africaines 54, 2018, 27–39.

Folch 2010
C. Folch, Stimulating Consumption. Yerba Mate Myths, Markets, and Meanings from Conquest to Present. Comparative Studies in Society and History 52.1, 2010, 6–36.

Fontanals-Coll et al. 2016
M. Fontanals-Coll/M. E. Subirà/M. Díaz-Zorita Bonilla/J. Gibaja, First Insight into the Neolithic Subsistence Economy in the North-East Iberian Peninsula. Paleodietary Reconstruction through Stable Isotopes. The Neolithic Subsistence in the NE Iberia. American Journal of Physical Anthropology 162.1, 2016, 36–50.

Förster 1883
B. Förster, Ein Deutschland der Zukunft. Bayreuther Blätter 6.1 (Januar/März), 1883, 44–56.

Förster 1886
B. Förster, Deutsche Colonien in dem oberen Laplata-Gebiete mit besonderer Berücksichtigung von Paraguay. Ergebnisse eingehender Prüfungen, praktischer Arbeiten und Reisen 1883–1885 (Naumburg 1886).

Förster/Bauch 2014
B. Förster/M. Bauch, Wasserinfrastrukturen und Macht von der Antike bis zu Gegenwart (Berlin 2014).

Förster-Nietzsche 1891
E. Förster-Nietzsche, Dr. Bernhard Förster´s Kolonie Neu-Germania in Paraguay (Berlin 1891).

Foster 1952
G. M. Foster, The Feixes of Ibiza. Geographical Review 42, 1952, 227–237. <https://doi.org/10.2307/211387>.

Fouache 2008
É. Fouache, Jiroft ii. Human Geography and Environment, last updated 2012, <https://www.iranicaonline.org/articles/jiroft-ii-human-geography-and-environment> (last access: 08.03.2024).

Fouache 2010
É. Fouache, L'approche géoarchéologique. In: H. Alarashi/M.-L. Chambrade/S. Gondet/A. Jouvenel/ C. Sauvage/H. Tronchère (eds.), Regards croisés sur l'étude archéologique des paysages anciens. Nouvelles recherches dans le Bassin méditerranéen, en Asie Centrale et au Proche et au Moyen-Orient (Actes des Rencontres internationales 'Broadening horizons-BH2', Lyon, Maison de l'Orient et

de la Méditerranée, 18–20 juin 2007). Travaux de la Maison de l'Orient et de la Méditerranée 56 (Lyon 2010) 17–30.

Fouache/Garçon 2007
É. Fouache/D. Garçon, Dynamiques géomorphologiques holocènes et contexte d'implantation des sites de l'âge du bronze dans la vallée de l'Halil Roud (bassin de Jiroft, SE de l'Iran) (Holocene Geomorphological Dynamics and Bronze Age Settlement Localisation along the Halil Roud Valley [Jiroft basin, SE of Iran]). Bulletin de l'Association de géographes français 84.2, 2007, 132–147.

Fouache et al. 2005
É. Fouache/D. Garçon/D. Rousset/G. Sénéchal/Y. Madjidzadeh, La vallée de l'Halil Roud (région de Jiroft, Iran). Étude géoarchéologique, méthodologie et résultats préliminaires. Paléorient 31.2, 2005, 107–122.

Foucault 1995
M. Foucault, Discipline & Punish. The Birth of the Prison. Translated by Alan Sheridan (New York 1995).

Foucault 2002
M. Foucault, The Archaeology of Knowledge, 2nd edition (London 2002).

Foucault 2005
M. Foucault, Analytik der Macht (Frankfurt am Main 2005).

Foucault [1966] 2021
M. Foucault, Die Heterotopien. Les héterotopies (Paris 1966, deutsche Übersetzung Frankfurt am Main 2021[5]).

Fragapane 1993
G. Fragapane, Lampedusa. Dalla preistoria al 1878 (Palermo 1993).

Francfort 2020
H. P. Francfort, The Grand'Route of Khorasan (Great Khorasan Road) during the Third Millennium BC and the "Dark Stone" Artefacts. In: J.-W. Meyer/E. Vila/M. Mashkour/M. Casanova/R. Vallet (eds.), The Iranian Plateau during the Bronze Age. Development of Urbanisation, Production and Trade. Archéologie(s) 1 (Lyon 2020) 247–266.

Franquesa 2019
J. Franquesa, Resources. Nature, Value and Time. In: J. Carrier (ed.), A Research Agenda for Economic Anthropology (Cheltenham 2019) 74–89.

Frauen 2024
W. Frauen, Staying at one Place while Leading a Life Shaped through Seasonal Migration. A Contradiction in Terms? In: S. Cvečk/B. Horejs (eds.), The Seasonal and the Material. Anthropology of Seasonal Practices. Anthropos Special Issue (Baden-Baden 2024) 89–100.

Frauen/Klocke-Daffa 2023
Valuation and Value Creation of Mineral Resources in Kermān, Southeastern Iran. In: R. Da Vela/M. Francheschini/F. Mazilli (eds.), Networks as Resources for Ancient Communities. Resourcen-Kulturen 22 (Tübingen 2023) 93–106.

Freeman 1968
L. G. Freeman, A Theoretical Framework for Interpreting Archaeological Materials. In: R. B. Lee/I. Devore (eds.), Man the Hunter (Chicago 1968) 262–267.

Frei 2005
N. Frei, 1945 und wir. Das Dritte Reich im Bewusstsein der Deutschen (München 2005).

Freud 1950
S. Freud, Totem and Taboo (London 1950).

Fried 1967
M. H. Fried, The Evolution of Political Society. An Essay in Political Anthropology. Studies in Anthropology (New York 1967).

Friedman 1982
J. Friedman, Catastrophe and Continuity on Social Evolution. In: C. Renfrew/M. Rowlands/B. Abott Seagraves (eds.), Theory and Explanation in Archaeology. The Southhampton Conference (New York 1982) 175–196.

Froehlich 2023
J. Froehlich, Im Kreis des Elefanten. Burgen als Ressourcen des Niederadels auf der Schwäbischen Alb 1250–1400. RessourcenKulturen 24 (Tübingen 2023).

Froehlich/Weidenbacher 2020
J. Froehlich/M. Weidenbacher, Kein Berg ohne Burg. Burgen und ihre Herren auf der Schwäbischen Alb. Mitteilungen der Deutschen Gesellschaft für Archäologie des Mittelalters und der Neuzeit 33, 2020, 111–122.

Fuchs 1956
H. Fuchs, Beiträge zur Anthropogeographie und Geschichte der Insel Linosa. Wiener Völkerkundliche Mitteilungen 4.1, 1956, 25–36.

Fullola et al. 2008
J. M. Fullola/V. Guerrero/M. A. Petit/M. Calvo/A. Malgosa/N. Armentano/P. Arnau/S. Cho/X. Esteve/T. Fadrique/I. Galtés/E. Garcia/J. Fornés/X. Jordana/M. Pedro/J. Riera/E. Sintes/M. Zubillaga, La Cova des Pas (Ferreries, Menorca). Un jaciment cabdal en la prehistòria de les Balears. Revista de l'Escola Superior de Conservació i Restauració de Béns Culturals de Catalunya. Unicum 7 (Barcelona 2008) 10–20.

Gäbel 2017
C. Gäbel, The Value of Ritual Feasting. Religious and Economic Considerations during the Renewal of the Deities and the Chariot Festival (Puri, Odisha). In: R. Hardenberg (ed.), Approaching Ritual Economy. Socio-Cosmic Fields in Globalised Contexts. RessourcenKulturen 4 (Tübingen 2017) 255–280.

Gardner/Osella 2003
K. Gardner/F. Osella, Migration, Modernity and Social Transformation in South Asia. An Overview. In: K. Gardner/F. Osella (eds.), Migration, Modernity and Social Transformation in South Asia. Contributions to Indian Sociology. Occasional Studies 11.1–2 (New Delhi 2003) v – xxviii.

Garrido/Orta 1978
J. P. Garrido/E. M. Orta, Excavaciones en la necrópolis de "La Joya" Huelva II (3ª, 4ª y 5ª campañas). Excavaciones Arqueológicas de España 96 (Madrid 1978).

Geddes 1983
G. D. Geddes, Neolithic Transhumance in the Mediterranean Pyrenees. World Archaeology 15, 1983, 52–66.

Geertz 1973
C. Geertz, The Interpretation of Cultures (New York 1973).

Geertz 1984
C. Geertz, Distinguished Lecture. Anti Anti-Relativism. American Anthropologist (NS) 86.2, 1984, 263–278.

Geist/McManus 2008
C. Geist/P. A. McManus, Geographical Mobility over the Life Course. Motivations and Implications. Population, Space and Place 14.4, 2008, 283–303.

Geiß 2018
M. Geiß, Kloster und Klientel. Fallstudien zum Verhältnis rittermäßigen Gefolges des Kuenringer zum Klarissenkloster Dürnstein mit einigen Bemerkungen zur Herrschaftsentwicklung in der Wachau, In: R. Zehetmeyer/C. Mochty-Weltin (eds.), Adel und Verfassung im hoch- und spätmittelalterlichen Reich (St. Pölten 2018) 187–245.

Gentelle 1989
P. Gentelle, Données paléogéographiques et fondements de l'irrigation. In: J.-C. Gardin (ed.), Prospections archéologiques en Bactriane orientale. 1974–1978 I. Mémoires de la mission archéologique française en Asie Centrale III (Paris 1989).

Giangiulio 1981
M. Giangiulio, Deformità eroiche e tradizioni di fondazione. Batto, Miscello e l'oracolo delfico. Annali della Scuola Normale Superiore di Pisa 9, 1981, 1–24.

Giangiulio 2001
M. Giangiulio, Constructing the Past. Colonial Traditions and the Writing of History. The Case of Cyrene. In: N. Luraghi (ed.), The Historian's Craft in the Age of Herodotus (Oxford 2001) 116–137.

Giangiulio 2011
M. Giangiulio, Greeks and Persians in Cyrenaica. The Campaigns Towards the Greek Cities. In: R. Rollinger/B. Truschnegg/R. Bichler (eds.), Herodot und das Persische Weltreich. Akten des 3. Internationalen Kolloquiums zum Thema "Vorderasien im Spannungsfeld klassischer und altorientalischer Überlieferungen", Innsbruck, 24.–28. November 2008. Classica et Orientalia 3 (Wiesbaden 2011) 705–716.

Giddens 1984
A. Giddens, The Constitution of Society. Outline of the Theory of Structuration (Cambridge 1984).

Giesen 1999
B. Giesen, Kollektive Identität. Die Intellektuellen und die Nation 2 (Frankfurt am Main 1999).

Gilfert 2023
J. Gilfert, Stören und Stören Lassen. Das Berghofgelände am Obersalzberg als Erinnerungsort der extremen Rechten. kuckuck – Notizen zur Alltagskultur 2/23 Jg. 38, 2023, 44–47.

Gilli et al. 2006
S. Gili/V. Lull/R. Micó/C. Rihuete/R. Risch, An Island Decides. Megalithic Burial Rites on Menorca. Antiquity 80, 2006, 829–842.

Gillich 2018
A. Gillich, Wasser als Ressource. Zur Erforschung des Wassersystems von Kloster Maulbronn. In: M. Krätschmer/K. Thode/C. Vossler-Wolf (eds.), Klöster und ihre Ressourcen. Räume und Reformen monastischer Gemeinschaften im Mittelalter. RessourcenKulturen 7 (Tübingen 2018) 117–126.

Gingrich 2011
A. Gingrich, Identität. In: F. Kreff/E.-M. Knoll/A. Gingrich (eds.), Lexikon der Globalisierung (Bielefeld 2011) 143–146.

Gingrich 2015
A. Gingrich, Tribe. In: J. Wright (ed.), International Encyclopedia of the Social & Behavioral Sciences. Second Edition (Amsterdam 2015) 645–647.

Girard 1972
R. Girard, Violence and the Sacred. Translated by Patrick Gregory (London 1972).

Giudice 2006
A. Giudice, Die Entwicklung der agorà von Kyrene als Beispiel für die Monumentalisierung des öffentlichen Raums. Forum Archaeologiae 40/IX/2006, <https://www.alpinehiking.eu/forum/forum0906/forum40giudice.pdf> (last access: 08.08.2024).

Giudice 2008
A. Giudice, La polis di Cirene in età adrianea. L'immagine dell'ideologia politica nella ricostruzione architettonica. Frankfurter elektronische Rundschau zur Altertumskunde 7, 2008, 14–26.

Giudice 2020
A. Giudice, Espace public et image du pouvoir à Cyrène à l'époque de Ptolémée Ier et Ptolémée III Évergète Ier. Göttinger Forum für Altertumswissenschaft 23, 2020, 33–43.

Gómez Gómez 2000
M. A. Gómez Gómez, Protocolos de Sancho Urtarte (Ayuntamiento de Güimar 2000).

Gómez Gómez 2016
M. A. Gómez Gómez, Estrategias y usos históricos del agua en Tenerife durante los siglos XVI y XVII (La Laguna 2016).

González/Marlovits 2020
R. J. González/J. Marlovits, Life under Lockdown. Notes on Covid-19 in Silicon Valley. Anthropology Today 36.3, 2020, 11–15.

González Wagner/Alvar Ezquerra 1989
C. G. González Wagner/J. Alvar Ezquerra, Fenicios en Occidente. La colonización agrícola. Rivista di Studi Fenici 17, 1989, 61–102.

Gornés/Gual 2018
J. S. Gornés/J. M. Gual, L'hipogeu XXI de la Necròpolis de Calescoves. Universitat de les Illes Balears. Series Monográficas Materialidades. Edicions UIB 2 (Barcelona 2018).

Govi 2014
E. Govi, Lo Studio delle stele felsinee. Approccio metodologico e analisi del linguaggio figurativo. In: G. M. Della Fina (ed.), Artisti, committenti e fruitori in Etruria tra VIII e V sec. a.C. Annali della Fondazione per il Museo Claudio Faina 21 (Rome 2014) 127–186.

Govi 2015
E. Govi, I segnacoli funerari di Bologna tra V e IV secolo a.C. In: G. Sassatelli/A. Russo Tagliente (eds.), Il viaggio oltre la vita. Gli Etruschi e l'Aldilà tra capolavori e realtà virtuale. Exhibition Catalogue Bologna (Bologna 2015) 111–119.

Graeber 2001
D. Graeber, Toward an Anthropological Theory of Value. The False Coin of Our Own Dreams (New York 2001).

Graf 1993
K. Graf, Literatur als adelige Hausüberlieferung? In: H. Joachim (ed.), Literarische Interessenbildung im Mittelalter. Germanistische Symposien. Berichtsbände 14 (Stuttgart 1993) 126–144.

Graf 2010
K. Graf, Der Mythos der Staufer – Eine schwäbische Königsdynastie wird erinnert und instrumentalisiert. Schwäbische Heimat 61.3, 2010, 296–306.

Gramsci 2010
A. Gramsci, Prison Notebooks. Translated by Joseph A. Buttigieg (New York 2010).

Granovetter 1985
M. Granovetter, Economic Action and Social Structure. The Problem of Embeddedness. American Journal of Sociology 91.3, 1985, 481–510.

Graves/Rechniewski 2010
M. Graves/E. Rechniewski, From Collective Memory to Transcultural Remembrance. PORTAL Journal of Multidisciplinary International Studies 7, 2010, 1–15.

Greco 2014
E. Greco, La 'tomba' del fondatore e le origini di Poseidonia (Paestum 2014).

Greco 2021
E. Greco, Mythical and Historical Heroic Founders. The Archaeological Evidence. In: K. Kalogeropoulos/D. Vassilikou/M. Tiverios (eds.), Sidelights on Greek Antiquity. Archaeological and Epigraphical Essays in Honour of Vasileios Petrakos (Berlin 2021) 299–319.

Greenwood/Hinings 1996
R. Greenwood/C. R. Hinings, Understanding Radical Organisational Change. Bringing Together the Old and the New Institutionalism. The Academy of Management Review 21.4, 1996, 1022–1054.

GSA: Goethe- und Schiller-Archiv. Klassik Stiftung Weimar, Jenaer Straße 1, 99425 Weimar, <https://www.klassik-stiftung.de/goethe-und-schiller-archiv/> (last access: 10.03.2025).

Gualtieri 2013
M. Gualtieri, Greeks, Lucanians and Romans at Poseidonia/Paestum (South Italy). In: J. DeRose Evans (ed.), A Companion to the Archaeology of the Roman Republic (Chichester 2013) 369–386.

Guarino/Brucato 2024
G. Guarino/A. Brucato, Spatial Archaeology. Remote Sensing for the Study and Preservation of Cultural Heritage through Open Data and FLOSS Tools. Proceedings 96.1, 2024, 1.

Gudeman 2001
S. Gudeman, The Anthropology of Economy. Community, Market, and Culture (Oxford 2001).

Gudeman 2005
S. Gudeman, Community and Economy. Economy's Base. In: J. Carrier (ed.), A Handbook of Economic Anthropology (Cheltenham 2005) 94–106.

Gudeman 2009
S. Gudeman, The Persuasion of Economics. Economic Persuasions (New York 2009) 62–80.

Guerra-Doce et al. 2023
E. Guerra-Doce/C. Rihuete-Herrada/R. Micó/R. Risch/V. Lull/H. M. Niemeyer, Direct Evidence of the Use of Multiple Drugs in Bronze Age Menorca (Western Mediterranean) from Human Hair Analysis. Scientific Reports 13.4782, 2023. DOI: <https://doi.org/10.1038/s41598-023-31064-2>.

Guichard 1977
P. Guichard, Structures sociales "orientales" et "occidentales" dans l'Espagne musulmane (Berlin 1977). DOI: <https://doi.org/10.1515/9783110816563>.

Guimerá Peraza 1960
M. Guimerá Peraza, Régimen jurídico de las aguas en Canarias (La Laguna 1960).

Halbwachs 1925
M. Halbwachs, Les cadres sociaux de la mémoire (Paris 1925).

Halbwachs [1925] 1966
M. Halbwachs, Das Gedächtnis und seine sozialen Bedingungen (Berlin [1925] 1966).

Halbwachs [1925] 2022
M. Halbwachs, Das Gedächtnis und seine sozialen Bedingungen (Frankfurt am Main [1925] 2022).

Halbwachs [1939] 1967
M. Halbwachs, Das kollektive Gedächtnis (Stuttgart [1939] 1967).

Halbwachs 1992
M. Halbwachs, On Collective Memory. Edited, Translated and with an Introduction by Lewis A. Coser. The Heritage of Sociology (Chicago 1992).

Hall 2008
J. M. Hall, Foundation Stories. In: G. R. Tsetskhladze (ed.), Greek Colonisation. An Account of Greek Colonies and Other Settlements Overseas II. Mnemosyne 193 (Leiden 2008) 383–426.

Halstead 1981
P. Halstead, Counting Sheep in Neolithic and Bronze Age Greece. In: I. Hodder/G. Isaac/N. Hammond (eds.), Patterns of the Past. Studies in Honour of David Clarke (Cambridge 1981) 307–339.

Hannam/Halewood 2006
K. Hannam/C. Halewood, European Viking Themed Festivals. An Expression of Identity. Journal of Heritage Tourism 1.1, 2006, 17–31.

Hansen/Müller 2011
S. Hansen/J. Müller (eds.), Sozialarchäologische Perspektiven. Gesellschaftlicher Wandel 5000–1500 v. Chr. zwischen Atlantik und Kaukasus. Internationale Tagung 15.–18. Oktober 2007 in Kiel. Archäologie in Eurasien 24 (Darmstadt 2011).

Harari 2010
M. Harari, Gli Etruschi nella valle del Po. In: G. Bonini/A. Brusa/R. Cervi, Il paesaggio agrario protostorico e antico. Storia e Didattica, Quaderni dell'Istituto Alcide Cervi 6 (Gattatico 2010) 51–57.

Haraway 1988
D. Haraway, Situated Knowledges. The Science Question in Feminism and the Privilege of Partial Perspective. Feminist Studies 14, 1988, 575–599.

Hardenberg 2008
R. Hardenberg, König ohne Reich. Rituale des Königtums in Orissa (Indien) (Berlin 2008).

Hardenberg 2011
R. Hardenberg, The Renewal of Jagannatha's Body. Ritual and Society in Coastal Orissa (New Delhi 2011).

Hardenberg 2016
R. Hardenberg, Beyond Economy and Religion. Resources and Socio-cosmic Fields in Odisha, India. Religion and Society. Advances in Research 7, 2016, 83–96.

Hardenberg 2017
R. Hardenberg, From Durkheim to Hocart. Sacred Resources and the Quest for 'Life'. Durkheimian Studies / Etudes Durkheimiennes 23.1, 2017, 40–56.

Hardenberg 2018a
R. Hardenberg, ‚Imperial Rice' and ‚Subaltern Millets'. Cereals as Resources in Odisha (India) and beyond. Paideuma 64, 2018, 265–283.

Hardenberg 2018b
R. Hardenberg, Children of the Earth Goddess. Society, Marriage, and Sacrifice in the Highlands of Odisha (India) (Boston 2018).

Hardenberg 2021a
R. Hardenberg, Dwelling in an Animated Landscape. Forms of Attachment Between Environment and People in Eastern India. In: M. Bartelheim/L. García Sanjuán/R. Hardenberg (eds.), Human-Made Environments. The Development of Landscapes as ResourceAssemblages. RessourcenKulturen 15 (Tübingen 2021) 23–34.

Hardenberg 2021b
R. Hardenberg, The Rise of a New Millet Assemblage. In: T. Schade/B. Schweizer/S. Teuber/R. Da Vela/W. Frauen/M. Karami/D. K. Ojha/K. Schmidt/R. Sieler/M. Toplak (eds.), Exploring Resources. On Cultural, Spatial and Temporal Dimensions of ResourceCultures. RessourcenKulturen 13 (Tübingen 2021) 141–156.

Hardenberg et al. 2017
R. Hardenberg/M. Bartelheim/J. Staecker, The ‚Resource Turn‘. A Sociocultural Perspective on Resources. In: A. K. Scholz/M. Bartelheim/R. Hardenberg/J. Staecker (eds.), ResourceCultures. Sociocultural Dynamics and the Use of Resources. Theories, Methods, Perspectives. RessourcenKulturen 5 (Tübingen 2017) 13–23.

Harris 1979
M. Harris, Cultural Materialism. The Struggle for a Science of Culture (New York 1979).

Harvey 1974
D. Harvey, Population, Resources, and the Ideology of Science. Economic Geography 50.3, 1974, 256–277.

Harvey 1989
D. Harvey, The Condition of Postmodernity. An Enquiry into the Origins of Cultural Change (Cambridge 1989).

Hausen 1954
H. M. Hausen, Hidrografía de las Islas Canarias. Rasgos generales y riego de los cultivos subtropicales (La Laguna 1954).

Hausmann/Perreaux 2018
D. Hausmann/N. Perreaux, Resources. A Historical and Conceptual Roadmap. In: I. Amelung/H. Leppin/C. A. Müller (eds.), Discourses of Weakness and Resource Regimes. Trajectories of a New Research Program (Frankfurt am Main 2018) 179–208.

Hayden 1981
B. Hayden, Subsistence and Ecological Adaptations of Modern Hunter-Gatherers. In: G. Teleki/R. Harding (eds.), Omnivorous Primates. Hunting and Gathering in Human Evolution (New York 1981) 344–422.

Hayter 1994
H. M. Hayter, Hunter-Gatherers and the Use of Ethnographic Analogy. Theoretical Perspectives. TOTEM 1.1, 1994, 39–49.

Hechberger 2007
W. Hechberger, Haus und Geschlecht. Anmerkungen zu den Welfen des 12. Jahrhunderts. Zeitschrift für Württembergische Landesgeschichte 66, 2007, 47–62.

Heintz 2009
M. Heintz, Introduction. In: M. Heintz (ed.), The Anthropology of Moralities (New York 2009) 1–19.

Henkner et al. 2017
J. Henkner/J. J. Ahlrichs/S. Downey/M. Fuchs/B. R. James/T. Knopf/T. Scholten/S. Teuber/P. Kühn, Archaeopedology and Chronostratigraphy of Colluvial Deposits as a Proxy for Regional Land Use History (Baar, southwest Germany). Catena 155, 2017, 93–113.

Henkner et al. 2018
J. Henkner/J. J. Ahlrichs/S. Downey/M. Fuchs/B. R. James/A. Junge/T. Knopf/T. Scholten/P. Kühn, Archaeopedological Analysis of Colluvial Deposits in Favourable and Unfavourable Areas. Reconstruction of Land Use Dynamics in SW Germany. Royal Society Open Science 5, 2018, 171624. DOI: <https://doi.org/10.1098/rsos.171624>.

Henning 2007
J. Henning, Kooperative Wertschöpfungsmodelle in der Asset Management und Wealth Management Industrie. Implikationen für die Wertschöpfung und für die Geschäftsmodelle von Banken (Bern 2007).

Hernández González 1986
M. Hernández González, La propiedad del agua y la conflictividad social en La Orotava. La contienda entre la justicia ordinaria y la alcaldía de aguas en el s. XVIII (La Laguna 1986).

Hernández González 1995
M. V. Hernández González, Canarias. La emigración. La emigración canaria a América a través de la historia (Santa Cruz 1995).

Hernández Ramos 2004
J. Hernández Ramos, Las Heredades de aguas de Gran Canaria (Santa Cruz 2004).

Herren 2021
X. Herren, Pandora, Pflug und Poesie. Immaterielle und materielle Ressourcen in Hesiods Werken und Tagen. Spudasmata 192 (Hildesheim 2021).

Herrmann 2019
V. Herrmann, The Reuse of Orthostats and Manipulation of Memories in the Iron Age Syro-Hittite Kingdoms. Semitica 61, 2019, 399–439.

Hesse 1997
B. Hesse, Cattle and Oxen. In: E. M. Meyers (ed.), The Oxford Encyclopedia of Archaeology in the Near East 1 (Oxford 1997) 442–443.

Hewitt 2007
G. Hewitt, Archaeologies of Utopia? Journal of Historical and European Studies 1, 2007, 120–131.

Hicks/McAtackney 2007
D. Hicks/L. McAtackney, Introduction. Landscapes as Standpoints. In: D. Hicks/L. McAtackney/G. Fairclough, Envisioning Landscape. Perspectives and Politics in Archaeology and Heritage (Walnut Creek 2007) 13–29.

Hidasi 2017
J. Hidasi, The Impact of Interconnectedness. Culture, Language, Communication. Language World 19, 2017, 62–69.

Higham et al. 2011
T. Higham/T. Compton/C. Stringer/R. Jacobi/B. Shapiro/E. Trinkaus/B. Chandler/F. Gröning/ C. Collins/S. Hillson/P. O'Higgins/C. FitzGerald/M. Fagan, The Earliest Evidence for Anatomically Modern Humans in Northwestern Europe. Nature 479.7374, 2011, 521–524. DOI: <https://doi.org/10.1038/nature10484>.

Higham et al. 2012
T. Higham/L. Basell/R. Jacobi/R. Wood/C. B. Ramsey/N. J. Conard, Testing Models for the Beginnings of the Aurignacian and the Advent of Figurative Art and Music. The Radiocarbon Chronology of Geißenklösterle. Journal of Human Evolution 62.6, 2012, 664–676. DOI: <http://dx.doi.org/10.1016/j.jhevol.2012.03.003>.

Higham et al. 2014

T. Higham/K. Douka/R. Wood/C. B. Ramsey/F. Brock/L. Basell/M. Camps/A. Arrizabalaga/J. Baena/C. Barroso-Ruiz/C. Bergman/C. Boitard/P. Boscato/M. Caparros/N. J. Conard/C. Draily/A. Froment/B. Galvan/P. Gambassini/A. Garcia-Moreno/S. Grimaldi/P. Haesaerts/B. Holt/M.-J. Iriarte-Chiapusso/A. Jelinek/J. F. Jorda Pardo/J.-M. Maillo-Fernandez/A. Marom/J. Maroto/M. Menendez/L. Metz/E. Morin/A. Moroni/F. Negrino/E. Panagopoulou/M. Peresani/S. Pirson/M. de la Rasilla/J. Riel-Salvatore/A. Ronchitelli/D. Santamaria/P. Semal/L. Slimak/J. Soler/N. Soler/ A. Villaluenga/R. Pinhasi/R. Jacobi, The Timing and Spatiotemporal Patterning of Neanderthal Disappearance. Nature 512.7514, 2014, 306–309. DOI: <https://doi.org/10.1038/nature13621>.

Hitlin/Piliavin 2004

S. Hitlin/J. A. Piliavin, Values. Reviving a Dormant Concept. Annual Review of Sociology 30, 2004, 359–393.

Hobson 2019

M. Hobson, Carthage after the Punic Wars and the Neo-Punic Legacy. In: C. López-Ruiz/B. R. Doak (eds.), The Oxford Handbook of the Phoenician and Punic Mediterranean (Oxford 2019) 183–196.

Hodder 1982

I. Hodder (ed.), Symbolic and Structural Archaeology. New Directions in Archaeology (Cambridge 1982).

Hodder 2014

I. Hodder, The Entanglements of Humans and Things. A Long-Term View. New Literary History 45.1, 2014, 19–36.

Hodos 2017

T. Hodos (ed.), The Routledge Handbook of Archaeology and Globalization (New York 2017).

Hoffmann 2000

G. Hoffmann, Brasidas ou le fait d'armes comme source d'héroïsation dans la Grèce classique. In: V. Pirenne-Delforge/E. Suárez de la Torre (eds.), Héros et héroïnes dans les mythes et les cultes grecs. Actes du colloque organisé à l'Université de Valladolid, du 26 au 29 mai 1999 (Liège 2000) 365–375.

Holck 2006

P. Holck, The Oseberg Ship Burial, Norway. New Thoughts on the Skeletons from the Grave Mound. European Journal of Archaeology 9.2–3, 2006, 185–210.

Hollis/Lukes 1982

M. Hollis/S. Lukes (eds.), Rationality and Relativism (Oxford 1982).

Hölscher 2019

T. Hölscher, Athen – die Polis als Raum der Erinnerung. In: E. Stein-Hölkeskamp/K.-J. Hölkeskamp (eds.), Die Griechische Welt. Erinnerungsorte der Antike (München 2019^{2}).

Hölzchen 2017

Y. M. Hölzchen, Religious Education and Cultural Change. The Case of Madrasas in Northern Kyrgyzstan. In: R. Hardenberg (ed.), Approaching Ritual Economy. Socio-Cosmic Fields in Globaliseised Contexts. RessourcenKulturen 4 (Tübingen 2017) 105–134.

Hölzchen 2023

Y. M. Hölzchen, Neue Moscheen braucht das Land. Religiöses Wissen ilim als Ressource in Nordost-Kirgistan. RessourcenKulturen 25 (Tübingen 2023).

Hornblower 1996

S. Hornblower, A Commentary on Thucydides II. Books IV–V. 24 (Oxford 1996).

Hubbert 1956
M. K. Hubbert, Nuclear Energy and the Fossil Fuels. Shell Development Company. Exploration and Production Research Division. Publication No. 95 (Houston 1956).

Hublin 2015
J. Hublin, The Modern Human Colonization of Western Eurasia. When and Where? Quaternary Science Reviews 118, 2015, 194–210.

Hublin et al. 2020
J. Hublin/N. Sirakov/V. Aldeias/S. Bailey/E. Bard/V. Delvigne/E. Endarova/Y. Fagault/H. Fewlass/M. Hajdinjak/B. Kromer/I. Krumov/J. Marreiros/N. L. Martisius/L. Paskulin/V. Sinet-Mathiot/M. Meye/S. Pääbo/V. Popov/Z. Rezek/S. Sirakova/M. M. Skinner/G. M. Smith/R. Spasov/S. Talamo/T. Tuna/L. Wacker/F. Welker/A. Wilcke/N. Zahariev/S. P. McPherron/T. Tsanova, Initial Upper Palaeolithic Homo Sapiens from Bacho Kiro Cave, Bulgaria. Nature 581.7808, 2020, 299–302. DOI: <https://doi.org/10.1038/s41586-020-2259-z>.

Hudson/Fernandez 2023
M. Hudson/I. M. M. Fernández, Bronze Fish. Marine Resources and the Bronze Age Economy. Old World. Journal of Ancient Africa and Eurasia 3.1, 2023, 1–44.

Hujoel/Lingström 2017
P. P. Hujoel/P. Lingström, Nutrition, Dental Caries and Periodontal Disease. A Narrative Review. Journal of Clinical Periodontology 44 Suppl. 18, 2017, 79–84.

Humm 2018
M. Humm, La 'barbarisation' de Poséidonia et la fin des cultes grecs à Paestum. Revue de l'histoire des religions 235, 2018, 353–372.

Hussain/Floss 2014
S. T. Hussain/H. Floss, The Role of River Courses in Organizing the Cultural Space of the Upper Paleolithic. Examples from the Rhine, Rhône, Danube and Garonne. In: M. Otte/F. Le Brun-Ricalens (eds.), Modes de contact et déplacements au Paléolithique eurasiatique. Actes du colloque international de la Commission 8 (Paléolithique Supérieur). Études et recherches archéologiques de l'Université de Liège 140 / Archéologiques 5 (Liège 2014) 307–320.

Ibrāhīmī/Ibrāhīmī-Pūr 2017
E. Ibrāhīmī/F. Ibrāhīmī-Pūr, Ǧoġrāfīā -ye ṭabīʿī-e Esfandaqeh (Kerman 2017).

Ingold 1993
T. Ingold, The Temporality of the Landscape. World Archaeology 25, 1993, 152–174.

Ingold 1995
T. Ingold, Building, Dwelling, Living. How Animals and People Make Themselves at Home in the World. In: M. Strathern (ed.), Shifting Contexts. Transformations in Anthropological Knowledge (New York 1995) 57–80.

Ingold 2020
T. Ingold, Hunting and Gathering as Ways of Perceiving the Environment. In: M. Susan/A. Giovanni (eds.), Posthumanism in Art and Science (New York 2020) 138–140.

Ingram/Silverman 2016
P. Ingram/B. S. Silverman, The Cultural Contingency of Structure. Evidence from Entry to the Slave Trade in and Around the Abolition Movement. American Journal of Sociology 122.3, 2016, 755–797.

Isler-Kerényi 2015
C. Isler-Kerényi, Dionysos in Classical Athens. An Understanding Through Images. Religions in the Graeco-Roman World 181 (Leiden 2015).

Itgenshorst 2021

T. Itgenshorst, Religion als Ressource in den frühen Apoikien (mit einem Kommentar zu Frank Bernstein). In: D. Delp/X. Herren (eds.), TextRessourcen. Agrarische, soziale und poetische Ressourcen in archaischer und hellenistischer Literatur. Spudasmata 187 (Hildesheim 2021) 87–121.

James et al. 2021

B. James/S. Teuber/J. J. Miera/S. Downey/J. Henkner/T. Knopf/F. Correa/B. Höpfer/S. Scherer/A. Michaelis/B. Wessel/K. Gibbons/P. Kuehn/T. Scholten, Soils, Landscapes, and Cultural Concepts of Favour and Disfavour within Complex Adaptive Systems and ResourceCultures. Human-Land Interactions during the Holocene. Ecology and Society 26.1, 2021, 6. DOI: <https://doi.org/10.5751/ES-12155-260106>.

Johansen/Bauer 2018

P. G. Johansen/A. M. Bauer, On the Matter of Resources and Techno-Politics. The Case of Water and Iron in the South Indian Iron Age. American Anthropologist 120.3, 2018, 412–428.

Jolivet 1995

L. Jolivet, La déformation des continents. Exemples régionaux, Collection Enseignement des sciences (Paris 1995).

Joshi 2017

R. Joshi, In Rebuilding Kedarnath, a New Disaster in the Making, last updated: 01.11.2017, <https://science.thewire.in/society/urban/kedarnath-temple-rebuilding-flood-lessons/> (last access: 17.04.2024).

Jussen 2009

B. Jussen, Perspektiven der Verwandtschaftsforschung fünfundzwanzig Jahre nach Jack Goodys "Entwicklung von Ehe und Familie in Europa". In: K.-H. Spieß (ed.), Die Familie in der Gesellschaft des Mittelalters. Vorträge und Forschungen 71 (Ostfildern 2009) 275–324.

Kamlah 2012

J. Kamlah, Temples of the Levant. Comparative Aspects. In: J. Kamlah (ed.), Temple Building and Temple Cult. Architecture and Cultic Paraphernalia of Temples in the Levant (2.-1. Mill. B.C.E.). Abhandlungen des Deutschen Palästina-Vereins 41 (Wiesbaden 2012) 507–534.

Kamlah 2016

J. Kamlah, Austausch als Ressource. Archäologische Zeugnisse für die Entwicklung des Fernhandels in Palästina während der späten Bronze- und frühen Eisenzeit. In: H. Jenni/M. Sauer (eds.), Nächstenliebe und Gottesfurcht. FS für H.-P. Mathys. Alter Orient und Altes Testament 439 (Münster 2016) 204–232.

Kamlah/Riehl 2021

J. Kamlah/S. Riehl, Agriculture in the Bronze Age Levant. In: D. Hollander/T. Howe (eds.), A Companion to Ancient Agriculture (Hoboken 2021) 193–209.

Kaneff 2018

D. Kaneff, Resources and their Re/valuation in Times of Political-Economic Reform. Max Planck Institute for Social Anthropology Working Papers 189 (Halle 2018).

Karami et al. 2020

M. Karami/H. Alimoradi/N. Alidadi Soleimani, Malekābād I. Development of an Occupation on the Western Slopes of Jebāl-e Bārez. Archäologische Mitteilungen aus Iran und Turan 49, 2017, 2020, 77–101.

Karami et al. 2021

M. Karami/W. Frauen/S. Herdt/S. Faust, On Meanings and Values. A (Re-) Construction from the Archaeological Record. In: T. Schade/B. Schweizer/S. Teuber/R. Da Vela/W. Frauen/M. Karami/

D. K. Ojha/K. Schmidt/R. Sieler/M. S. Toplak (eds.), Exploring Resources. On Cultural, Spatial and Temporal Dimensions of ResourceCultures. RessourcenKulturen 13 (Tübingen 2021) 261–294.

Kelly 1995
R. L. Kelly, The Foraging Spectrum. Diversity in Hunter-Gatherer Lifeways (Washington 1995).

Kenoyer 1998
J. M. Kenoyer, Ancient Cities of the Indus Valley Civilization (New York 1998).

Kerry/Oliver 2011
R. Kerry/M. A. Oliver, Soil Geomorphology. Identifying Relations Between the Scale of Spatial Variation and Soil Processes Using the Variogram. Geomorphology 130, 2011, 40–54. DOI: <https://doi.org/10.1016/j.geomorph.2010.10.002>.

Kertzer 1988
D. Kertzer, Ritual, Politics and Power (New Haven 1988).

Khazanov 1984
A. M. Khazanov, Nomads and the Outside World (Cambridge 1984).

Kienlin 2010
T. L. Kienlin, Traditions and Transformations. Approaches to Eneolithic, Copper Age, and Bronze Age Metalworking and Society in Eastern Central Europe and the Carpathian Basin. BAR International Series 2184 (Oxford 2010).

Kienlin 2012
T. L. Kienlin, Beyond Elites. An Introduction. In: T. L. Kienlin/A. Zimmermann (eds.), Beyond Elites. Alternatives to Hierarchical Systems in Modelling Social Formations; International Conference at the Ruhr-Universität Bochum, Germany, October 22–24, 2009. Universitätsforschungen zur prähistorischen Archäologie aus dem Institut für Archäologische Wissenschaften an der Universität Bochum, Fach Ur- und Frühgeschichte 215 (Bonn 2012) 15–32.

Kienlin/Zimmermann 2012
T. L. Kienlin/A. Zimmermann (eds.), Beyond Elites. Alternatives to Hierarchical Systems in Modelling Social Formations. International Conference at the Ruhr-Universität Bochum, Germany, October 22–24, 2009. Universitätsforschungen zur prähistorischen Archäologie aus dem Institut für Archäologische Wissenschaften an der Universität Bochum, Fach Ur- und Frühgeschichte 215 (Bonn 2012).

Kienzle 2016
M. Kienzle, Burg und Herrschaft Blankenstein. Grundzüge einer hochmittelalterlichen Adelsherrschaft im Gebiet des Großen Lautertals. Münsinger Jahrbuch 7/8.2014/2015, 2016, 35–89.

Kirchner 2009
H. Kirchner, Original Design, Tribal Management and Modifications in Medieval Hydraulic Systems in the Balearic Islands (Spain). World Archaeology 41.1, 2009, 151–168.

Kirchner 2024
H. Kirchner, The Archaeology of Field Systems in Al-Andalus. Agronomy 14.1, 2024, 196. DOI: <https://doi.org/10.3390/agronomy14010196>.

Kirchner/Retamero 2016
H. Kirchner/F. Retamero, Becoming Islanders. Migration and Settlement in the Balearic Islands (10th–13th Centuries). In: F. Retamero/I. Schjellerup/A. Davies (eds.), Agricultural and Pastoral Landscapes in Pre-Industrial Society. Choices, Stability and Change (Oxford 2016) 57–78.

Kirchner/Sabaté 2022
H. Kirchner/F. Sabaté, Agricultural Landscapes of Al-Andalus, and the Aftermath of the Feudal Conquest (Turnhout 2022).

Kitagawa/Conard 2020
K. Kitagawa/N. J. Conard, Split-Based Points from the Swabian Jura Highlight Aurignacian Regional Signatures. PLoS ONE 15.11, 2020, e0239865.

Klingbeil 1889
J. Klingbeil, Enthüllungen über die Dr. Bernhard Förster'sche Ansiedelung Neu-Germanien in Paraguay. Ein Beitrag zur Geschichte unserer gegenwärtigen colonialen Bestrebungen. Nach eigenen Erfahrungen mitgetheilt von Julius Klingbeil (Leipzig 1889).

Klocke-Daffa 2017
S. Klocke-Daffa, ResourceComplexes, Networks, and Frames. The Sambatra in Madagascar. In: A. K. Scholz/M. Bartelheim/R. Hardenberg/J. Staecker (eds.), ResourceCultures. Sociocultural Dynamics and the Use of Resources. Theories, Methods, Perspectives. RessourcenKulturen 5 (Tübingen 2017) 253–268.

Kluckhohn 1951
C. Kluckhohn, Values and Value-Orientations in the Theory of Action. In: T. Parsons/E. Shils (eds.), Toward a General Theory of Action (New York 1951) 388–433.

Knoblauch 2020
H. Knoblauch, Wissen. In: J. Hinrichsen/R. Johler/S. Ratt (eds.), Katastrophen/Kultur. Beiträge zu einer interdisziplinären Begriffswerkstatt (Tübingen 2020) 167–182.

Knopf 2017
T. Knopf, Ressourcennutzung und Umweltverhalten prähistorischer Bauern. Eine Analyse archäologischer und ethnographischer Untersuchungen. RessourcenKulturen 3 (Tübingen 2017).

Knopf et al. 2021
T. Knopf/T. Scholten/P. Kühn, Der RessourcenKomplex 'Boden' als dynamisches Element in Gunst- und Ungunsträumen. In: T. Schade/B. Schweizer/S. Teuber/R. Da Vela/W. Frauen/M. Karami/D. K. Ojha/K. Schmidt/R. Sieler/M. Toplak (eds.), Exploring Resources. On Cultural, Spatial and Temporal Dimensions of ResourceCultures. RessourcenKulturen 13 (Tübingen 2021) 365–380.

Kohl 1975
P. L. Kohl, Carved Chlorite Vessels. A Trade in Finished Commodities in the Mid-Third Millennium. Expedition 18.1, 1975, 18.

Kohl 1978
P. L. Kohl, The Balance of Trade in Southwestern Asia in the Mid-Third Millennium BC [and Comments and Reply]. Current Anthropology 19.3, 1978, 463–492.

Kohl 1981
P. L. Kohl, Materialist Approaches in Prehistory. Annual Review of Anthropology 10.1, 1981, 89–118.

Kohl 1987
P. L. Kohl, The Ancient Economy, Transferable Technologies and the Bronze Age World-System. A View from the Northeastern Frontier of the Ancient Near East. In: M. Rowlands/M. Larsen/K. Kristiansen (eds.), Centre and Periphery in the Ancient World. New Directions in Archaeology (Cambridge 1987) 13–24.

Kohl 2003
K. H. Kohl, Die Macht der Dinge. Geschichte und Theorie sakraler Objekte (München 2003).

Kohring/Wynne-Jones 2007
S. Kohring/S. Wynne-Jones (eds.), Socialising Complexity. Structure, Interaction and Power in Archaeological Discourse (Oxford 2007).

Kollewe 2007
C. Kollewe, Wie ‚böse Steine' zu ‚wertvollen Stücken' werden. Die Produktion kulturellen Erbes, der Aufbau eines Gemeindemuseums und die Konstruktion lokaler Identität in Südmexiko. In: D. Hemme/M. Tauschek/R. Bendix (eds.), Prädikat Heritage. Wertschöpfungen aus kulturellen Ressourcen (Münster 2007) 253–276.

König 1938
E. König, Historia Welforum. Schwäbische Chroniken der Stauferzeit 1 (Stuttgart 1938).

Kopytoff 1986
I. Kopytoff, The Cultural Biography of Things. Commoditisation as Process. In: A. Appadurai (ed.), The Social Life of Things. Commodities in Cultural Perspective (Cambridge 1986) 64–94.

Koster 2021
M. Koster, Political Anthropology. In: L. Pedersen/L. Cliggett (eds.), The SAGE Handbook of Cultural Anthropology (Los Angeles 2021) 330–347.

Koukouli-Chrysanthaki 2002
C. Koukouli-Chrysanthaki, Excavating Classical Amphipolis. In: M. Stamatopoulou/M. Yeroulanou (eds.), Excavating Classical Culture. Recent Archaeological Discoveries in Greece (Oxford 2002) 57–73.

Koukouli-Chrysanthaki 2011
C. Koukouli-Chrysanthaki, Amphipolis. In: R. J. L. Fox (ed.), Brill's Companion to Ancient Macedon. Studies in the Archaeology and History of Macedon, 650 BC–300 AD (Leiden 2011) 409–436.

Kozakavich 2017
S. Kozakavich, The Archaeology of Utopian and Intentional Communities (Gainesville 2017).

Krämer 2015
R. P. Krämer, Non di questo mondo? Riflessioni sul significato dei fregi animalistici etruschi con figure antropomorfe nel VII e VI sec. a.C. In: M. C. Biella/E. Giovanelli (eds.), Nuovi studi sul bestiario fantastico orientalizzante nella penisola italiana. Quaderni di Aristonothos 5 (Trento 2015) 187–220.

Krätschmer et al. 2018
M. Krätschmer/K. Thode/C. Vossler-Wolf (eds.), Klöster und ihre Ressourcen. Räume und Reformen monastischer Gemeinschaften im Mittelalter. RessourcenKulturen 7 (Tübingen 2018).

Kraus 1999
D. Kraus, Bernhard und Elisabeth Försters Nueva Germania in Paraguay. Eine antisemitische Utopie (Diss. Universität Wien 1999).

Kraus 2008
D. Kraus, Bernhard Förster und seine Siedlung Nueva Germania in Paraguay. Jahrbuch für Antisemitismusforschung 17, 2008, 171–194.

Kraus 2009
D. Kraus, Förster, Bernhard. In: W. Benz (ed.), Handbuch des Antisemitismus. Judenfeindschaft in Geschichte und Gegenwart. Band 2 (Berlin 2009) 236–237.

Krause et al. 2012
K. Krause/G. Alex/D. Parkin, Medical Knowledge, Therapeutic Practice and Processes of Diversification. MMG Working Papers 12–11, 2012, 1–30.

Krentz/Wheeler 1994
P. Krentz/E. L. Wheeler, Polyaenus. Stratagems of War. Edited and Translated by Peter Krentz and Everett L. Wheeler (Chicago 1994).

Kreutz 2016
N. Kreutz, Two Tumuli for Battus in the agora of Cyrene. In: O. Henry/U. Kelp (eds.), Tumulus as Sema. Space, Politics, Culture and Religion in the First Millennium B.C., Topoi. Berlin Studies of the Ancient World 27 (Berlin 2016) 35–42.

Krieb 2001
S. Krieb, Erinnerungskultur und adeliges Selbstverständnis im Spätmittelalter. Zeitschrift für Württembergische Landesgeschichte 60, 2001, 59–76.

Krins/Scholkmann 1992
H. Krins/B. Scholkmann (eds.), Pfullingen. Zeugen der Geschichte. Bedrohung – Erforschung – Erhaltung. Archäologische Informationen aus Baden-Württemberg 24 (Stuttgart 1992).

Kristiansen 1998
K. Kristiansen, Europe Before History (Cambridge 1998).

Kristiansen 2011
K. Kristiansen, Constructing Social and Cultural Identities in the Bronze Age. In: B. W. Roberts/M. Vander Linden (eds.), Investigating Archaeological Cultures. Material Culture, Variability and Transmission (New York 2011) 201–210.

Kristiansen/Larsson 2005
K. Kristiansen/T. B. Larsson, The Rise of Bronze Age Society. Travels, Transmissions and Transformations (Cambridge 2005).

Kristiansen/Suchowska-Ducke 2015
K. Kristiansen/P. Suchowska-Ducke, Connected Histories the Dynamics of Bronze Age Interaction and Trade 1500–1100. Proceedings of the Prehistoric Society 81, 2015, 361–392.

Krobath 2009
H. T. Krobath, Werte. Ein Streifzug durch Philosophie und Wissenschaft (Würzburg 2009).

Kroeber 1952
A. L. Kroeber, The Nature of Culture (Chicago 1952).

Kuada/Gullestrup 1999
J. Kuada/H. Gullestrup, Cultural Categories and Profiles. A Framework for Studying Individuals' Behaviour in Organizations. In: M. Lauristin/L. Rahnu (eds.), Intercultural Communication and Changing National Identities. Selected Papers Presented at Nordic Network for Intercultural Communication IV Annual Symposium, November 6–9, 1997 (Tartu 1999) 155–174.

Kuhlwilm et al. 2016
M. Kuhlwilm/I. Gronau/M. J. Hubisz/C. de Filippo/J. Prado-Martinez/M. Kircher/Q. Fu/H. A. Burbano/C. Lalueza-Fox/M. de la Rasilla/A. Rosas/P. Rudan/D. Brajkovic/Ž. Kucan/I. Gušic/T. Marques-Bonet/A. M. Andrés/B. Viola/S. Pääbo/M. Meyer/A. Siepel/S. Castellano, Ancient Gene Flow from Early Modern Humans into Eastern Neanderthals. Nature 530/7591, 2016, 429–433. DOI: <https://doi.org/10.1038/nature16544>.

Kukahn/Blanco 1959
E. Kukahn/A. Blanco, El tesoro de El Carambolo. Archivo Español de Arqueología 32, 1959, 38–49.

Kumar 2015
A. Kumar, When Officials and Media Failed. The Response to the Uttarakhand Floods, 2013. In: R. Jeffrey/R. Sen (eds.), Media at Work in China and India. Discovering and Dissecting (New Delhi 2015) 271–294.

Kurzwelly 2019
J. Kurzwelly, Being German, Paraguayan and Germanino. Exploring the Relation between Social and Personal Identity. Identity 19.2, 2019, 144–156.

Kurzwelly 2024
J. Kurzwelly, People and Identities in Nueva Germania. Göttingen Studies in Cultural Anthropology/European Ethnology 18 (Göttingen 2024).

Lal et al. 2021
R. Lal/J. Bouma/E. Brevik/L. Dawson/D. J. Field/B. Glaser/R. Hatano/A. E. Hartemink/T. Kosaki/B. Lascelles/C. Monger/C. Muggler/G. M. Ndzana/S. Norra/X. Pan/R. Paradelo/L. B. Reyes-Sánchez/T. Sandén/B. R. Singh/H. Spiegel/J. Zhang, Soils and Sustainable Development Goals of the United Nations. An International Union of Soil Sciences Perspective. Geoderma Regional 25, 2021, e00398. DOI: <https://doi.org/10.1016/j.geodrs.2021.e00398>.

Laland/O'Brien 2010
K. N. Laland/M. J. O'Brien, Niche Construction Theory and Archaeology. Journal of Archaeological Method and Theory 17, 2010, 303–322.

Laland et al. 2000
K. N. Laland/J. Odling-Smee/M. W. Feldman, Niche Construction, Biological Evolution, and Cultural Change. Behavioral Brain Science 23, 2000, 131–175.

Lamberg-Karlovsky 1975
C. C. Lamberg-Karlovsky, Third Millennium Modes of Exchange and Modes of Production. In: J. A. Sabloff/C. C. Lamberg-Karlovsky (eds.), Ancient Civilization and Trade (Albuquerque 1975) 341–368.

Lamberg-Karlovsky 1988
C. C. Lamberg-Karlovsky, The "Intercultural Style" Carved Vessels. Iranica Antiqua 23, 1988, 45.

Lamberg-Karlovsky/Tosi 1973
C. C. Lamberg-Karlovsky/M. Tosi, Shahr-i Sokhta and Tepe Yahya. Tracks on the Earliest History of the Iranian Plateau. East and West 23.1/2, 1973, 21–57.

Langer 2021
C. Langer, Egyptian Deportations of the Late Bronze Age. A Study in Political Economy. Zeitschrift für ägyptische Sprache und Altertumskunde, Beiheft 13 (Berlin 2021).

Lansing 2003
J. S. Lansing, Complex Adaptive Systems. Annual Review of Anthropology 32, 2003, 183–204.

Laronde 1988
A. Laronde, La Cyrénaïque romaine, des origines à la fin des Sévères (96 av. J.-C.-235 ap. J.–C.). In: H. Temporini (ed.), Aufstieg und Niedergang der römischen Welt II. Principat. 10.1. Politische Geschichte. Provinzen und Randvölker. Afrika und Ägypten (Berlin 1988) 1006–1064.

Laronde 1990
A. Laronde, Cyrène sous la derniers Battiades. In: B. Gentili (ed.), Cirene. Storia, mito e letteratura. Atti del Convegno della S.I.S.A.C., Urbino, 3 luglio 1988. Atti del Convegni 4 (Urbino 1990) 35–50.

Latour 1996
B. Latour, On Actor-Network Theory. A few Clarifications. Soziale Welt 47.4, 1996, 369–381.

Latour 2005
B. Latour, Reassembling the Social. An Introduction to Actor-Network-Theory (Oxford 2005).

Lawrence et al. 2020
D. Lawrence/G. Philip/M. Altaweel, New Agendas in Remote Sensing and Landscape Archaeology in the Near East. Studies in Honour of Tony J. Wilkinson (Oxford 2020).

Lazaridis 1997
D. Lazaridis, Amphipolis. Guides of the Archaeological Receipts Fund (Athens 1997).

Leschhorn 1984
W. Leschhorn, Gründer der Stadt. Studien zu einem politisch-religiösen Phänomen der griechischen Geschichte (Stuttgart 1984).

Lévi-Strauss [1962] 1968
C. Lévi-Strauss, Das wilde Denken (Frankfurt am Main [1962] 1968).

Lewis et al. 2015
S. L. Lewis/D. P. Edwards/D. Galbraith, Increasing Human Dominance of Tropical Forests. Science 349, 2015, 827–832.

Lewontin 1983
R. C. Lewontin, Gene, Organism, and Environment. In: D. S. Bendall (ed.), Evolution from Molecules to Men (Cambridge 1983) 273–285.

Liesching 1984
W. P. Liesching, Die Wappengruppe mit der Kirchenfahne. Der Herold für Geschlechter-, Wappen- und Siegelkunde. Neue Folge 11, 1984, 1–5.

Liesching 1989
W. P. Liesching, Das Stammwappen der Pfalzgrafen von Tübingen. Bemerkungen zu einer Wappentradition. Zeitschrift für Württembergische Landesgeschichte 48, 1989, 69–86.

Linkenbach 2007
A. Linkenbach, Forest Futures. Global Representations and Ground Realities in the Himalayas (Bangalore 2007).

Lipner 1994
J. Lipner, Hindus. Their Religious Beliefs and Practices (London 1994).

Lipo et al. 2021
C. P. Lipo/R. J. DiNapoli/M. E. Madsen/T. L. Hunt, Population Structure Drives Cultural Diversity in Finite Populations. A Hypothesis for Localized Community Patterns on Rapa Nui (Easter Island, Chile). PLoS ONE 16.5, 2021, e0250690. DOI: <https://doi.org/10.1371/journal.pone.0250690>.

Litt et al. 2021
T. Litt/J. Richter/F. Schäbitz (eds.), The Journey of Modern Humans from Africa to Europe. Culture-Environmental Interaction and Mobility (Stuttgart 2021).

Longo 2012
F. Longo, Agorai di Magna Grecia. In: C. Ampolo (ed.), Agora Greca e Agorai di Sicilia (Pisa 2012) 329–345.

Longo 2014
F. Longo, Poseidonia/Paestum – La città lucana e romana. Continuità e trasformazioni. In: C. Rescigno/F. Sirano (eds). Immaginando Città. Racconti di Fondazioni Mitiche, Forma e Funzioni delle Città Campane (Napoli 2014) 254–259.

López Mayol et al. 2017
T. López Mayol/V. Homar/C. Ramis/J. A. Guijarro, PREGRIDBAL 1.0. Towards a High-Resolution Rainfall Atlas for the Balearic Islands (1950–2009), <https://nhess.copernicus.org/preprints/nhess-2016-330/nhess-2016-330-manuscript-version4.pdf> (last access: 28.08.2024).

López Padilla et al. 2006
J. A. López Padilla/D. Belmonte Mas/M. Paz de Miguel Ibáñez, Los enterramientos argáricos de la Illeta dels Banyets de El Campello. Prácticas Funerarias en la frontera oriental de El Argar. In: J. A. Soler Díaz (ed.), La ocupación prehistórica de la Illeta dels Banyets (Alicante 2006) 119–171.

López-Ruiz/Doak 2019
C. López-Ruiz/B. R. Doak (eds.), The Oxford Handbook of the Phoenician and Punic Mediterranean (Oxford 2019).

Lorenz 1999
S. Lorenz, Herrenbergs Gründer und Herren. Die Pfalzgrafen von Tübingen (11.–14. Jahrhundert). In: J. Janssen/O. Auge (eds.), Herrenberger Persönlichkeiten aus acht Jahrhunderten. Herrenberger Historische Schriften 6 (Herrenberg 1999) 11–32.

Lowenthal 1985
D. Lowenthal, The Past is a Foreign Country (Cambridge 1985).

Lucas 2012
G. Lucas, Understanding the Archaeological Record (Cambridge 2012).

Luhmann 2021
I. Luhmann, Die Staufer in der populären Geschichtskultur. Ein Rezeptionspanorama seit den 1970er Jahren. Historische Lebenswelten in populären Wissenskulturen 20 (Bielefeld 2021).

Lull et al. 1999a
V. Lull/R. Micó/C. Rihuete/R. Risch, La Cova des Mussol. Un lugar de culto en la Menorca Prehistórica. Consell Insular de Menorca. Sa Nostra Obra Social i Cultural (Barcelona 1999).

Lull et al. 1999b
V. Lull/R. Micó/C. Rihuete/R. Risch, Rituales de vida y muerte en la prehistoria de Menorca. La Cova des Càrritx. Consell Insular de Menorca. Sa Nostra Obra Social i Cultural (Barcelona 1999).

Lull et al. 2013
V. Lull/R. Micó/C. Rihuete/R. Risch, Taphonomy and Funerary Practices in Collective Cemeteries. A Prehistoric Case from Menorca (Balearic Islands, Spain). In: M. C. Lozada/B. O'Donnabhain (eds.), The Dead Tell Tales. Essays in Honor of Jane E. Buikstra (Los Angeles 2013) 154–161.

Lüning 2005
J. Lüning, Bandkeramische Hofplätze und die absolute Chronologie der Bandkeramik. In: J. Lüning/C. Frirdich/A. Zimmermann (eds.), Die Bandkeramik im 21. Jahrhundert. Symposium in der Abtei Brauweiler bei Köln vom 16.9.–19.9.2002 (Rahden 2005) 49–74.

Lüning 2012
J. Lüning, Bandkeramische Hofplätze und Erbregeln. In: T. Kienlin/A. Zimmermann (eds.), Metals and Societies. Alternatives to Hierarchical Systems in Modelling Social Formations. International Conference at the Ruhr-Universität Bochum, Germany, October 22–24, 2009. Universitätsforschungen zur Prähistorischen Archäologie (Bonn 2012) 197–202.

de Luxán Meléndez/Viña Brito 2006
S. de Luxán Meléndez/A. Viña Brito, El azúcar y el mundo atlántico. Economía y hacienda; patrimonio cultural y geobotánico. 16 Coloquio de Historia Canario-Americana (Las Palmas de Gran Canaria 2006).

Lyons 2010
D. Lyons, Heroines, Heroes, and Apotheosis. Where the Bodies are Buried. In: M. Meyer/R. von den Hoff (eds.), Helden wie sie. Übermensch – Vorbild – Kultfigur in der griechischen Antike, Beiträge zu einem altertumswissenschaftlichen Kolloquium in Wien, 2.–4. Februar 2007 (Freiburg 2010) 71–84.

Macías Hernández 2005
A. Macías Hernández, Los Heredamientos de Aguas de Canarias y la cuestión señorial. In: L. Aguir de Luque (ed.), Constitución, estado de las autonomías y justicia constitucional. Libro homenaje al profesor Gurmesindo Trujillo (Valencia 2005) 1327–1340.

Macintyre 1992
B. Macintyre, Forgotten Fatherland. The Search for Elisabeth Nietzsche (New York 1992).

Mackowiak 2018
K. Mackowiak, Hagnon et Brasidas à Amphipolis. Chronique d'une ‚fin de culte' annoncée? Revue de l'histoire des religions 235.2, 2018, 311–328.

Madjidzadeh 2008
Y. Madjidzadeh, Excavations at Konar Sandal in the Region of Jiroft in the Halil Basin. First Preliminary Report (2002–2008). With a contribution by H. Pittman 2008, Iran 46, 2008, 69–103.

Maeir 2023
A. M. Maeir, Did Trade Stop in the Early Iron Age? The Evidence from Philistia and Beyond. Diacritica 37, 2023, 82–90.

Maggiani 1996
A. Maggiani, Appunti sulle magistrature etrusche, Studi Etruschi 62, 1996, 95–138.

Maggiani 1997
A. Maggiani, Modello etico o antenato eroico? Sul motivo di Aiace suicida nelle stele felsinee, Studi Etruschi 63, 1997, 149–165.

Maggiani 2018
A. Maggiani, Cippi pisani. In: S. Steingräber (ed.), Cippi, Stele, Statue-Stele e Semata. Testimonianze in Etruria, nel mondo italico e in Magna Grecia dalla prima Età del Ferro all'Ellenismo. Proceedings Sutri 24th–25th April 2015. Mousai 8 (Pisa 2018) 81–98.

Maggiani 2010–2013
A. Maggiani in REE Parte II Padania. Rubiera. Studi Etruschi 76, 2010–2013, 276–278.

Maggiani 2024
A. Maggiani, La regalità arcaica e i cippi di Rubiera. In: G. Sassatelli (ed.), Gli Etruschi nella valle del Po. Atti del XXX Convegno di Studi Etruschi e Italici. Bologna 23–25 giugno 2022 (Rome 2024) 147–164.

Malinowski 1926
B. Malinowski, Myth in Primitive Psychology. Psyche Miniatures, General Series 6 (London 1926).

Malkin 1987
I. Malkin, Religion and Colonization in Ancient Greece. Studies in Greek and Roman Religion 3 (Leiden 1987).

Malkin 2003
I. Malkin, 'Tradition' in Herodotus. The Foundation of Cyrene. In: P. Derow/R. Parker (eds.), Herodotus and his World. Essays from a Conference in Memory of George Forrest (Oxford 2003) 153–170.

Malnati/Bermond Montanari 1989
L. Malnati/G. Bermond Montanari, Nuove iscrizioni etrusche da Rubiera (Reggio Emilia). In: G. Maetzke (ed.), Atti del Secondo Congresso Internazionale Etrusco. Supplemento di Studi Etruschi (Rome 1989) 1567–1578.

Mandolesi 2020
A. Mandolesi, Grandi tumuli etruschi (Florence 2020).

von Mangoldt 2013
B. von Mangoldt, Griechische Heroenkultstätten in klassischer und hellenistischer Zeit. Untersuchungen zu ihrer äußeren Gestaltung, Ausstattung und Funktion (Tübingen 2013).

Manhart 2001
C. Manhart, The Afghan Cultural Heritage Crisis. UNESCO's Response to the Destruction of Statues in Afghanistan. American Journal of Archaeology 105.3, 2001, 387–388.

Manhart 2015
C. Manhart, The Intentional Destruction of Heritage. Bamiyan and Timbuktu. In: W. Logan/M. N. Craith/U. Kockel (eds.), A Companion to Heritage Studies. Blackwell Companions to Anthropology 28 (Oxford 2015) 280–294.

Mannheim 1952
K. Mannheim, Essays on the Sociology of Knowledge (London 1952)

Männlein-Robert 2021
I. Männlein-Robert, Mythos als Ressource. Prometheus und Epimetheus als Reflexionsfiguren bei Hesiod. In: D. Delp/X. Herren (eds.), TextRessourcen. Agrarische, soziale und poetische Ressourcen in archaischer und hellenistischer Zeit. Spudasmata 187 (Hildesheim 2021) 123–145.

Maras 2014
D. F. Maras, Appendice 1. Note epigrafiche sulla stele di via Saffi. In: G. M. Della Fina (ed.), Artisti, Committenti e fruitori in Etruria tra VIII e V sec. a.C. Annali della Fondazione per il Museo Claudio Faina 21 (Rome 2014) 336–342.

Maras 2015
D. F. Maras, Stele e corredo della tomba 11 della necropoli di via Saffi. Nota epigrafica. In: G. Sassatelli/A. Russo Tagliente (eds.), Il viaggio oltre la vita. Gli Etruschi e l'Aldilà tra capolavori e realtà virtuale. Exhibition Catalogue Bologna (Bologna 2015) 216–217.

Maravall/Baten 2019
L. Maravall/J. Baten, Valkyries: Was Gender Equality High in the Scandinavian Periphery Since Viking Times? Evidence from Enamel Hypoplasia and Height Ratios. Economics and Human Biology 34, 2019, 181–193.

Marchesi 2000
M. Marchesi, La scultura di età orientalizzante in Etruria padana. In: G. Bartoloni/C. Morigi Govi (eds.), Principi Etruschi tra Mediterraneo e Europa, Exhibition Catalogue Bologna (Venezia 2000) 336–337.

Marchesi 2011
M. Marchesi, Le sculture di età orientalizzante in Etruria padana (Bologna 2011).

Marett 1909
R. R. Marett, The Threshold of Religion (London 1909).

Mari 2010
M. Mari, Atene, l'impero e le apoikiai. Riflessioni sulla breve vita di Anfipoli 'ateniese'. Annuario della Scuola Archeologica di Atene e delle Missioni Italiane in Oriente 88=Ser. 3.10, 2010 [2012], 391–413.

Mari 2012
M. Mari, Amphipolis between Athens and Sparta. A Philological and Historical Commentary on Thuc. V 11,1. Mediterraneo Antico 15, 2012, 327–353.

Martín Martín 2003
V. Martín Martín, Agua y agricultura en Canarias. El sur de Tenerife (Santa Cruz 2003).

Martín Medina/Gouzévitch 2008
A. Martín Medina/M. Gouzévitch (eds.), Los secretos de 'La presión del vapor de agua', de A. de Betancourt, desvelados más de dos siglos después (Santa Cruz 2008).

Marzol Jaén 2005
M. V. Marzol Jaén, La captación del agua de la niebla en la isla de Tenerife (Santa Cruz 2005).

Massa/Palmisano 2018
M. Massa/A. Palmisano, Change and Continuity in the Long-Distance Exchange Networks between Western/Central Anatolia, Northern Levant and Northern Mesopotamia, c. 3200–1600 BCE. Journal of Anthropological Archaeology 49, 2018, 65–87.

Massey 2004
D. Massey, The Responsibilities of Place. Local Economy 19.2, 2004, 97–101.

Mauersberg 2022
M. Mauersberg, Antike Identitätswahrnehmungen im Kontext der ‚griechischen Kolonisation' und ihr Nutzen für die moderne Forschung. In: K. Riehle/V. Sossau (eds.), Mistaken Identity. Identitäten als Ressourcen im zentralen Mittelmeerraum, RessourcenKulturen 19 (Tübingen 2022) 37–51.

McNaspy 1987
C. J. McNaspy, The Archaeology of the Paraguay Reductions (1609–1767). World Archaeology 18.3, 1987, 398–410.

Meadows et al. 1972
D. H. Meadows/D. L. Meadows/J. Randers/W. W. Behrens III, The Limits to Growth. A Report for The Club of Rome's Project on the Predicament of Mankind (New York 1972).

Mehler 2009
N. Mehler, The Perception and Identification of Hanseatic Material Culture in the North Atlantic. Problems and Suggestions. Journal of the North Atlantic, Special Volume 1: Historical Archaeology in the North Atlantic, 2009, 89–108.

Mehler/Dézsi 2024
N. Mehler/A. Dézsi, Nueva Germania (Paraguay): Siedlungsgeographie einer utopischen Kolonie. In: C. Theune/B. Hausmair/C. Schmid (eds.), Das 19. und 20. Jahrhundert im Fokus der Historischen Archäologie. The Historical Archaeology of the 19th and 20th Centuries. Beiträge zur Mittelalter- und Neuzeitarchäologie in Österreich 15 (Wien 2024) 219–229.

Meier 2009
T. Meier, Das Kloster im See. Überlegungen zu einem mittelalterlichen Lagetyp. Siedlungsforschung 27, 2009, 113–161.

Meinzer et al. 2019
N. Meinzer/R. Steckel/J. Baten, Agricultural Specialization, Urbanization, Workload and Stature. In: R. Steckel/C. S. Larsen/C. A. Roberts/J. Baten (eds.), The Backbone of Europe. Health, Diet, Work and Violence over Two Millennia (Cambridge 2019) 231–252.

Mele 1996a
A. Mele, Storia di Poseidonia tra VI e V secolo a.C. In: M. Cipriani/F. Longo (eds.), I Greci in Occidente. Poseidonia e i Lucani (Napoli 1996) 17–20.

Mele 1996b
A. Mele, Le fonte storiche. In: M. Cipriani/F. Longo (eds.), I Greci in Occidente. Poseidonia e i Lucani (Napoli 1996) 67–70.

Mellars 2011
P. Mellars, The Earliest Modern Humans in Europe. Nature 479.7374, 2011, 483–485.

Melles 2021
M. Melles, The Representation of the Dehesa Landscape in Spanish Local Museums. In: M. Bartelheim/L. García Sanjuán/R. Hardenberg (eds.), Human-Made Environments. The Development of Landscapes as Resource Assemblages. RessourcenKulturen 15 (Tübingen 2021) 35–51.

Menichetti/Pellegrino 2020
M. Menichetti/C. Pellegrino, Le aristocrazie arcaiche. Gestione della tradizione e della memoria. In: G. M. Della Fina (ed.), Ascesa e crisi delle aristocrazie arcaiche. Annali della Fondazione per il Museo Claudio Faina di Orvieto 27, 2020, 103–127.

Mertens 2006
D. Mertens, Städte und Bauten der Westgriechen. Von der Kolonisationszeit bis zur Krise um 400 vor Christus (München 2006).

Mertens 2019
D. Mertens, Il tempio di Nettuno alla luce di un nuovo rilievo. In: F. Mangone/V. Russo/G. Zuchtriegel (eds.), »L'emblema dell'eternità«. Il tempio di Nettuno a Paestum tra archeologia, architettura e restauro. Argonautica 1 (Pisa 2019) 27–57.

Messerli 1958
B. Messerli, Linosa. Geographica Helvetica 13, 1958, 232–240.

Miera 2020
J. J. Miera, Ur- und frühgeschichtliche Siedlungsdynamiken zwischen Gunst- und Ungunsträumen in Südwestdeutschland. Landschaftsarchäologische Untersuchungen zur Baar und den angrenzenden Naturräumen des Schwarzwaldes und der Schwäbischen Alb. RessourcenKulturen 10 (Tübingen 2020).

Miera et al. 2019
J. J. Miera/J. Henkner/K. Schmidt/M. Fuchs/T. Scholten/P. Kühn/T. Knopf, Neolithic Settlement Dynamics Derived from Archaeological Data and Colluvial Deposits between the Baar and Adjacent Low Mountain Ranges, Southwest Germany. E&G Quaternary Science Journal 68, 2019, 75–93. DOI: <https://doi.org/10.5194/egqsj-68-75-2019>.

Millek 2017
J. M. Millek, Sea Peoples, Philistines, and the Destruction of Cities. A Critical Examination of Destruction Layers 'Caused' by the 'Sea Peoples'. In: P. M. Fischer/T. Bürge (eds.), Sea Peoples Up-to-Date. New Research on the Transformations in the Eastern Mediterranean in the 13th–11th Centuries BCE. Österreichische Academie der Wissenschaften 81 / Contributions to the Chronology of the Eastern Mediterranean 35 (Vienna 2017) 113–140.

Millek 2018
J. M. Millek, Destruction and the Fall of Egyptian Hegemony over the Southern Levant. Journal of Ancient Egyptian Interconnections 19, 2018, 1–21.

Millek 2019
J. M. Millek, Exchange, Destruction, and a Transitioning Society. Interregional Exchange in the Southern Levant from the Late Bronze Age to the Iron I. RessourcenKulturen 9 (Tübingen 2019).

Millek 2022
J. M. Millek, The Impact of Destruction on Trade at the End of the Late Bronze Age in the Southern Levant. In: F. Hagemeyer (ed.), Jerusalem and the West. Perspectives from Archaeology (Tübingen 2022).

Millek 2023
J. M. Millek, Destruction and Its Impact on Ancient Societies at the End of the Late Bronze Age (Columbus 2023).

Millennium Ecosystem Assessment 2005
Millennium Ecosystem Assessment, Ecosystems and Human Well-being. Synthesis (Washington 2005).

Miller 1997
T. Miller, Die griechische Kolonisation im Spiegel literarischer Zeugnisse (Tübingen 1997).

Mir Ahmadi 2008
M. Mir Ahmadi, Marco Polo in Iran. Quaderni di Oriente Moderno 6, 2008, 1–13.

de Miroschedji 1973
P. de Miroschedji, Vases et objets en stéatite susiens du musée du Louvre. Cahiers de la Délégation Archéologique Française en Iran 3, 1973, 9–80.

de Miroschedji 2011
P. de Miroschedji, At the Origin of Canaanite Cult and Religion. The Early Bronze Age Fertility Ritual in Palestine. Eretz Israel 30, 2011, 74–103.

Montgomery 2007
D. R. Montgomery, Soil Erosion and Agricultural Sustainability. Proceedings of the National Academy of Sciences of the United States of America 104.33, August 14, 2007, 13268–13272.

Monumenta 1868
Gisleberti chronicon Hanoniense a. 1070–1195. In: G. H. Pertz (ed.), Monumenta Germaniae Historica. Scriptores 21 (Hannover 1868).

Monumenta 1872
Lamberti Ardensis historia comitum Ghisnensium. In: G. H. Pertz (ed.), Monumenta Germaniae Historica. Scriptores 24 (Hannover 1872).

Moreno Onorato et al. 2019
A. Moreno Onorato/E. Alarcón García/M. Díaz-Zorita Bonilla/L. Arboledas Martínez, El descubrimiento de la cueva de Biniedrís (Alaior, Menorca, España). Revista de Menorca 98, 2019, 271–286.

Morpurgo 2013
G. Morpurgo, Sulla cronologia delle stele felsinee. Qualche novità dal sepolcreto De Luca. In: G. M. Della Fina (ed.), Artisti, committenti e fruitori in Etruria tra VIII e V sec. a.C., Annali della fondazione per il Museo Claudio Faina 21 (Rome 2013) 243–290.

Morpurgo 2014
G. Morpurgo, Sulla cronologia delle stele felsinee. Qualche novità dal sepolcreto De Luca. In: G. M. Della Fina (ed.), Artisti, committenti e fruitori in Etruria tra VIII e V sec. a.C., Annali della Fondazione per il Museo Claudio Faina 21, 2014 (Rome 2014) 243–289.

Morpurgo 2015a
G. Morpurgo, Per un quadro aggiornato della cronologia. In: E. Govi (ed.), Studi sulle stele etrusche di Bologna tra V e IV sec. a.C. (Rome 2015) 73–100.

Morpurgo 2015b
G. Morpurgo, L'ideologia funeraria attraverso i corredi di Bologna tra VI e V sec. a.C. In: G. Sassatelli/A. Russo Tagliente (eds.), Il viaggio oltre la vita. Gli Etruschi e l'Aldilà tra capolavori e realtà virtuale, Exhibition Catalogue Bologna (Bologna 2015) 121–129.

Morpurgo 2018
G. Morpurgo, I sepolcreti etruschi di Bologna nei terreni De Luca e Battistini (fine VI – inizi IV sec. a.C.), Dipartimento di storia culture e civiltà Archeologia 20 Felsina 1 (Bologna 2018).

Mouffe 2013
C. Mouffe, Agonistics. Thinking the World Politically (London 2013).

Mozdzen 2024
N. Mozdzen, Was würde Solon tun? Vergangenheitswissen als Ressource im politischen Diskurs des 4. Jhd. v. Chr. in Athen (Unpubl. Diss. Eberhard Karls Universität Tübingen 2024).

Müller 2012

J. Müller, Wie entsteht Ungleichheit? In: T. Kienlin/A. Zimmermann (eds.), Metals and Societies. Alternatives to Hierarchical Systems in Modelling Social Formations. International Conference at the Ruhr-Universität Bochum, Germany, October 22–24, 2009. Universitätsforschungen zur Prähistorischen Archäologie (Bonn 2012) 203–212.

Munro 2024

A. Munro, Art. power. The Encyclopaedia Britannica online, last updated 19.01.2024, <https://www.britannica.com/topic/power-political-and-social-science> (last access: 20.03.2024).

Munslow 2007

A. Munslow, Narrative and History (Hampshire 2007).

Murdock 1967

G. P. Murdock, Ethnographic Atlas. A Summary. Ethnology 6.2, 1967, 109–236.

Murray 1993

O. Murray, Early Greece (London 1993).

Murrieta-Flores 2010

P. Murrieta-Flores, Travelling in a Prehistoric Landscape. Exploring the Influences that Shape Human Movement. In: B. Frischer/J. Webb Crawford/D. Koller (eds.), Making History Interactive. Computer Applications and Quantitative Methods in Archaeology (CAA). Proceedings of the 37th International Conference, Williamsburg, Virginia, USA, March 22–26, 2009. BAR International Series 2079 (Oxford 2010) 249–267.

Nail 2018

T. Nail, The Ontology of Motion. Qui Parle 27, 2018, 47–76.

Neer/Kurke 2019

R. Neer/L. Kurke, Pindar, Song, and Space. Towards a Lyric Archaeology (Baltimore 2019).

Nelson 1996

M. Nelson, Technological Strategies Responsive to Subsistence Stress. In: J. Tainter/B. Tainter (eds.), Evolving Complexity and Environmental Risk in the Prehistoric Southwest, Proceedings of the Workshop "Resource Stress, Economic Uncertainty, and Human Response in the Prehistoric Southwest," Held February 25–29, 1992, in Santa Fe, NM (Reading 1996) 107–144.

Nesbitt/Samuel 1996

M. Nesbitt/D. Samuel, From Staple Crop to Extinction? The Archaeology and History of the Hulled Wheats. In: S. Paludosi/K. Hammer/J. Heller (eds.), Hulled Wheats. Promoting the Conservation and Use of Underutilized and Neglected Crops, Workshop on Hulled Wheats, 21–22 July 1995, Castelvecchio Pascoli, Tuscany, IPGRI (International Plant Genetic Resources Institute) (Rom 1996) 41–100.

Neuburger 2017

M. Neuburger, Geographical Approaches on Territorialities, Resources and Frontiers. In: A. Scholz/M. Bartelheim/R. Hardenberg/J. Staecker (eds.), ResourceCultures. Sociocultural Dynamics and the Use of Resources. Theories, Methods, Perspectives (Tübingen 2017) 179–193.

Neuburger 2022

M. Neuburger, Marginalräume? Von der gesellschaftlichen Herstellung räumlicher Hierarchien. In: J. J. Miera/T. Knopf/T. Scholten/P. Kühn (eds.), Gunst/Ungunst. Nutzung und Wahrnehmung von (Marginal-)Räumen. RessourcenKulturen 20 (Tübingen 2022) 119–130.

Neumann 2003

B. Neumann, Literatur als Medium (der Inszenierung) kollektiver Erinnerungen und Identitäten. In: A. Erll/M. Gymnich/A. Nünning (eds.), Literatur – Erinnerung – Identität. Theoriekonzeptionen und Fallstudien (Trier 2003) 49–78.

Nicolì 2024
M. Nicolì, Agriculture in Transition. An Archaeobotanical Study of Bronze Age Sites in the Southern Levant (Diss. Eberhard Karls Universität Tübingen 2024). DOI: <http://dx.doi.org/10.15496/publikation-99040>.

Niebuhr 2023
F. A. Niebuhr, Dhārī Devī, Goddess of the Floods. Development, Disaster and the Transitions of a Place of Worship (Heidelberg 2023).

Nieto-Espinet et al. 2021
A. Nieto-Espinet/A. Trentacoste/S. Guimarães/S. Valenzuela-Lamas, Cattle from the East, Cattle from the West. Diversity of Bos Morphotypes in the Iberian Peninsula During Late Prehistoric and Roman Times. In: U. Albarella/C. Detry/S. Gabriel/C. Ginja/A. E. Pires/J. P. Tereso (eds.), Themes in Old World Zooarchaeology. From the Mediterranean to the Atlantic (Oxford 2021) 129–143.

Nippa 2011
A. Nippa, Kleines abc des Nomadismus (Hamburg 2011).

Nisbet et al. 1991
R. Nisbet/R. Maggi/G. Barker (eds.), Archeologia Della Pastorizia Nell'Europa Meridionale. Rivista di Studi Liguri 56/57, 1991, 1–4.

Nora 1984–1994
P. Nora (ed.), Les Lieux de Mémoire, 7 Bände, Paris 1984–1994.

Nora 1989
P. Nora, Between Memory and History. Les Lieux de Mémoire. Representations 26, 1989, 7–24.

Nora 2009
P. Nora, Rethinking France. Les Lieux de Mémoire 3 (Chicago 2009).

Nowak 2014
C. Nowak, Bestattungsrituale in Unteritalien vom 5. bis 4. Jh. v. Chr. Überlegungen zur sogenannten Samnitisierung Kampaniens. Italikà 3 (Wiesbaden 2014).

Nüllen 2021
H. Nüllen, Narratives of Insular Transformation in 8th Century CE Historio- and Hagiography. In: L. Dierksmeier/F. Schön/A. Kouremenos/A. Condit/V. Palmowski (eds.), European Islands Between Isolated and Interconnected Life Worlds. RessourcenKulturen 16 (Tübingen 2021) 227–241.

Obbink 1993
D. Obbink, Dionysus Poured Out. Ancient and Modern Theories of Sacrifice and Cultural Formation. In: T. H. Carpenter/C. A. Faraone (eds.), Masks of Dionysus (Ithaca 1993) 65–86.

O'Connor 2010
B. O'Connor, From Dorchester to Dieskau. Some Aspects of Relations Between Britain and Central Europe During the Early Bronze Age. In: H. Meller/F. Bertemes (eds.), Der Griff nach den Sternen. Internationales Symposium in Halle (Saale) 16.-21. Februar 2005. Tagungen des Landesmuseums für Vorgeschichte Halle (Saale) (Halle 2010) 591–602.

Odling-Smee et al. 2003
F. J. Odling-Smee/K. N. Laland/M. W. Feldman, Niche Construction. The Neglected Process in Evolution, Monographs in Population Biology (Princeton 2003).

Oexle 1986
O. G. Oexle, Adliges Selbstverständnis und seine Verknüpfung mit dem liturgischen Gedenken. Das Beispiel der Welfen. Zeitschrift für die Geschichte des Oberrheins 134, 1986, 47–75.

Oexle 1995
O. G. Oexle, Memoria als Kultur. In: O. G. Oexle (ed.), Memoria als Kultur. Veröffentlichung des Max-Planck-Instituts für Geschichte 121 (Göttingen 1995) 9–78.

O'Flanagan et al. 2011
P. O'Flanagan/T. Lasanta Martínez/M. Paz Errea Abad, Restoration of Sheep Transhumance in the Ebro Valley, Aragon, Spain. Geographical Review 101.4, 2011, 556–575.

Ohri 2019a
L. Ohri, Till Kingdom Come. Medieval Hinduism in the Modern Himalaya (Delhi 2019).

Ohri 2019b
L. Ohri, Multiple Realities, Realpolitik, and Anti-heroes Duryodhana and Karna of the Himalayas in a Secularising India. In: M. Kaushal/S. P. Kumar (eds.), Aesthetic Textures. Living Traditions of Mahabharata (New Delhi 2019) 259–276.

Ohri/Bishnoi 2016
L. Ohri/A. Bishnoi, A Guide to Losing Yourself to Shiva. The Kedarnath Foot Pilgrimage (New Delhi 2016)

Olías/Nieto 2015
M. Olías/J. M. Nieto, Background Conditions and Mining Pollution throughout History in the Río Tinto (SW Spain). Environments 2015.2, 2015, 295–316.

Oliver Martin 1957
W. Oliver Martin, The Order and Integration of Knowledge (Ann Arbor 1957).

Orendi 2020
A. Orendi, Entwicklung und Bedeutung agrarischer Ressourcen in der Südlevante der Bronze- und Eisenzeit (Diss. Eberhard Karls Universität Tübingen 2020). DOI: <http://dx.doi.org/10.15496/publikation-40980>.

Orsingher et al. 2021
A. Orsingher/J. Kamlah/H. Sader/A. Schmitt/S. Amicone/C. Berthold, Making, Trading and Consuming Phoenician Wine. Ancient Near East Today IX.9, September 2021.

Ortalli 2011
J. Ortalli, Tumuli a Felsina? In: A. Naso (ed.), Tumuli e sepolture monumentali nella protostoria europea. Atti del convegno internazionale, Celano 21–24 settembre 2000. Römisches Germanisches Zentralmuseum Mainz, Tagungen 5 (Mainz 2011) 57–70.

Osborne 1996
R. Osborne, Greece in the Making, 1200–479 BC (London 1996).

Osborne 1997
R. Osborne, The Ecstasy and the Tragedy. Varieties of Religious Experience in Art, Drama, and Society. In: C. B. R. Pelling (ed.), Greek Tragedy and the Historian (Oxford 1997) 187–211.

Otto 1917
R. Otto, Das Heilige. Über das Irrationale in der Idee des Göttlichen und sein Verhältnis zum Rationalen (Breslau 1917).

Padgham 2012
M. Padgham, Human Movement Is Both Diffusive and Directed. PLoS ONE 7.5, 2012, e37754.

Pajuelo Pando/López Aldana 2014
A. Pajuelo Pando/P. M. López Aldana, Las ofrendas animales. In: Á. Fernández Flores/A. Rodríguez Azogue/M. Casado Ariza/E. Prados Pérez (eds.), La necrópolis de época tartésica de la Angorrilla, Alcalá del Río, Sevilla. Historia y Geografía 271 (Sevilla 2014) 535–553.

Palumbo 1997
B. Palumbo, Identità nel tempo. Saggi di antropologia della parentela. Mnemosyne 17 (Lecce 1997).

Papadimitriou 2019
N. Papadimitriou, From Habitus to Visual Memory. Changing Mnemonic Processes in Early Mycenean Greece. In: E. Borgna/I. Caloi/F. M. Carinci/R. Laffineur (eds.), MNHMH/MNEME. Past and Memory in the Aegean Bronze Age. Proceedings of the 17th International Aegean Conference Udine-Venice 17th–21st April 2018 (Leuven 2019).

Parisi Presicce 2007
C. Parisi Presicce, La casa di Batto, il culto del eroe fondatore e il regime democratico a Cirene. In: C. Dobias-lalou (ed.), Questions de religion cyrénéenne. Actes du colloque de Dijon 21–23 mars 2002. Karthago 27 (Paris 2007) 245–264.

Pathak 2021
S. Pathak, The Chipko Movement. A People's History (Ranikhet 2021).

Patze 2002
H. Patze, Adel und Stifterchronik. Frühformen territorialer Geschichtsschreibung im hochmittelalterlichen Reich. Ausgewählte Aufsätze von Hans Patze. Vorträge und Forschungen 50 (Stuttgart 2002).

Pérez et al. 2018
G. Pérez/L. Peña/L. Picornell/Y. Carron, Agriculture between the Third and First Millennium BC in the Balearic Islands. The Archaeobotanical Data. Vegetation History and Archaeobotany 27.1, 2018, 253–154.

Pérez Herrero 1961
E. Pérez Herrero, Plan de obras hidráulicas de la isla de Tenerife. Cabildo Insular de Tenerife (Santa Cruz 1961).

Pérez Herrero 2006
E. Pérez Herrero, Reflexiones en torno al azúcar, agua, tabaco y plátano en Canarias (Las Palmas de Gran Canaria 2006).

Pérez Jordà/Peña Chocarro 2024
G. Pérez Jordà/L. Peña Chocarro, New Products, New Tastes? Agricultural Innovations and Continuities Between the Ninth and Fourth Centuries BC in Mediterranean Iberia. Oxford Journal of Archaeology 43.1, 2024, 43–60. DOI: <https://doi.org/10.1111/ojoa.12288>.

Pérez Macías 1996
J. A. Pérez Macías, Metalurgia extractiva prerromana en Huelva (Huelva 1996).

Pérez Ripoll 1999
M. Pérez Ripoll, La explotación ganadera durante el III milenio a. C. en la Península Ibérica. Sagvntvm Extra 2, 1999, 95–103.

Perrot 2008
J. Perrot, JIROFT iv. ICONOGRAPHY OF CHLORITE ARTIFACTS, last updated 2012, <https://iranicaonline.org/articles/jiroft-iv-iconography-of-chlorite-artifacts> (last access: 28.03.2022).

Petzold et al. 2023
J. Petzold/E. T. Joe/I. Kelman/A. K. Magnan/C. Mirbach/G. Nagle Alverio/P. D. Nunn/B. M. W. Ratter, The Global Adaptation Mapping Initiative Team, 2023. Between Tinkering and Transformation. A Contemporary Appraisal of Climate Change Adaptation Research on the World's Islands. Frontiers in Climate 4, 2023, 1072231. DOI: <https://doi.org/10.3389/fclim.2022.1072231>.

Pfälzner 2010

P. Pfälzner, Introduction and Synthesis. Urban Development and Ecology at Tell Mozan. In: K. Deckers/M. Doll/P. Pfälzner/S. Riehl, Development of the Environment, Subsistence and Settlement of the City of Urkeš and Its Region. Studien zur Urbanisierung Nordmesopotamiens, Serie A. Ausgrabungen 1998–2001 in der zentralen Oberstadt von Tall Mozan/Urkeš, Volume 3 (SUN A3) (Wiesbaden 2010) 1–12.

Pfälzner/Alidadi Soleimani 2017

P. Pfälzner/N. Alidadi Soleimani, The ICAR – University of Tübingen South-of-Jiroft Archaeological Survey. Results of the First Season 2015, AMIT 47, 2017, 105–141.

Pfälzner/Faist 2020

P. Pfälzner/B. Faist, Eine Geschichte der Stadt Mardama(n). In: J. Baldwin/J. Matuszak (eds.), mu-zu an-za3-še3 kur-ur2-še3 ḫe2-ĝal2. Altorientalische Studien zu Ehren von Konrad Volk. dubsar 17 (Münster 2020) 347–389.

Pfälzner/Qasim 2017

P. Pfälzner/H. A. Qasim, The First and Second Season of German-Kurdish Excavations at Bassetki in 2015 and 2016. Zeitschrift für Orient-Archäologie 10, 2017, 10–43.

Pfälzner/Qasim 2020

P. Pfälzner/H. A. Qasim, From Akkadian Maridaban to Middle Assyrian Mardama. Excavations at Bassetki in 2018 and 2019 (with contributions by H. Arnhold/B. Faist/O. Hence/S. Herdt/P. Sconzo). Zeitschrift für Orient-Archäologie 13, 2020, 12–89.

Pfälzner/Sconzo 2015

P. Pfälzner/P. Sconzo, First Results of the Eastern Ḫabur Archaeological Survey in the Dohuk Region of Iraqi Kurdistan (with a contribution by I. Puljiz). The Season of 2013. Zeitschrift für Orient-Archäologie 8, 2015, 90–122.

Pfälzner/Sconzo 2016a

P. Pfälzner/P. Sconzo, From the Banks of the Upper Tigris River to the Zagros Highlands. The First Season (2013) of the Tübingen Eastern Ḫabur Archaeological Survey. In: K. Kopanias/J. MacGinnis (eds.), The Archaeology of the Kurdistan Region of Iraq (Oxford 2016) 311–319.

Pfälzner/Sconzo 2016b

P. Pfälzner/P. Sconzo, The Eastern Ḫabur Archaeological Survey in Iraqi Kurdistan. A Preliminary Report on the 2014 Season (with contributions by R. Beutelschieß, A. J. Edmonds, B. Glissmann, S. Herdt, J. Hermann, S. Heydari-Guran, J. Köhler, M. Müller-Wiener, I. Puljiz and M. Sharp). Zeitschrift für Orient-Archäologie 9, 2016, 10–69.

Pfälzner et al. 2018

P. Pfälzner/H. A. Qasim/I. Puljiz/P. Sconzo/B. Faist, Urban Developments in Northeastern Mesopotamia from the Ninevite V to the Neo-Assyrian Periods. Excavations at Bassetki 2017. Zeitschrift für Orient-Archäologie 11, 42–87.

Pfälzner et al. 2019

P. Pfälzner/N. Alidadi Soleimani/M. Karami, SOJAS 2015–2018. A Résumé of Four Seasons of Archaeological Survey South of Jiroft, Archaeology. Journal of the Iranian Center for Archaeological Research 2.2, 2019, 107–124.

Picornell 2012

L. Picornell, Paisaje vegetal y comunidades prehistóricas y protohistóricas en Mallorca y Menorca (Illes Balears). Una aproximación desde la antrocología (PhD-Thesis University of Barcelona 2012).

Pietsch/Kühn 2017
D. Pietsch/P. Kühn, Buried Soils in the Context of Geoarchaeological Research. Two Examples from Germany and Ethiopia. Archaeological and Anthropological Sciences 9, 2017, 1571–1583.

Pizzirani 2015
C. Pizzirani, Il mare nell'immaginario funebre degli Etruschi. In: G. Sassatelli/A. Russo Tagliente (eds.), Il viaggio oltre la vita. Gli Etruschi e l'Aldilà tra capolavori e realtà virtuale. Exhibition Catalogue Bologna (Bologna 2015) 71–79.

de Planhol 2000
X. de Planhol, Garmsīr and Sardsīr (warm zones and cold zones), last updated 2012, <https://www.iranicaonline.org/articles/garmsir-and-sardsir-> (last access: 24.02.2023).

Plant 2023
I. Plant, Brasidas and Thucydides. Hero and His Historian. Dialogues d'histoire ancienne 49.2, 2023, 11–36.

Platenkamp 2003
J. Platenkamp, Strukturalismus in der Ethnologie. In: B. Beer/H. Fischer (eds.), Ethnologie. Einführung und Überblick (Berlin 2003) 295–308.

Poggesi 2014
G. Poggesi, Le "pietre fiesolane". In: G. Poggesi/L. Sarti (eds.), Passaggi a Nord-Ovest. Interventi di archeologia preventiva nell'area fiorentina (Mezzana-Perfetti Ricasoli) tra preistoria ed età romana, Millenni. Studi di Archeologia preistorica 10 (Florence 2014) 100–104.

Polit 2011
K. Polit, Heritage, Art and Elections. Popular Culture, Ritual and Political Competition in Uttarakhand. In: C. Brosius/K. Polit (eds.) Ritual, Heritage and Identity. The Politics of Culture and Performance in a Globalised World (New Delhi 2011) 149–164.

Polit 2012a
K. Polit, Wenn Rituale Kulturerbe werden. Zur Konstruktion und Ästhetik kulturellen Erbes in Uttarakhand, Nordindien. In: J. Karolewski/N. Miczek/C. Zotter (eds.), Ritualdesign. Zur kultur- und ritualwissenschaftlichen Analyse »neuer« Rituale (Bielefeld 2012) 175–200.

Polit 2012b
K. Polit, Women of Honour. Gender and Agency among Dalit Women in the Central Himalayas (Hyderabad 2012).

Polit 2024
K. M. Polit, Subjectivities, Agency and Ritual Performance in the Garhwal Himalayas. In: S. Thies/S. Goumegou/G. Cebey (eds.), The Routledge Handbook of Global South Studies on Subjectivities (New York 2024) 338–351.

Pomar et al. 2002
L. Pomar/A. Obrador/H. Westphal, Sub-Wavebase Cross-Bedded Grainstone on a Distally Steepened Carbonate Ramp, Upper Miocene, Menorca, Spain. Sedimentology 49, 2002, 139–169.

Porres Caballero 2013
S. Porres Caballero, Maenadic Ecstasy in Greece. Fact or Fiction? In: A. Bernabé/M. Herrero de Jáuregui/A. I. Jiménez San Cristóbal/R. Martín Hernández (eds.), Redefining Dionysus (Berlin 2013) 159–184.

Potts 1982
D. Potts, The Road to Meluhha. Journal of Near Eastern Studies 41.4, 1982, 279–288.

Potts 1989

T. F. Potts, Foreign Stone Vessels of the Late Third Millennium BC from Southern Mesopotamia. Their Origins and Mechanisms of Exchange. Iraq 51, 1989, 123–164.

Potts 1993

T. F. Potts, Patterns of Trade in Third-Millennium BC Mesopotamia and Iran. World Archaeology 24.3, 1993, 379–402.

Powell 1985

M. L. Powell, Analysis of Dental Wear and Caries for Dietary Reconstruction. In: R. I. Gilbert/J. H. Mielke (eds.), Analysis of Prehistoric Diets (San Diego 1985) 307–338.

Proietti 2012

G. Proietti, Prospettive socio-antropologiche sull'arcaismo Greco. La storiografia erodotea tra tradizione orale e 'storia intenzionale'. In: E. Franchi/G. Proietti (eds.), Forme della memoria e dinamiche identitarie nell'antichità greco-romana (Trento 2012) 181–206.

Proietti 2014

G. Proietti, Annual Games for War Dead and Founders in Classical Times. Between Hero-Cult and Civic Honours. Nikephoros 27, 2014, 199–213.

Puljiz 2021

I. Puljiz, Gold von Königen und Göttern. Zur Bedeutung von Goldobjekten in den syrisch-nordlevantinischen Königtümern der Mittleren und Späten Bronzezeit ausgehend von den Funden aus den Grüften von Qaṭna. Qaṭna-Studien 10 (Wiesbaden 2021).

Puljiz/Qasim 2019

I. Puljiz/H. A. Qasim, A New Mittani Centre on the Middle Tigris (Kurdistan Region). Report on the 2018 Excavations at Kemune (with contributions by O. Hense and B. Faist). Zeitschrift für Orient-Archäologie 12, 2019, 10–43.

Puljiz/Qasim 2020

I. Puljiz/H. A. Qasim, The Middle Assyrian Rural Complex of Muqable (with contributions by O. Hense/B. Faist). Zeitschrift für Orient-Archäologie 13, 2020, 90–118.

Puljiz/Schwab 2021

I. Puljiz/R. Schwab, Die Korrelation zwischen chemischen und typologischen Gruppen der Goldobjekte von Qaṭna. In: I. Puljiz (ed.), Gold von Königen und Göttern. Zur Bedeutung von Goldobjekten für die syrisch-nordlevantinischen Königtümer der Mittleren und Späten Bronzezeit ausgehend von den Funden aus den Grüften von Qaṭna. Qaṭna Studien 10 (Wiesbaden 2021).

Purcaro 2001

V. Purcaro, L'Agorà di Cirene. II 3 L'area meridionale del lato ovest dell'Agorà. Monografie di Archeologia Libica 24 (Rome 2001).

Purohit 1993

D. R. Purohit, Medieval English Folk Drama and Garhwali Folk Theatre. A Comparative Study (PhD Diss. H. N. B. Garhwal University Srinagar 1993).

Pyke et al. 1977

G. H. Pyke/H. R. Pulliam/E. L. Charnov, Optimal Foraging. A Selective Review of Theory and Tests. Quarterly Review of Biology 52, 1977, 137–154.

Quet 2014

M. Quet, La circulation des savoirs. Interdisciplinarité, concepts nomades, analogies, métaphores. F. Darbellay (ed.), P. Lang, 2012. Revue d'anthropologie des connaissances 8.1, 2014, 221–224.

Rachkov 2022
Y. Rachkov, Destruction, Preservation, and Rethinking of Ukraine's Urban Cultural Heritage during the Russo-Ukrainian War. The Journal of VN Karazin Kharkiv National University. Series History 62, 2022, 12–48.

Ramis 2010
D. Ramis, From Colonisation to Habitation. Early Cultural Adaptations in the Balearic Bronze Age. In: P. van Dommelen/A. B. Knapp (eds.), Material Connections in the Ancient Mediterranean. Mobility, Materiality and Identity (Oxon 2010) 64–84.

Ramis 2017
D. Ramis, Evidènces de contactes exteriors al món talaiòtic a partir de l'estudi del registre faunístic. In: F. Praos/H. Jimenez/J. J. Martinez (eds.), Menorca entre fenicis i púnics. Centro de Estudios de Próximo Oriente y Antiguedad Tardía (CEPOAT) (Menorca 2017) 201–217.

Ramos Soldado 2021
J. L. Ramos Soldado, Culto, Sacrificio y Simbolismo animal en Tartessos. Análisis arqueológico y paleobiológico de los registros de santuarios y necrópolis (Sevilla 2021).

Rappaport 1999
R. A. Rappaport, Ritual and Religion in the Making of Humanity (Cambridge 1999).

Rausch 2000
M. Rausch, Das Hypogäum auf der Agora von Poseidonia. Ein Kultort der Tritopatores? Kernos 13, 2000, 107–116.

Reber 1988
V. B. Reber, The Demographics of Paraguay. A Reinterpretation of the Great War, 1864–1870. Hispanic American Historical Review 68.2, 1988, 289–319.

Reckwitz 2006
A. Reckwitz, Die Transformation der Kulturtheorien. Zur Entwicklung eines Theorieprogramms (Weilerswist 2006).

Renfrew 1978
C. Renfrew, Trajectory, Discontinuity, and Morphogenesis. The Implications of Catastrophe Theory for Archaeology. American Antiquity 43, 1978, 203–222.

Respondek 2025
M. Respondek, From Entanglement to ResourceCulture. Rise of Mining as a Resource in the Riotinto Mining Basin. In: E. Revert Francés/M. Respondek/M. Bartelheim/R. Hardenberg (eds.), Entangled Economies. Mining and Agriculture from Antiquity to Modern Ages in Southern Spain (Tübingen forthcoming).

Richard et al. 2019
M. Richard/C. Falguères/H. Valladas/B. Ghaleb/E. Pons-Branchu/N. Mercier/D. Richter/N. J. Conard, New Electron Spin Resonance (ESR) Ages from Geißenklösterle Cave. A Chronological Study of the Middle and Early Upper Paleolithic Layers. Journal of Human Evolution 133, 2019, 133–145.

Richardson/Weszkalnys 2014
T. Richardson/G. Weszkalnys, Introduction. Resource Materialities. Anthropological Quarterly 87.1, 2014, 5–30.

Riehl 2009
S. Riehl, Archaeobotanical Evidence for the Interrelationship of Agricultural Decision-Making and Climate Change in the Ancient Near East. Quaternary International 197, 2009, 93–114.

Riehl 2024
S. Riehl, Archaeobotany. Crop Husbandry. Encyclopedia of Archaeology (Second Edition) Volume 2B (Elsevier 2024) 733–742.

Riera Rullan 2014
M. Riera Rullan (ed.), El Monestir de Cabrera. Segles V–VIII dC. Exposition Catalogue 15 March to 15 June 2014, Castell de Bellver (Palma 2014).

Riera Rullan 2017
M. Riera Rullan, El Monacat Insular de la Mediterrània Occidental. El Monestir de Cabrera (Balears, Segles V–VIII) (Barcelona 2017).

Rihuete Herrada 2003
C. Rihuete Herrada, Bio-Arqueología de las prácticas funerarias. Análisis de la comunidad enterrada en el cementerio prehistórico de la Cova des Càrritx (Cuitadella, Menorca), ca. 1450–800 cal ANE. BAR International Series 1161 (Cambridge 2003).

Rix 1982
H. Rix, Felsina. Rivista di Epigrafia Etrusca 82. Studi Etruschi 50, 1982, 304–320.

Robbins 2007
J. Robbins, Between Reproduction and Freedom. Morality, Value and Radical Cultural Change. Ethnos 72.3, 2007, 293–314.

Robbins 2015
J. Robbins, On Happiness, Values, and Time. The Long and the Short of it. Hau. Journal of Ethnographic Theory 5.3, 2015, 215–233.

Robbins/Sommerschuh 2016
J. Robbins/J. Sommerschuh, Values. In: F. Stein/S. Lazar/M. Candea/H. Diemberger/J. Robbins/A. Sanchez/R. Stasch (eds.), The Cambridge Encyclopedia of Anthropology (Cambridge 2016).

Roberts/Manchester 2010
C. Roberts/K. Manchester, The Archaeology of Disease (Stroud 2010).

Rodrigues et al. 2023
J. Rodrigues/K. Body/H. Carel, The Pandemic Body. The Lived Body during the COVID-19 Pandemic. Medical Humanities 49.4, 2023, 725–734.

Rodriguez Morales 2010
C. Rodriguez Morales, Tres cartas de emigrantes canarios en indias (Siglo 17). Cartas Diferentes. Revista Canaria De Patrimonio Documental (Islas Canarias) 06, 2010, 1–54.

Rogers 2015
D. Rogers, Oil and Anthropology. Annual Review of Anthropology 44, 2015, 365–380.

Rokeach 1973
M. Rokeach, The Nature of Human Values (New York 1973).

Roncalli 2014
G. Roncalli, Eredità geometrica e linguaggio figurativo. Il caso delle stele felsinee. In: G. M. Della Fina (ed.), Artisti, committenti e fruitori in Etruria tra VIII e V sec. a.C. Annali della Fondazione per il Museo Claudio Faina 21, 2014 (Rome 2014) 221–241.

Roncalli 2015
G. Roncalli, L'Aldilà. Dall'idea al passaggio. In: G. Sassatelli/A. Russo Tagliente (eds.), Il viaggio oltre la vita. Gli Etruschi e l'Aldilà tra capolavori e realtà virtuale, Exhibition Catalogue Bologna (Bologna 2015) 53–59.

Roscoe 2000
P. Roscoe, New Guinea Leadership as Ethnographic Analogy. A Critical Review. Journal of Archaeological Method and Theory 7, 2000, 79–126.

Rowlands 1971
M. J. Rowlands, The Archaeological Interpretation of Prehistoric Metalworking. World Archaeology 3, 1971, 210–224.

Ruiz Ballesteros 1998
E. Ruiz Ballesteros, Minería y poder. Antropología política en Riotinto (Huelva 1998).

Ruiz Ballesteros 2011
E. Ruiz Ballesteros, "Ingleses", minería y cultura. Un ensayo a contracorriente. In: A. Galán García (ed.), La presencia "inglesa" en Huelva. Entre la seducción y el abandono (Sevilla 2011) 17–40.

Rüpke 2018
J. Rüpke, Religious Agency, Sacralisation, Tradition. Istrazinvanja. Journal of Historical Researches 29, 2018, 22–38.

Sahlins 1963
M. D. Sahlins, Poor Man, Rich Man, Big-Man, Chief. Political Types in Melanesia and Polynesia. Comparative Studies in Society and History 5, 1963, 285–303.

Sahlins 1985
M. Sahlins, Island of History (London 1985).

Sahlins 1999
M. Sahlins, Two or Three Things That I Know about Culture. Journal of the Royal Anthropological Institute 5.3, 1999, 399–421.

Sahlins 2000
M. Sahlins, Culture in Practice. Selected Essays (New York 2000).

Sahlins 2013
M. Sahlins, On the Culture of Material Value and the Cosmography of Riches. HAU. Journal of Ethnographic Theory 3.2, 2013, 161–195.

Salazar García et al. 2018
D. C. Salazar-García/M. Fontanals-Coll/G. Goude/M. E. Subirà, "To 'Seafood' or not to 'Seafood'?" An Isotopic Perspective on Dietary Preferences at the Mesolithic-Neolithic Transition in the Western Mediterranean. Quaternary International 470 (Part B), 2018, 497–510.

Salzman 2004
P. C. Salzman, Pastoralists. Equality, Hierarchy and the State (Oxford 2004).

Samuel 1993
D. Samuel, Ancient Egyptian Cereal Processing. Beyond the Artistic Record. Cambridge Archaeological Journal 3, 1993, 276–283.

Samuelsen/Toé 2021
H. Samuelsen/L. P. Toé, COVID-19 Temporalities. Ruptures of Everyday Life in Urban Burkina Faso. Medicine Anthropology Theory 8.1, 2021, 1–21.

Sánchez García 2007
I. Sánchez García, Garoé. Patrimonio cultural en el ámbito rural. El Hierro (Islas Canarias) (Gran Canaria 2007).

Santocchini Gerg 2015
S. Santocchini Gerg, Felsina villanoviana. La "città visibile". Strategie insediative tra Bronzo Finale e Primo Ferro. In: M. Rendeli (ed.), Le città visibili. Archeologia dei processi di formazione urbana,

Proceedings in Honour of Gilda Bartoloni and Alberto Moravetti, Alghero 31st Januar–1st Februar 2014. Officina Etruscologia 11 (Rome 2015) 13–59.

Santocchini Gerg 2021
S. Santocchini Gerg, L'orientalizzante nel Bolognese. In: S. Bourdin/O. Dally/A. Naso/C. Smith (eds.), The Orientalizing Cultures in the Mediterranean, 8th-6th cent. BC. Origins, Cultural Contacts and Local Developments. The Case of Italy, Mediterranea. Studi e Ricerche sul Mediterraneo Antico Suppl. N.s. 1 (Rome 2021) 63–92.

Sanvisente 1849
B. Sanvisente, L'isola di Lampedusa eretta a colonia dal munificentissimo nostro sovrano Ferdinando II (Neapel 1849).

Sassatelli 1988
G. Sassatelli, Topografia e "sistemazione monumentale" delle necropoli felsinee. In: La formazione della città preromana in Emilia-Romagna. Proceedings Bologna/Marzabotto 1985 (Bologna 1988) 197–259.

Sassatelli 1990
G. Sassatelli, La situazione in Etruria padana. In: École Française de Rome (ed.), Crise et transformation des sociétés archaïques de l'Italie antique au Ve siècle Av. J.-C. Proceedings Rome 19–21 november 1987 (Rome 1990) 51–100.

Sassatelli 2009
G. Sassatelli, Riflessioni sulla 'stele della nave' di Bologna. In: S. Bruni (ed.), Etruria e Italia preromana. Studi in onore di Giovannangelo Camporeale (Pisa 2009) 833–840.

Sassatelli 2015a
G. Sassatelli, L'area padana e Bologna etrusca tra Grecia ed Etruria. In: G. Sassatelli/A. Russo Tagliente (eds.), Il viaggio oltre la vita. Gli Etruschi e l'Aldilà tra capolavori e realtà virtuale. Exhibition Catalogue Bologna (Bologna 2015) 99–109.

Sassatelli 2015b
G. Sassatelli, Bologna etrusca. Stele e società. In: G. Sassatelli/L. Cerchiai, Riflessioni sull'immaginario delle stele tra committenza e artigiani. In: E. Govi (ed.), Studi sulle stele etrusche di Bologna tra V e IV sec. a.C. (Rome 2015) 43–58.

Sassatelli 2015c
G. Sassatelli, Noterelle su Felsina, Archeologia Classica 66, 2015, 407–416.

Sassatelli 2018
G. Sassatelli, Etruschi e Italici in Italia settentrionale. Rapporti culturali e mobilità individuale. In: L. Aigner-Foresti/P. Amann, Beiträge zur Sozialgeschichte der Etrusker, Proceedings Wien 8th–10th June 2016 (Wien 2018) 355–370.

Sassatelli/Cerchiai 2014
G. Sassatelli/L. Cerchiai, Le stele felsinee. Invenzione e uso dell'immaginario tra artigiani e committenza. In: G. M. Della Fina (ed.), Artisti, committenti e fruitori in Etruria tra VIII e V sec. a.C. Annali della Fondazione per il Museo Claudio Faina 21, 2014 (Rome 2014) 187–219.

Sassatelli/Govi 2007
G. Sassatelli/E. Govi, Ideologia funeraria e celebrazione del defunto nelle stele etrusche di Bologna. Studi Etruschi 73, 2007, 67–92.

Sassatelli/Mancuso 2018
G. Sassatelli/G. Mancuso, Marmi d'Etruria. Verso un quadro aggiornato del problema. In: S. Steingräber (ed.), Cippi, Stele, Statue-Stele e Semata. Testimonianze in Etruria, nel mondo italico e in Magna Grecia dalla prima Età del Ferro all'Ellenismo. Proceedings Sutri 24th–25th April 2015. Mousai 8 (Pisa 2018) 103–121.

Schade 2020
T. Schade, Das ‚Nydamboot' im Museum. Inwertsetzungen und Präsentationen im Wandel der Zeit. Arkæologi i Slesvig/Archäologie in Schleswig 18, 2020, 139–157.

Schade 2021
T. Schade, Vom Fund zum Exponat. Inwertsetzungsprozesse und Authentizitätsvorstellungen am Beispiel der ‚Bremer Kogge'. In: T. Schade/B. Schweizer/S. Teuber/R. Da Vela/W. Frauen/ M. Karami/D. K. Ojha/K. Schmidt/R. Sieler/M. S. Toplak (eds.), Exploring Resources. On Cultural, Spatial and Temporal Dimensions of ResourceCultures. RessourcenKulturen 13 (Tübingen 2021) 233–260.

Schade 2022
T. Schade, Musealisierung und Museumsleben. Wertbildungen am Beispiel der Kon-Tiki. In: M. Farrenkopf/A. Filippidou/T. Meyer/S. Przigoda/A. Saupe/T. Schade (eds.), Alte Dinge – Neue Werte. Musealisierung und Inwertsetzung von Objekten (Göttingen 2022) 163–180.

Schäfer 2019
T. Schäfer, Dal dominio cartaginese a quello romano un complesso gioco di interscambio. In: A. Russo/ F. Guarneri/P. Xella/J. A. Zamora López (eds.), Carthago. Il mito immortale. La mostra (Mailand 2019) 138–145.

Scharl 2017
S. Scharl, Organisationsformen von Tausch im Neolithikum. Eine Fallstudie aus Nordwestbayern. In: P. Eisenach/T. Stöllner/A. Windler (eds.), The RITaK conferences 2013–2014. Raw Materials, Innovation, Technology of Ancient Cultures. RITaK 1 (Bochum 2017) 111–124.

Schattner 2022
T. G. Schattner, Zu den Megalographien in der tartessischen Vasenmalerei. Madrider Mitteilungen 63, 2022, 154–187.

Schattner 2025
T. G. Schattner (ed.) MULVA VIII, Die wirtschaftlichen Grundlagen von Munigua. Madrider Beiträge 43 (Wiesbaden 2025).

Scheibelreiter 2006a
G. Scheibelreiter, Wappen und adeliges Selbstverständnis. Das Mittelalter. Perspektiven mediävistischer Forschung 11.2, 2006, 7–27.

Scheibelreiter 2006b
G. Scheibelreiter, Heraldik. Oldenbourg Historische Hilfswissenschaften 1 (München 2006).

Scheid/Svenbro 1985
J. Scheid/J. Svenbro, La ruse d'Elyssa et la fondation de Carthage. Annales Économies, Sociétés, Civilisations 40.2, 1985, 328–342.

Scherer et al. 2021a
S. Scherer/K. Deckers/J. Dietel/M. Fuchs/J. Henkner/B. Höpfer/A. Junge/E. Kandeler/E. Lehndorff/P. Leinweber/J. Lomax/J. J. Miera/C. Poll/M. B. Toffolo/T. Knopf/T. Scholten/P. Kühn, What's in a Colluvial Deposit? Perspectives from Archaeopedology. Catena 198, 2021, 105040.

Scherer et al. 2021b
S. Scherer/B. Höpfer/K. Deckers/E. Fischer/M. Fuchs/E. Kandeler/J. Lechterbeck/E. Lehndorff/ J. Lomax/S. Marhan/E. Marinova/J. Meister/C. Poll/H. Rahimova/M. Rösch/K. Wroth/ J. Zastrow/T. Knopf/T. Scholten/P. Kühn, Middle Bronze Age Land Use Practices in the Northwestern Alpine Foreland. A Multi-Proxy Study of Colluvial Deposits, Archaeological Features and Peat Bogs. SOIL 7, 2021, 269–304.

Schiek et al. 1989
S. Schiek/H. G. Brand/H. Krins, Die Grabdenkmale im Kloster Bebenhausen. In: S. Schiek/ H. G. Brand/H. Krins (eds.), Beiträge zur Tübinger Geschichte 2 (Stuttgart 1989) 25–126.

Schlanger et al. 2016
N. Schlanger/L. Nespoulous/J.-P. Demoule, Year 5 at Fukushima. A 'Disaster-Led' Archaeology of the Contemporary Future. Antiquity 90.350, 2016, 409–424.

Schmid 1853
L. Schmid, Geschichte der Pfalzgrafen von Tübingen, nach meist ungedruckten Quellen, nebst Urkundenbuch. Ein Beitrag zur schwäbischen und deutschen Geschichte (Tübingen 1853).

Schneidmüller 2000
B. Schneidmüller, Die Welfen. Herrschaft und Erinnerung 819–1252 (Stuttgart 2000).

Schneppen 2001
H. Schneppen, Nietzsche und Paraguay. Der Philosoph als Bauer? Nietzscheforschung 8, 2001, 249–265.

Schofield 2005
J. Schofield, Discordant Landscapes. Managing Modern Heritage at Twyford Down, Hampshire (England). International Journal of Heritage Studies 11.2, 2005, 143–159.

Scholten et al. 2002
T. Scholten/M. Szibalski/P. Felix-Henningsen, Multifunktionalität von Mittelgebirgslandschaften. Auf den Standort kommt es an. Berichte über Landwirtschaft 80.4, 2002, 509–539.

Schön 2014
F. Schön, Insulare Wasserversorgung. Antike Regenwassersammel- und -speicheranlagen auf Pantelleria und Linosa (Italien). In: T. Schäfer/F. Schön/J. Heinrichs/A. Gerdes (eds.), Antike und moderne Wasserspeicherung. Internationaler Workshop vom 11.–14.05.2011 in Pantelleria (Italien). Tübinger Archäologische Forschungen 12 (Rahden 2014) 103–118.

Schön 2020a
F. Schön, Wasserressourcen in Karthago und den Punischen Städten des Mittelmeerraumes. In: J. Bergemann (ed.), Kultur und Natur in der antiken Mittelmeerwelt. Italienische und deutsche Forschungen in Archäologie und Geschichte und ihre Bedeutung für die moderne Diskussion über Klimawandel, Villa Vigoni Gespräch vom 03.–06. April 2018. Göttinger Studien zur Mediterranen Archäologie 9 (Rahden 2020) 45–55.

Schön 2020b
F. Schön, Isolation as a Resource for an Island Community in the Strait of Sicily. The Case of Linosa Island (Italy) in Late Antiquity. In: S. Teuber/A. K. Scholz/T. Scholten/M. Bartelheim (eds.), Waters. Conference Proceedings for 'Waters as a Resource' of the SFB 1070 ResourceCultures and DEGUWA (Deutsche Gesellschaft zur Förderung der Unterwasserarchäologie e.V.). RessourcenKulturen 11 (Tübingen 2020) 49–63.

Schön 2025
F. Schön, Aethusa. Diachrone Untersuchungen zur Besiedlungsgeschichte eines insularen ‚Marginalraumes' im zentralen Mittelmeerraum. Die Ergebnisse des Tübinger Linosa Survey Projektes. RessourcenKulturen 29 (Tübingen 2025).

Schön/Dierksmeier 2021
F. Schön/L. Dierksmeier, Water Scarcity at Sea. Historical and Archaeological Perspectives on the Preservation of Freshwater on Volcanic Islands. In: T. Schade/B. Schweizer/S. Teuber/R. Da Vela/ W. Frauen/M. Karami/D. K. Ojha/K. Schmidt/R. Sieler/M. Toplak (eds.), Exploring Resources. On Cultural, Spatial and Temporal Dimensions of ResourceCultures. RessourcenKulturen 13 (Tübingen 2021) 157–182.

Schön/Schäfer 2021

F. Schön/T. Schäfer, Römer werden, punisch bleiben? Kult und Vergangenheit als Ressourcen kollektiver Identität auf Pantelleria (Cossyra) im 3. Jahrhundert v. Chr. bis in das 3. Jahrhundert n. Chr. In: J. Bergemann/O. Belvedere (eds.), Imperium Romanum. Romanization between Colonization and Globalization, Villa Vigoni, 04. –07.11.2019. Studi e Materiali 2 (Palermo 2021) 145–163.

Schön/Töpfer 2016

F. Schön/H. Töpfer (eds.), Karthago Dialoge. Karthago und der punische Mittelmeerraum. Kulturkontakte und Kulturtransfers im 1. Jahrtausend v. Chr. RessourcenKulturen 2 (Tübingen 2016).

Schörner 2007

H. Schörner, Sepulturae graecae intra urbem. Untersuchungen zum Phänomen der intraurbanen Bestattungen bei den Griechen (Möhnesee 2007).

Schritt 2016

J. Schritt, From Nuclear Imperialism to Petro-Democracy? Resource Assemblages and the Emergence of a New Political Configuration in Niger. Canadian Journal of African Studies 50.2, 2016, 229–254.

Schubert/Klein 2020

K. Schubert/M. Klein, Das Politiklexikon. 7. aktual. u. erw. Aufl. (Bonn 2020).

Schulte 2014

P. Schulte, Defining Water Scarcity, Water Stress, and Water Risk, Pacific Institute, February 2014, <https://pacinst.org/water-definitions> (last access: 15.12.2023).

Schwab/Pernicka 2021

R. Schwab/E. Pernicka, Metallanalysen an den Goldfunden aus der Gruft VII und der Königsgruft von Qaṭna. In: I. Puljiz (ed.), Gold von Königen und Göttern. Zur Bedeutung von Goldobjekten für die syrisch-nordlevantinischen Königtümer der Mittleren und Späten Bronzezeit ausgehend von den Funden aus den Grüften von Qaṭna. Qaṭna Studien 10 (Wiesbaden 2021) 565–589.

Schwarmaier/Schaab 1995

H. Schwarmaier/M. Schaab (eds.), Handbuch der baden-württembergischen Geschichte 2. Die Territorien im Alten Reich (Stuttgart 1995).

Schwartz 1994

S. H. Schwartz, Are there Universal Aspects in the Structure and Content of Human Values? Journal of Social Issues 50.4, 1994, 19–45.

Schwartz/Bilsky 1987

S. H. Schwartz/W. Bilsky, Towards a Psychological Structure of Human Values. Journal of Personality and Social Psychology 53, 1987, 550–562.

Schweizer 2020

B. Schweizer, Knowledge-scapes as Resources. An Archaeological Approach to the Construction of Cultural and Social Identities. Archaeological Review from Cambridge 35.2 (Knowledge-scapes), 2020, 45–59.

Schweizer 2021

B. Schweizer, Wine as a Key Resource in the Fifth Century BC. Substances and Things, Bodies and Emotions in Dionysian Practices. In: T. Schade/B. Schweizer/S. Teuber/R. Da Vela/W. Frauen/M. Karami/D. K. Ojha/K. Schmidt/R. Sieler/M. S. Toplak (eds.), Exploring Resources. On Cultural, Spatial and Temporal Dimensions of ResourceCultures. RessourcenKulturen 13 (Tübingen 2021) 307–328.

Schweizer 2022
B. Schweizer, Dinge und Monumente als Ressourcen kollektiver Identitäten. Zur Medialität kultureller Gedächtnisse. In: V. Sossau/K. Riehle (eds.), Mistaken Identity. Identitäten als Ressourcen im zentralen Mittelmeerraum. RessourcenKulturen 19 (Tübingen 2022) 53–79.

Schweizer/Schön 2018
B. Schweizer/F. Schön, Settlement and Mobility. In: A.-M. Wittke/J. N. Dillon (eds.), The Early Mediterranean World, 1200–600 BC. Brill's New Pauly Suppl. II 9 (Leiden 2018).

SCI 2006
Statistical Centre of Iran, Census of the Islamic Republic of Iran (Teheran 2006).

SCI 2014
Statistical Centre of Iran, Census of the Islamic Republic of Iran (Teheran 2014).

Scifo 1989
O. Scifo, Linosa. Un'isola hawaiana nel Mediterraneo (Agrigent 1989).

Sconzo/Qasim 2021
P. Sconzo/H. A. Qasim, Investigating Jubaniyah, a Late Chalcolithic Site on the Upper Tigris River, Kurdistan Region – Iraq. Preliminary Report (with contributions by M. al Besso/B. Glissmann/ A. Titolo). Mesopotamia LVI, 2021, 1–59.

Scott 2013
M. Scott, Space and Society in the Greek and Roman Worlds (Cambridge 2013).

Service 1962
E. R. Service, Primitive Social Organization. An Evolutionary Perspective (New York 1962).

Service 1975
E. R. Service, Origins of the State and Civilization. The Process of Cultural Evolution (New York 1975).

Setzler 2003
W. Setzler, Bebenhausen. In: N. Priesching/W. Zimmermann (eds.), Württembergisches Klosterbuch. Klöster, Stifte und Ordensgemeinschaften von den Anfängen bis zur Gegenwart (Ostfildern 2003) 184–187.

Shennan et al. 2013
S. Shennan/S. S. Downey/A. Timpson/K. Edinborough/S. Colledge/T. Kerig/K. Manning/ M. G. Thomas, Regional Population Collapse Followed Initial Agriculture Booms in Mid-Holocene Europe. Nature Communications 4.2486, 2013. DOI: <https://doi.org/10.1038/ncomms3486>.

Sherratt 1981
A. Sherratt, Plough and Pastoralism. Aspects of the Secondary Products Revolution. In: I. Hodder/ G. Isaac/N. Hammond (eds.), Patterns of the Past. Studies in Honour of David Clarke (Cambridge 1981) 261–306.

Sherratt 1983
A. Sherratt, The Secondary Exploitation of Animals in the Old World. World Archaeology 15.1, 1983, 90–104.

Sicking 2014
L. Sicking, The Dichotomy of Insularity. Islands between Isolation and Connectivity in Medieval and Early Modern Europe, and Beyond. International Journal of Maritime History 26.3, 2014, 494–511.

Sieg 2019
U. Sieg, Die Macht des Willens. Elisabeth Förster-Nietzsche und ihre Welt (München 2019).

Sieler 2022
R. Sieler, Religion, Alchemy, and Medicine in South India. Tamil Siddha Traditions and Siddha Medicine. Asian Medicine 17.1, 2022, 1–15.

Simonton 2018
M. Simonton, The Burial of Brasidas and the Politics of Commemoration in the Classical Period. American Journal of Philology 139.1=553, 2018, 1–30.

Slimak et al. 2022
L. Slimak/C. Zanolli/T. Higham/M. Frouin/J. Schwenninger/L. J. Arnold/M. Demuro/K. Douka/N. Mercier/G. Guérin/H. Valladas/P. Yvorra/Y. Giraud/A. Seguin-Orlando/L. Orlando/J. E. Lewis /X. Muth/H. Camus/S. Vandevelde/M. Buckley/C. Mallol/C. Stringer/L. Metz, Modern Human Incursion into Neanderthal Territories 54,000 Years ago at Mandrin, France. Science Advances 8/eabj9496, 2022, 1–16.

Smelser 1992
N. J. Smelser, External and Internal Factors in Theories of Social Change. In: H. Haferkamp/N. J. Smelser (eds.), Social Change and Modernity (Berkeley 1992) 370–394.

Smith 1919–1923
Thucydides, History of the Peloponnesian War. With an English Translation by Charles Forster Smith, in Four Volumes. I (Cambridge 1919); II (Cambridge 1920); III (Cambridge 1921); IV (Cambridge 1923).

Sommer 1971
I. K. Sommer, Die Chronik des Stuttgarter Ratsherrn Sebastian Küng. Edition und Kommentar. Veröffentlichungen des Archivs der Stadt Stuttgart 24 (Stuttgart 1971).

Sørensen/Carman 2009
M. L. S. Sørensen/J. Carman (eds.), Heritage Studies. Methods and Approaches (Abingdon 2009).

Soressi 2010
M. Soressi, Les Cottés (Vienne). Nouveaux travaux sur l'un des gisements de référence pour la transition Paléolithique moyen/supérieur. In: J. Bouisson-Catil/J. Primault (eds.), Préhistoire entre Vienne et Charente. Hommes et sociétés du Paléolithique, Ministère de la Culture et de la Communication (Chauvigny 2010) 221–234.

Soressi/Roussel 2014
M. Soressi/M. Roussel, European Middle to Upper Paleolithic Transitional Industries. Châtelperronian. In: C. Smith (ed.), Encyclopedia of Global Archaeology (New York 2014) 2679–2693.

Soto Company/Mas Forners 2015
R. Soto Company/A. Mas Forners, Feudal Colonisation and Socio-Ecological Transition in Mayûrqa (Muslim Majorca) in the Thirteenth Century. Continuity and Change 30, 2015, 341–366. DOI: <https://doi.org/10.1017/S0268416015000375>.

Spangenberg et al. 2008
J. E. Spangenberg/I. Matuschik/S. Jacomet/J. Schibler, Direct Evidence for the Existence of Dairying Farms in Prehistoric Central Europe (4th millennium BC). Isotopes in Environmental and Health Studies 44.2, 2008, 189–200.

Spanish Ministry of Agriculture, Fisheries and Food 2023
<https://www.mapa.gob.es/es/ganaderia/temas/zootecnia/razas-ganaderas/razas/catalogo-razas/> (last access: 05.09.2023).

Spieß 1993
K.-H. Spieß, Familie und Verwandtschaft im deutschen Hochadel des Spätmittelalters. 13. bis Anfang des 16. Jahrhunderts. Vierteljahrschrift für Sozial- und Wirtschaftsgeschichte Beihefte 111, 1993, 490–492.

Staecker 2005a
J. Staecker, The Concepts of Imitatio and Translatio. Perceptions of a Viking-Age Past. Norwegian Archaeological Review, 2005, 3–28.

Staecker 2005b
J. Staecker, The Making of a Viking-Age Past. Norwegian Archaeological Review, 2005, 116–121.

Stähli 1999
A. Stähli, Die Verweigerung der Lüste. Erotische Gruppen in der antiken Plastik (Berlin 1999).

Stalnaker 1998
R. Stalnaker, On the Representation of Context. Journal of Logic, Language and Information 7, 1998, 3–19.

Starbuck 2004
D. R. Starbuck, Neither Plain nor Simple. New Perspectives on the Canterbury Shakers (Lebanon 2004).

Stehle 2012
E. Stehle, Women and Religion in Greece. In: S. L. James/S. Dillon (eds.), A Companion to Women in the Ancient World (Malden 2012) 191–203.

Steingräber 2018
S. Steingräber (ed.), Cippi, Stele, Statue-Stele e Semata. Testimonianze in Etruria, nel mondo italico e in Magna Grecia dalla prima Età del Ferro all'Ellenismo. Proceedings Sutri 24th–25th April 2015. Mousai 8 (Pisa 2018).

Stevenson/Harrison 1992
A. C. Stevenson/R. J. Harrison, Ancient Forest in Spain. A Model for Land-Use and Dry Forest Management in South-West Spain from 4000 BC to 1900 AD. Proceedings of the Prehistoric Society 58, 1992, 227–247.

Stöber 1978
G. Stöber, Die Afshār. Nomadismus im Raum Kermān/Zentralirān (Marburg 1978).

Stöber 2002
G. Stöber, The Nomads of Kermān. In: R. Tapper/J. Thompson (eds.), The Nomadic Peoples of Iran (London 2002) 252–259.

Stöllner et al. 2006
T. Stöllner/J. Cierny/C. Eibner/N. Boenke/R. Herd/A. Maass/K. Röttger/T. Sormaz/G. Steffens/P. Thomas, Der bronzezeitliche Bergbau im Südrevier des Mitterberggebietes. Bericht zu den Forschungen der Jahre 2002 bis 2006. Archaeologia Austriaca 90, 2006, 87–137.

Störmer 1980
W. Störmer, Die Hausklöster der Wittelsbacher. Wittelsbach und Bayern 1.1. Die Zeit der frühen Herzöge. Von Otto I. zu Ludwig dem Bayern (München 1980).

Stratford 2016
E. Stratford, Island Geographies. Essays and Conversations (Florence 2016).

Straub 2004
J. Straub, Identität. In: F. Jäger/B. Liebsch (eds.), Handbuch der Kulturwissenschaften. Grundlagen und Schlüsselbegriffe, Band 1 (Stuttgart 2004) 277–303.

Strawson 1950
P. F. Strawson, On Referring. Mind 59, 1950, 320–344.

Strezewski 2015a
M. Strezewski, Excavations at the Harmonist Dormitory #1 Complex, New Harmony, Indiana. University of Southern Indiana, Archaeology Laboratory Reports of Investigations, 14–02 (Evansville 2015).

Strezewski 2015b
M. Strezewski, Harmonist Demography and Town Planning in New Harmony, Indiana. Communal Societies 35.1, 2015, 1–28.

Stringer 2011
C. Stringer, The Origin of our Species (München 2011).

Stroszeck 2020
J. Stroszeck, Grabrelief und Kanalbau. Zum neuen Fund einer Familienstele im Kerameikos. In: A. Delivorrias/E. Vikela/A. Zarkadas/N. Kaltsas/I. Triante (eds.), ΣΠΟΥΔΗ. Αφιέρωμα στη μνήμη του Γιώργου Δεσπίνη. Studies in Memory of Giorgos Despinis (Athen 2020) 403–416.

Stucchi 1965
S. Stucchi, L'agora di Cyrene I. I lati nord ed est della platea inferiore. Monografie di Archeologia Libica 7 (Rome 1965).

Stucchi 1967
S. Stucchi, Cirene 1957–1966. Un decennio di attività della missione archeologica italiana a Cirene (Tripoli 1967).

Suárez Grimón 1987
V. Suárez Grimón, La propiedad pública, vinculada y eclesiástica en Gran Canaria, en la crisis del antiguo régimen (Las Palmas de Gran Canaria 1987).

Sutton et al. 2011
M. A. Sutton/C. M. Howard/J. W. Erisman/G. Billen/A. Bleeker/P. Grennfelt/H. van Grinsven/B. Grizzetti (eds.), The European Nitrogen Assessment. Sources, Effects and Policy Perspectives (Cambridge 2011).

Svoboda-Baas 2019
D. F. Svoboda-Baas, Kultlandschaften. Räumliche Organisation in Heiligtümern Lukaniens des 4. und 3. Jhs. v. Chr. Tübinger Archäologische Forschungen 27 (Rahden 2019).

Sydow 1984
J. Sydow, Das Bistum Konstanz 2. Die Zisterzienserabtei Bebenhausen. Germania Sacra Neue Folge 16 (Berlin 1984).

Tanganelli et al. 2015
V. Tanganelli/G. Bocchini/M. G. Belcastro, Appendice 3. Studio antropologico dei resti ossei provenienti dalla tomba 11 della necropoli di via Saffi (Bologna). In: E. Govi (ed.), Studi sulle stele etrusche di Bologna tra V e IV sec. a.C. (Rome 2015) 133–138.

Tapper 1997
R. Tapper, Frontier Nomads of Iran. A Political and Social History of the Shahsevan (Cambridge 1997).

Tarlow 2002
S. Tarlow, Excavating Utopia. Why Archaeologists Should Study "Ideal" Communities of the Nineteenth Century. International Journal of Historical Archaeology 6.4, 2002, 299–323.

Tarlow 2006
S. Tarlow, Representing Utopia. The Case of Cyrus Teed's Koreshan Unity Settlement. Historical Archaeology 40.1, 2006, 89–99.

Taylor 2004
C. Taylor, Modern Social Imaginaries (Durham 2004).

Teuber/Schweizer 2020
S. Teuber/B. Schweizer, Resources Redefined. Resources and ResourceComplexes. In: S. Teuber/A. Scholz/T. Scholten/M. Bartelheim (eds.), Waters. Conference Proceedings for 'Waters as a Resource' of the SFB 1070 ResourceCultures and DEGUWA (Deutsche Gesellschaft zur Förderung der Unterwasserarchäologie e.V.). RessourcenKulturen 11 (Tübingen 2020) 9–19.

Teuber et al. 2017
S. Teuber/J. J. Ahlrichs/J. Henkner/P. Kühn/T. Scholten, SoilCultures. The Adaptive Cycles of Agrarian Soil Use in Central Europe. An Interdisciplinary Study Using Soil Scientific and Archaeological Research. Ecology and Society 22.4:13, 2017. DOI: <https://doi.org/10.5751/ES-09729-220413>.

Teuber et al. 2020
S. Teuber/A. K. Scholz/T. Scholten/M. Bartelheim (eds.), Waters. Conference Proceedings for 'Waters as a Resource' of the SFB 1070 ResourceCultures and DEGUWA (Deutsche Gesellschaft zur Förderung der Unterwasserarchäologie e.V.). RessourcenKulturen 11 (Tübingen 2020).

Teuber et al. 2022
S. Teuber/M. Bartelheim/R. Hardenberg/M. Knopf/T. Knopf/P. Kühn/T. Schade/K. Schmidt/T. Scholten, Why Do We Need Interdisciplinary Cooperation with Anthropologists and Archaeologists in Soil Science. Journal of Plant Nutrition and Soil Science 185.6, 2022, 752–765.

Theodossopoulos 2013
D. Theodossopoulos, Laying Claim to Authenticity. Five Anthropological Dilemmas. Anthropological Quarterly 86.2, 2013, 337–360.

Thesing 1977
R. Thesing, Die Größenentwicklung des Haushuhns in vor- und frühgeschichtlicher Zeit (München 1977).

Thiemeyer 2021
T. Thiemeyer, Identität. In: T. Schade/B. Schweizer/S. Teuber/R. Da Vela/W. Frauen/M. Karami/D. K. Ojha/K. Schmidt/R. Sieler/M. Toplak (eds.), Exploring Resources. On Cultural, Spatial and Temporal Dimensions of ResourceCultures. RessourcenKulturen 13 (Tübingen 2021) 81–89.

Thomas 2018
R. Thomas, Truth and Authority in Herodotus' Narrative. False Stories and True Stories. In: E. Bowie (ed.), Herodotus. Narrator, Scientist, Historian (Berlin 2018) 265–284.

Thompson 1979
M. Thompson, Rubbish Theory. The Creation and Destruction of Value (Oxford 1979).

Tilley 1994
C. Tilley, A Phenomenology of Landscape. Places, Paths and Monuments. Explorations in Anthropology (Oxford 1994).

Tilley 2015
L. Tilley, Theory and Practice in the Bioarchaeology of Care. Bioarchaeology and Social Theory (Cham 2015).

Timmermann 2020
A. Timmermann, Quantifying the Potential Causes of Neanderthal Extinction. Abrupt Climate Change versus Competition and Interbreeding. Quaternary Science Reviews 238.106331, 2020, 1–14.

Tobler 1993
W. Tobler, Three Presentations on Geographical Analysis and Modeling. Non- Isotropic Geographic Modeling. Speculations on the Geometry of Geography and Global Spatial Analysis. Technical Reports 93.1, 1993. DOI: <https://escholarship.org/uc/item/05r820mz>.

Tomaso et al. 2006
M. Tomaso/R. Veit/C. DeRooy/S. Walling, Social Status and Landscape in a Nineteenth-Century Planned Industrial Alternative Community. Archaeology and Geography of Feltville, New Jersey. Historical Archaeology 40.1, 2006, 20–36.

Torelli 1999a
M. Torelli, Paestum Romana (Rome 1999).

Torelli 1999b
M. Torelli, Roman Paestum. In: M. Torelli (ed.), Tota Italia. Essays in the Cultural Formation of Roman Italy (Oxford 1999) 43–88.

Trias 2004
M. Trias, El paisatge subterrani meridional de Menorca. In: J. J. Fornós/A. Obrador/V. M. Roselló (eds.), Història Natural del Migjorn de Menorca. El medi fisic i l'influx humà. Societat Historia Natural Balears 11 (Menorca 2004) 275–290.

Tsetskhladze 2006
G. Tsetskhladze, Revisiting Ancient Greek Colonisation. In: G. Tsetskhladze (ed.), Greek Colonisation. An Account of Greek Colonies and other Settlements overseas, Volume I (Leiden 2006) xxiii-lxxxiii.

Tsing 2004
A. L. Tsing, Friction. An Ethnography of Global Connection (Princeton 2004).

Tsing 2015
A. L. Tsing, The Mushroom at the End of the World. On the Possibility of Life in Capitalist Ruins (Princeton 2015).

Turck et al. 2019
R. Turck/T. Stöllner/G. Goldenberg (eds.), Alpine Copper II. New Results and Perspectives on Copper Production. Der Anschnitt, Beiheft 42 (Bochum 2019).

Turner 1967
V. Turner, The Forest of Symbols. Aspects of Ndembu Ritual (Ithaca 1967).

Turner 2005
N. J. Turner, The Earth's Blanket. Traditional Teachings for Sustainable Living. Culture, Place, and Nature (Seattle 2005).

Tvedt 2015
T. Tvedt (ed.), A History of Water. Series I–II (London 2015).

Tvedt 2016
T. Tvedt, Water and Society. Changing Perceptions of Societal and Historical Development (London 2016).

Uchibori 2011
M. Uchibori, Theoretical Themes for an Anthropology of Resources. Social Science Information 50.1, 2011, 142–153.

Ulmansky 1918
S. Ulmansky, Die andalusische Rinderrasse. Mitteilungen der landwirtschaftlichen Lehrkanzeln der Hochschule für Bodenkultur in Wien 3.3, 1918, 457–547.

UNESCO 2009

Ramman, Religious Festival and Ritual Theatre of the Garhwal Himalayas, India, <https://ich.unesco.org/en/RL/ramman-religious-festival-and-ritual-theatre-of-the-garhwal-himalayas-india-00281> (last access: 23.07.2024).

UNESCO 2020

UNESCO, Practices and Traditional Knowledge of Terere in the Culture of Pohã Ñana, Guaraní Ancestral Drink in Paraguay. Inscribed in 2020 (15.COM) on the Representative List of the Intangible Cultural Heritage of Humanity (2020), <https://ich.unesco.org/en/RL/practices-and-traditional-knowledge-of-terere-in-the-culture-of-poha-nana-guarani-ancestral-drink-in-paraguay-01603> (last access: 14.08.2024).

UNESCO 2023

Talayotic Menorca, <https://whc.unesco.org/en/list/1528/> (last access: 08.11.2023).

United Nations 2015

United Nations 2015. Resolution Adopted by the General Assembly on 25 September 2015. Transforming our World. The 2030 Agenda for Sustainable Development.

Ur 2003

J. Ur, CORONA Satellite Photography and Ancient Road Networks. A Northern Mesopotamian Case Study. Antiquity 77, 2003, 102–115.

Uttaranchal Tourism 2002

Uttaranchal Tourism, The Inner Soul of Uttaranchal. Myths and Traditions of a Living State (Dehra Dun 2002).

Vaesen et al. 2019

K. Vaesen/F. Scherjon/L. Hemerik/A. Verpoorte, Inbreeding, Allee Effects and Stochasticity Might Be Sufficient to Account for Neanderthal Extinction. PLoS ONE 14/11, 2019, e0225117. DOI: <https://doi.org/10.1371/journal.pone.0225117>.

Van Bueren 2006

T. Van Bueren, Between Vision and Practice. Archaeological Perspectives on the Llano del Rio Cooperative. Historical Archaeology 40.1, 2006, 133–151.

Van Bueren/Tarlow 2006

T. Van Bueren/S. Tarlow, The Interpretive Potential of Utopian Settlements. Historical Archaeology 40.1, 2006, 1–5.

Van Strydonck 2014

M. Van Strydonck, From Myotragus to Metellus. A Journey through the Pre- and Early-History of Majorca and Minorca (Hochwald 2014).

Van Strydonck et al. 2005

M. Van Strydonck/M. Boudin/A. Ervynck/J. Orvay/H. Borms, Spatial and Temporal Variation of Dietary Habits during the Prehistory of the Balearic Islands as Reflected by 14C, d15N and d13C Analyses on Human and Animal Bones. Mayurqa 30, 2005, 523–554.

Varjāvand 1998

P. Varjāvand, EMĀMZĀDA iii. Number, Distribution, and Important Examples, last updated 2011, <https://iranicaonline.org/articles/emamzada-iii> (last access: 01.08.2023).

Veblen 1899

T. Veblen, The Theory of the Leisure Class. An Economic Study in the Evolution of Institutions (New York 1899).

Vega Toscano et al. 1998
L. G. Vega Toscano/M. L. Cerdeño Serrano/B. Córdoba de Oya, El origen de los mastines ibéricos. La trashumancia entre los pueblos prerromanos de la meseta. Complutum 9, 1998, 117–135.

Vermeersch et al. 2021
S. Vermeersch/S. Riehl/B. M. Starkovich/J. Kamlah, Developments in Subsistence during the Early Bronze Age through the Iron Age in the Southern and Central Levant. Integration of Faunal and Botanical Remains Using Multivariate Statistics. Quaternary Science Reviews 253, 2021, 106776.

Vidale 2015
M. Vidale, Searching for Mythological Themes on the 'Jiroft' Chlorite Artefacts. Iranica Antiqua 50, 2015, 15–58.

Voigt 2014
F. Voigt, Die Aktualität von Halbwachs' Kritik an Bergson und Durkheim. Zu einigen Problemen in den Halbwachs-Lektüren bei Aleida und Jan Assmann. Zeitschrift für Religions- und Geistesgeschichte 66, 2014, 243–269.

Vok 1997
F. V. Vok, Wappenbuch, -rolle. Lexikon des Mittelalters 8 (München 1997) 2035.

von deutschen Freunden 1889
von deutschen Freunden, Eine Schrift zum Andenken und zur Rechtfertigung. Schriften über Bernhard Förster (Leipzig 1889).

Vossler-Wolf 2016
C. Vossler-Wolf, Kloster & UmWelt. Ressourcennutzung im Kontext mittelalterlicher Klosterlandschaften. Mitteilungen der Deutschen Gesellschaft für Archäologie des Mittelalters und der Neuzeit 29 (Paderborn 2016) 55–68.

Vossler-Wolf 2020
C. Vossler-Wolf, Aqua Viva and the Monk in the Pond. Multi-Dimensional Water Use in Medieval Monasteries. In: S. Teuber/A. K. Scholz/T. Scholten/M. Bartelheim (eds.), Waters. Conference Proceedings for 'Waters as a Resource' of the SFB 1070 ResourceCultures and DEGUWA (Deutsche Gesellschaft zur Förderung der Unterwasserarchäologie e.V.). RessourcenKulturen 11 (Tübingen 2020) 65–76.

Wang et al. 2024
S. Wang/Q. Hu/S. Wang/M. Ai/P. Zhao, Archaeological Site Segmentation of Ancient City Walls Based on Deep Learning and LiDAR Remote Sensing. Journal of Cultural Heritage 66, 2024, 117–131.

Warner 2002
M. Warner, Publics and Counterpublics (New York 2002).

Washington 1908
H. S. Washington, Linosa and its Rocks. Journal of Geology 16, 1908, 1–35.

Watts 2005
M. Watts, Resource Curse? Governmentality, Oil and Power in the Niger Delta, Nigeria. In: P. Le Billon (ed.), Geopolitics of Resource Wars. Resource Dependence, Governance and Violence. Case Studies in Geopolitics (London 2005) 50–80.

Weber 1978
M. Weber, Economy and Society. An Outline of Interpretive Sociology (Berkeley 1978).

Wehling 2006
H-G. Wehling, Baden-Württemberg. Zur Geschichte eines jungen Bundeslandes. Baden-Württemberg. Gesellschaft, Geschichte, Politik. Schriften zur politischen Landeskunde Baden-Württembergs 34 (Stuttgart 2006).

Weiler 2005
N. Weiler, Wissen und Macht in einer Welt der Konflikte. Zur Politik der Wissensproduktion. In: K. Gerlof/A. Ulrich (eds.), Die Verfasstheit der Wissensgesellschaft (Münster 2005).

Wein 1979
G. Wein, Die Ausgrabung der „Königswart" bei Baiersbronn, Landkreis Freudenstadt. Forschungen und Berichte der Archäologie des Mittelalters in Baden-Württemberg 6 (Stuttgart 1979) 77–96.

Wentink 2020
K. Wentink, Stereotype. The Role of Grave Sets in Corded Ware and Bell Beaker Funerary Practices (Leiden 2020).

White 2003
D. R. White, Network Analysis, Social Dynamics and Feedback in Social Systems. Cybernetics and Systems 35.2–3, 2003, 173–192.

White 2015
D. A. White, The Basics of Least Cost Analysis for Archaeological Applications. Advances in Archaeological Practice 3.4, 2015, 407–414.

White/Surface-Evans 2012
D. A. White/S. L. Surface-Evans (eds.), Least Cost Analysis of Social Landscapes. Archaeological Case Studies (Salt Lake City 2012).

Whitmore 2018
L. Whitmore, Mountain, Water, Rock, God. Understanding Kedarnath in the Twenty-First Century (Oakland 2018).

WHO 1986
World Health Organization, Ottawa Charter for Health Promotion, 1986 (No. WHO/EURO: 1986-4044-43803-61677). World Health Organization. Regional Office for Europe. DOI: <https://iris.who.int/handle/10665/349652>.

Wiesehöfer 2013
J. Wiesehöfer, Iranian Empires. In: P. Fibiger Bang/W. Scheidel (eds.), The Oxford Handbook of the State in the Ancient Near East and Mediterranean (Oxford 2013) 199–232.

Wilkinson 1993
T. J. Wilkinson, Linear Hollows in the Jazira, Upper Mesopotamia. Antiquity 67.256, 1993, 548–562.

Wilkinson 2014
T. C. Wilkinson, Tying the Treads of Eurasia. Trans-Regional Routes and Material Flows in Transcaucasia Eastern Anatolia and Western Central Asia, c. 3000–1500 BC (Leiden 2014).

Wilkinson et al. 2010
T. J. Wilkinson/C. French/J. A. Ur/M. Semple, The Geoarchaeology of Route Systems in Northern Syria. Geoarchaeology 25.6, 2010, 745–771.

Williams et al. 2001
J. Williams/D. Kraus/H. Knowles, Flights from Modernity. German and Australian Utopian Colonies in Paraguay 1886–1896. Journal of Australian Studies 25, 2001, 49–62.

Williamson/Carr 2009
D. L. Williamson/J. Carr, Health as a Resource for Everyday Life. Advancing the Conceptualization. Critical Public Health 19.1, 2009, 107–122.

Williams-Thorpe 1995
O. Williams-Thorpe, Obsidian in the Mediterranean and Near East. A Provenancing Success Story. Archaeometry 37, 1995, 217–248.

Wilson 1979
B. Wilson, Rationality (Oxford 1979).

Wilson 2007
L. Wilson, Understanding Prehistoric Lithic Raw Material Selection. Application of a Gravity Model. Journal of Archaeological Method and Theory 14, 2007, 388–411. DOI: <https://doi.org/10.1007/s10816-007-9042-4>.

Wilson et al. 2017
S. Wilson/A. Carlson/I. Szeman (eds.), Petrocultures. Oil, Politics, Culture (Montreal 2017).

Wissing 2017
A. Wissing, Die Bestattungen der Frühen und Mittleren Bronzezeit in der zentralen Oberstadt von Tall Mozan/Urkeš. Eine vergleichende Analyse zu den Bestattungspraktiken des Oberen Ḫābūr-gebietes. Mit einem Beitrag von Katja Zipp. Studien zur Urbanisierung Nordmesopotamiens, Serie A: Ausgrabungen 1998–2001 in der zentralen Oberstadt von Tall Mozan/Urkeš, Band 5 (Wiesbaden 2017).

Wittfogel 1981
K. A. Wittfogel, Die orientalische Despotie. Eine vergleichende Untersuchung totaler Macht (Frankfurt am Main 1981).

Wittwer et al. 2021
R. A. Wittwer/S. F. Bender/K. Hartman/S. Hydbom/R. A. A. Lima/V. Loaiza/T. Nemecek/F. Oehl/P. A. Olsson/O. Petchey/U. E. Prechsl/K. Schlaeppi/T. Scholten/S. Seitz/J. Six/M. G. A. Van der Heijden, Organic and Conservation Agriculture Promote Ecosystem Multifunctionality. Science Advances 7.34, 2021, eabg6995. DOI: <https://doi.org/10.1126/sciadv.abg6995>.

Wolf 2015
S. Wolf, Schmuckstücke. Die Elfenbeinbearbeitung im schwäbischen Aurignacien (Tübingen 2015).

Wozniak 2021
T. Wozniak, Eine kleine Geschichte Mallorcas (Regensburg 2021).

WUB
Württembergisches Urkundenbuch. Band 1–11, Stuttgart 1849–1913, <http://www.wubonline.de> (last access: 23.09.2024).

Wuepper et al. 2020
D. Wuepper/P. Borrelli/R. Finger, Countries and the Global Rate of Soil Erosion. Nature Sustainability 3, 2020, 51–55.

Wynn et al. 2016
T. Wynn/K. A. Overmann/F. K. Coolidge, The False Dichotomy. A Refutation of the Neandertal Indistinguishability Claim. Journal of Anthropological Sciences 94, 2016, 1–22. DOI: <https://doi.org/10.4436/jass.94022>.

Yadav et al. 2018
A. Yadav/G. Sharma/R. Gangwar, Determining Factors for Migration in Uttarakhand. International Journal of Agricultural Science and Research (IJASR) ISSN (P): 2250–0057, ISSN (E): 2321–0087, Vol. 8:1, 2018, 85–90, <https://ssrn.com/abstract=3172063> (last access: 11.10.2024).

Yoffee 1993
N. Yoffee, Too Many Chiefs? (or, Safe Texts for the '90s). In: N. Yoffee/A. G. Sherratt (eds.), Archaeological Theory. Who Sets the Agenda? New Directions in Archaeology (Cambridge 1993) 60–78.

Yonge 1854
C. D. Yonge, Athenaeus of Naucratis. The Deipnosophists, or Banquet of the Learned of Athenæus (London 1854).

Zádorová et al. 2023

T. Zádorová/V. Penížek/L. Lisá/M. Koubová/D. Žížala/V. Tejnecký/O. Drábek/R. Kodešová/M. Fér/A. Klement/A. Nikodem/J. Reyes Rojas/P. Vokurková/L. Pavlů/A. Vaněk/P. Moska, Formation of Colluvisols in Different Soil Regions and Slope Positions (Czechia). Stratification and Upbuilding of Colluvial Profiles. Catena 221, 2023, 106755.

Zagolla 2002

R. Zagolla, Bebenhäuser Annalen. Textkritische Untersuchung und Neuedition. Tübinger Bausteine zur Landesgeschichte, Band 2 (Leinfelden-Echterdingen 2002).

Zamboni 2018

L. Zamboni, Sepolture arcaiche della pianura emiliana. Il riconoscimento di una società di frontiera. Reditus. Riflessioni di archeologia 1 (Rome 2018).

Zanjani/Nejatian 2017

H. Zanjani/M. Nejatian, Kerman. Population, last updated 2017, <http://www.iranicaonline.org/articles/kerman-03-population> (last access: 09.03.2022).

Zeune 1996

J. Zeune, Burgen. Symbole der Macht (Regensburg 1996).

Zeuske 2019

M. Zeuske, Handbuch Geschichte der Sklaverei. Eine Globalgeschichte von den Anfängen bis zur Gegenwart (Berlin 2019).

Zimmer-Dauphinee et al. 2024

J. Zimmer-Dauphinee/P. VanValkenburgh/S. A. Wernke, Eyes of the Machine. AI-Assisted Satellite Archaeological Survey in the Andes. Antiquity 98.397, 2024, 245–259.

Zimmerman 1933

E. W. Zimmerman, World Resources and Industries. A Functional Appraisal of the Availability of Agricultural and Industrial Resources (New York 1933).

Zimmermann 2012

A. Zimmermann, Cultural Cycles in Central Europe between the Neolithic and the Iron Age. In: T. Kienlin/A. Zimmermann (eds.), Metals and Societies. Alternatives to Hierarchical Systems in Modelling Social Formations. International Conference at the Ruhr-Universität Bochum, Germany, October 22–24, 2009. Universitätsforschungen zur Prähistorischen Archäologie (Bonn 2012) 137–146.

Zohar/Artzy 2019

I. Zohar/M. Artzy, The Role of Preserved Fish. Evidence of Fish Exploitation, Processing and Long-Term Preservation in the Eastern Mediterranean during the Late Bronze Age (14th–13th Century BCE). Journal of Archaeological Science (Reports) 23, 2019, 900–909.

Zuchtriegel 2018

G. Zuchtriegel (ed.), The Invisible Image. The Tomb of the Diver in the Fiftieth Anniversary of its Discovery (Napoli 2018).

Zuchtriegel 2022

G. Zuchtriegel, Paestum (Rome 2022).

Züfle 2017

L. Züfle, The two Faces of Mahaprasada. Ritual and Business in the Kitchen of Sri Jagannath. In: R. Hardenberg (ed.), Approaching Ritual Economy. Socio-Cosmic Fields in Globalised Contexts. RessourcenKulturen 4 (Tübingen 2017) 281–308.

Zürcher Wappenrolle
Schweizerisches Nationalmuseum (Zürich), AG 2760: Zürcher Wappenrolle, <http://www.e-codices.unifr.ch/de/list/one/snm/AG002760> (last access: 05.03.2025).

Zvelebil/Lillie 2000
M. Zvelebil/M. Lillie, The Transition to Agriculture in Eastern Europe. In: T. D. Price (ed.), Europe's First Farmers (Cambridge 2000) 57–92.